AF787090

Advanced Biomaterials, Coatings and Techniques: Applications in Medicine and Dentistry

Advanced Biomaterials, Coatings and Techniques: Applications in Medicine and Dentistry

Guest Editors

Lavinia Cosmina Ardelean
Laura-Cristina Rusu

MDPI

Basel • Beijing • Wuhan • Barcelona • Belgrade • Novi Sad • Cluj • Manchester

Guest Editors
Lavinia Cosmina Ardelean
Academic Department of
Technology of Materials and
Devices in Dental Medicine
"Victor Babes" University of
Medicine and Pharmacy
Timișoara
Romania

Laura-Cristina Rusu
University Clinic of Oral
Pathology
"Victor Babes" University of
Medicine and Pharmacy
Timișoara
Romania

Editorial Office
MDPI AG
Grosspeteranlage 5
4052 Basel, Switzerland

This is a reprint of the Special Issue, published open access by the journal *Coatings* (ISSN 2079-6412), freely accessible at: www.mdpi.com/journal/coatings/special_issues/Biomaterials_Medicine.

For citation purposes, cite each article independently as indicated on the article page online and as indicated below:

Lastname, A.A.; Lastname, B.B. Article Title. *Journal Name* **Year**, *Volume Number*, Page Range.

ISBN 978-3-7258-4064-9 (Hbk)
ISBN 978-3-7258-4063-2 (PDF)
https://doi.org/10.3390/books978-3-7258-4063-2

Contents

About the Editors

Lavinia Cosmina Ardelean

Professor Lavinia Cosmina Ardelean, DMD, PhD, is a professor and head of the Department of Technology of Materials and Devices in Dental Medicine, "Victor Babes" University of Medicine and Pharmacy, Timisoara, Romania. She has authored/co-authored 20 books, 15 book chapters, and over 120 peer-reviewed papers. With a current H-index of 13, she is an Editorial Board Member of numerous journals, including *Scientific Reports* and a member of reviewer/topic boards of the following MDPI journals: *Materials, Prosthesis, Metals, Coatings, Polymers, Dentistry,* and *Biomedicines.* She has guest-edited 10 Special Issues in different journals and seven books, and she currently has one ongoing book editing project. Being an active reviewer with more than 300 reviews to her credit so far, she was awarded with the Top reviewers in Cross-Field award 2019 and the Top reviewers in Materials Science award 2019. She currently holds three patents. Her research interests include most areas of dentistry, with a focus on dental materials/biomaterials, dental alloys, resins, ceramics/bioceramics, CAD/CAM milling, 3D printing/bioprinting in dentistry, welding, scanning, prosthodontics, and oral health care.

Laura-Cristina Rusu

Professor Laura Cristina Rusu, DMD, PhD, is a professor and head of the Oral Pathology Department, Faculty of Dental Medicine, "Victor Babes" University of Medicine and Pharmacy, Timisoara, Romania. Her PhD thesis was centered on allergens in dental materials. In 2017, she obtained a Dr. Habil and was given the role of PhD coordinator in the field of dental medicine. She has taken part in 11 research projects, including FP7 COST Action MP 1005, and has authored over 140 peer-reviewed papers. She has published 10 books and book chapters as an author and co-author. She has guest-edited seven Special Issues in different journals and edited or co-edited six books. She currently holds three patents. With a current H-index of 15, she is a member of the Editorial Board of the *Journal of Science and Arts* and *Medicine in Evolution* and is a topic editor for the journal *Materials.* Her main scientific interests are oral pathology and oral diagnosis in dental medicine, with a focus on oral cancer.

coatings

MDPI

Editorial

Advanced Biomaterials, Coatings, and Techniques: Applications in Medicine and Dentistry

Lavinia Cosmina Ardelean [1,*] and Laura-Cristina Rusu [2]

1 Department of Technology of Materials and Devices in Dental Medicine, Multidisciplinary Center for Research, Evaluation, Diagnosis and Therapies in Oral Medicine, "Victor Babes" University of Medicine and Pharmacy Timisoara, 2 Eftimie Murgu Sq., 300041 Timisoara, Romania

2 Department of Oral Pathology, Multidisciplinary Center for Research, Evaluation, Diagnosis and Therapies in Oral Medicine, "Victor Babes" University of Medicine and Pharmacy Timisoara, 2 Eftimie Murgu Sq., 300041 Timisoara, Romania; laura.rusu@umft.ro

* Correspondence: lavinia_ardelean@umft.ro

The field of biomaterials is very extensive, encompassing both the materials themselves and the manufacturing methods, which are constantly developing. Biomaterials, natural or synthetic, alive or lifeless, due to their biological interactions, are frequently used in medical or oral applications to augment or replace a natural function [1].

A biomaterial for medical or oral applications has been defined as a natural or synthetic material that can be inserted into live tissues without developing an immune reaction [2]. As a consequence of the close proximity with human tissues, their use implies specific issues related to properties such as biocompatibility, bio-integration, antimicrobial action, corrosion resistance, and long-term performance. Based on biocompatibility, they are classified as bioactive, biotolerant, biodegradable, or bioinert [3].

The wide range of biomaterial applications in medicine and dentistry include both hard and soft tissue regeneration [4,5]. The basic characteristics of biomaterials for tissue engineering have constantly improved due to the advancements in the field, being characterized by corrosion resistance, non-toxicity and non-carcinogenic properties, bioactivity, and proper mechanical strength, depending on the surrounding tissue type [2,6].

Starting with medical devices or grafts, regenerative medicine has improved the field of tissue engineering, with the aid of biomaterials. Natural or synthetic biomaterial scaffolds have been developed to induce replacement of the missing tissue by means of generating specific regenerative cell responses, through bioactive molecules. Once placed into a specific tissue, the biomaterial surface initiates the interaction with the surrounding cells, inducing the charging of its surface energy and resulting in an adequate matrix for biomolecule adhesion [2,7–9].

Scaffolds promote cell growth and differentiation, resulting in tissue healing. Based on their type, scaffold biomaterials can be categorized in natural-based and synthetic-based polymers, ceramics, hydrogels, and bioactive glasses [9,10].

The latest generation of scaffolds can induce specific cellular responses: adhesion, differentiation, and proliferation. In order to improve tissue response and intensify the regenerative capability scaffolds were combined with growth factors and bioactive molecules. They are used to provide an extracellular matrix, an attachment site, or 3D support for regenerative cells, as well as a template for tissue regeneration [1,2,11].

Smart scaffolds, which incorporate bioactive molecules and nanoparticles, with tailored physical and chemical properties, aim to improve the interactions with cells by enhancing the osteogenic differentiation and generate a better response to the surrounding environment. Providing a proper microenvironment that ensures cell adhesion and differentiation is the ultimate goal [12–14].

Platelet-rich fibrin is a biomaterial scaffold with trapped platelets and leukocytes aimed to accelerate musculoskeletal tissue recovery by providing a binding site for platelets and growth factors [15,16]. It promotes tissue regeneration and reduces healing time by increasing the local concentration of growth factors, and has been frequently used in combination with bone graft materials in maxillofacial and orthopaedic surgery and sports-related injuries [17–23]. It has excellent handling characteristics, and can be firmly sutured in an anatomically desired location during open surgery [24–26].

The additive manufacturing (3D printing) of biomaterials launched a new perspective for the field of biomedical engineering, considering its patient-specific clinical applications. Scaffolds are now being fabricated using 3D bioprinting methods and progress has been made in 3D printing of biocompatible materials, seed cells, and supporting components into functional living tissue [1,2,27].

Based on layer-by-layer precise positioning of biological constituents, biochemicals and living cells, this novel technology facilitates the printing of cells, tissues, and organs for regenerative medicine purposes, enabling the manufacturing of tissue-engineered constructs with tailored structures and properties [1,28–32].

Coatings play an important role in achieving the most crucial properties of biomaterials by surface modification, making them suitable for medical and oral applications. The application of coatings onto medical devices is quite vast, ranging from implantable to non-implantable medical devices, from orthopedic prostheses to dental implants, including hydroxyapatite (which enhances cell attachment onto orthopaedic implants), antimicrobial silver coatings on catheters, drug-eluting coatings on stents, and blood-compatible coatings (such as heparin). The protective medical coating of prosthetic device surfaces results in healing stimulation; meanwhile, porous bioactive coatings make implants far better suited to bone tissue interaction. The coating of scaffolds with stem or differentiated cells is a complex and novel method used in the field of regenerative medicine [2,33].

Among the different methods used to deposit coatings, plasma spraying, dipping, and spin coating are quite usual. Meanwhile, recently developed techniques such as laser, low-temperature atmospheric plasmas, and microblasting have been used for the deposition of bioactive coatings [34–36].

This Special Issue aims to provide a forum for researchers to share current research findings to promote further research and provide an updated outlook on the applications of biomaterials and coatings in medicine and dentistry, as well as presenting innovative manufacturing technologies.

Funding: This research received no external funding.

Conflicts of Interest: The authors declare no conflict of interest.

References

1. Matichescu, A.; Ardelean, L.C.; Rusu, L.-C.; Craciun, D.; Bratu, E.A.; Babucea, M.; Leretter, M. Advanced Biomaterials and Techniques for Oral Tissue Engineering and Regeneration—A Review. *Materials* **2020**, *13*, 5303. [CrossRef] [PubMed]
2. Roi, A.; Ardelean, L.C.; Roi, C.I.; Boia, E.-R.; Boia, S.; Rusu, L.-C. Oral Bone Tissue Engineering: Advanced Biomaterials for Cell Adhesion, Proliferation and Differentiation. *Materials* **2019**, *12*, 2296. [CrossRef] [PubMed]
3. Marin, E.; Boschetto, F.; Pezzotti, G. Biomaterials and biocompatibility: An historical overview. *J. Biomed. Mater. Res.* **2020**, *108*, 1617–1633. [CrossRef] [PubMed]
4. Huang, G.T. Pulp and dentin tissue engineering and regeneration: Current progress. *Regen. Med.* **2009**, *4*, 697–707. [CrossRef]
5. Ohara, T.; Itaya, T.; Usami, K.; Ando, Y.; Sakurai, H.; Honda, M.J.; Ueda, M.; Kagami, H. Evaluation of scaffold materials for tooth tissue engineering. *J. Biomed. Mater. Res. A* **2010**, *94*, 800–805. [CrossRef]
6. Anderson, J.M. The future of biomedical materials. *J. Mater. Sci. Mater. Med.* **2006**, *17*, 1025–1028. [CrossRef]
7. Bhat, S.; Kumar, A. Biomaterials and bioengineering tomorrow's healthcare. *Biomatter* **2013**, *3*, e24717. [CrossRef]
8. Gilbert Triplett, R.; Budinskaya, O. New Frontiers in Biomaterials. *Oral Maxillofac. Surg. Clin.* **2017**, *29*, 105–115. [CrossRef]
9. Jafari, M.; Paknejad, Z.; Rezai Rad, M.; Motamedian, S.R.; Eghbal, M.J.; Nadjmi, N.; Khojasteh, A. Polymeric scaffolds in tissue engineering: A literature review. *J. Biomed. Mater. Res. Part B* **2017**, *105*, 431–459. [CrossRef]
10. Rusu, L.-C.; Ardelean, L.; Negrutiu, M.-L.; Dragomirescu, A.-O.; Albu, M.G.; Ghica, M.V.; Topala, F.I.; Podoleanu, A.; Sinescu, C. SEM for the General Structural Features Assesing of the Synthetic Polymer Scaffolds. *Rev. Chim.* **2011**, *62*, 841–845.
11. Baranova, J.; Büchner, D.; Götz, W.; Schulze, M.; Tobiasch, E. Tooth Formation: Are the Hardest Tissues of Human Body Hard to Regenerate? *Int. J. Mol. Sci.* **2020**, *21*, 4031. [CrossRef] [PubMed]

12. Zhang, K.; Wang, S.; Zhou, C.; Cheng, L.; Gao, X.; Xie, X.; Sun, J.; Wang, H.; Weir, M.D.; Reynolds, M.A.; et al. Advanced smart biomaterials and constructs for hard tissue engineering and regeneration. *Bone Res.* **2018**, *6*, 31. [CrossRef] [PubMed]
13. Motamedian, S.R.; Hosseinpour, S.; Ahsaie, M.G.; Khojasteh, A. Smart scaffolds in bone tissue engineering: A systematic review of literature. *World J. Stem Cells* **2015**, *7*, 657–668. [CrossRef] [PubMed]
14. Khan, F.; Tanaka, M. Designing smart biomaterials for tissue engineering. *Int. J. Mol. Sci.* **2018**, *19*, 17. [CrossRef] [PubMed]
15. Grecu, A.F.; Reclaru, L.; Ardelean, L.C.; Nica, O.; Ciucă, E.M.; Ciurea, M.E. Platelet-Rich Fibrin and Its Emerging Therapeutic Benefits for Musculoskeletal Injury Treatment. *Medicina* **2019**, *55*, 141. [CrossRef]
16. Sánchez, A.R.; Sheridan, P.J.; Kupp, L.I. Is platelet-rich plasma the perfect enhancement factor? A current review. *Int. J. Oral Maxillofac. Implant.* **2003**, *18*, 93–103.
17. Alsousou, J.; Thompson, M.; Hulley, P.; Noble, A.; Willett, K. The biology of platelet-rich plasma and its application in trauma and orthopaedic surgery: A review of the literature. *J. Bone Jt. Surg. Br.* **2009**, *91*, 987–996. [CrossRef]
18. Zumstein, M.A.; Berger, S.; Schober, M.; Boileau, P.; Nyffeler, R.W.; Horn, M.; Dahinden, C.A. Leukocyteand platelet-rich fibrin (L-PRF) for long-term delivery of growth factor in rotator cuff repair: Review, preliminary results and future directions. *Curr. Pharm. Biotechnol.* **2012**, *13*, 1196–1206. [CrossRef]
19. Naik, B.; Karunakar, P.; Jayadev, M.; Marshal, V.R. Role of Platelet rich fibrin in wound healing: A critical review. *J. Conserv. Dent.* **2013**, *16*, 284–293. [CrossRef]
20. Ali, S.; Bakry, S.A.; Abd-Elhakam, H. Platelet-Rich Fibrin in Maxillary Sinus Augmentation: A Systematic Review. *J. Oral. Implantol.* **2015**, *41*, 746–753. [CrossRef]
21. Borie, E.; Oliví, D.G.; Orsi, I.A.; Garlet, K.; Weber, B.; Beltrán, V.; Fuentes, R. Platelet-rich fibrin application in dentistry: A literature review. *Int. J. Clin. Exp. Med.* **2015**, *8*, 7922–7929. [PubMed]
22. Asutay, F.; Yolcu, Ü.; Geçör, O.; Acar, A.H.; Öztürk, S.A.; Malkoç, S. An evaluation of effects of platelet rich-fibrin on postoperative morbidities after lower third molar surgery. *Niger. J. Clin. Pract.* **2017**, *20*, 1531–1536. [CrossRef] [PubMed]
23. Ozgul, O.; Senses, F.; Er, N.; Tekin, U.; Tuz, H.H.; Alkan, A.; Kocyigit, I.D.; Atil, F. Efficacy of platelet rich fibrin in the reduction of the pain and swelling after impacted third molar surgery: Randomized multicenter split-mouth clinical trial. *Head Face Med.* **2015**, *11*, 37. [CrossRef]
24. Guinot, A.; Arnaud, A.; Azzis, O.; Habonimana, E.; Jasienski, S.; Frémond, B. Preliminary experience with the use of an autologous platelet-rich fibrin membrane for urethroplasty coverage in distal hypospadias surgery. *J. Pediatr. Urol.* **2014**, *10*, 300–305. [CrossRef] [PubMed]
25. Garin, P.; Peerbaccus, Y.; Mullier, F.; Gheldof, D.; Dogne, J.M.; Putz, L.; Van Damme, J.P. Platelet-rich fibrin (PRF): An autologous packing material for middle ear microsurgery. *B-ENT* **2014**, *10*, 27–34.
26. Niedzielska, I.; Ciapiński, D.; Bąk, M.; Niedzielski, D. The Assessment of the Usefulness of Platelet-Rich Fibrin in the Healing Process Bone Resorption. *Coatings* **2022**, *12*, 247. [CrossRef]
27. Ma, Y.; Xie, L.; Yang, B.; Tian, W. Three-dimensional printing biotechnology for the regeneration of the tooth and tooth-supporting tissues. *Biotechnol. Bioeng.* **2019**, *116*, 452–468. [CrossRef]
28. Zadpoor, A.A.; Malda, J. Additive manufacturing of biomaterials, tissues, and organs. *Ann. Biomed. Eng.* **2017**, *45*, 1–11. [CrossRef]
29. Murphy, S.V.; Atala, A. 3D bioprinting of tissues and organs. *Nat. Biotechnol.* **2014**, *32*, 773–785. [CrossRef]
30. Wang, X.; Ao, Q.; Tian, X.; Fan, J.; Wei, Y.; Hou, W.; Tong, H.; Bai, S. 3D bioprinting technologies for hard tissue and organ engineering. *Materials* **2016**, *9*, 802. [CrossRef]
31. Ji, X.; Zhu, H.; Zhao, L.; Xiao, J. Recent advances in 3D bioprinting for the regeneration of functional cartilage. *Regen. Med.* **2018**, *13*, 73–87. [CrossRef]
32. Nakamura, M.; Kobayashi, A.; Takagi, F.; Watanabe, A.; Hiruma, Y.; Ohuchi, K.; Iwasaki, Y.; Horie, M.; Morita, I.; Takatani, S. Biocompatible inkjet printing technique for designed seeding of individual living cells. *Tissue Eng.* **2005**, *11*, 1658–1666. [CrossRef] [PubMed]
33. Taranu, B.-O.; Ianasi, P.; Rus, S.F.; Bucur, A.I. Simultaneous Precipitation and Electrodeposition of Hydroxyapatite Coatings at Different Temperatures on Various Metal Substrates. *Coatings* **2022**, *12*, 288. [CrossRef]
34. Wu, X.; Liu, C.; Chen, H.; Zhang, Y.; Li, L.; Tang, N. Layer-by-Layer Deposition of Hyaluronan and Quercetin-Loaded Chitosan Nanoparticles onto Titanium for Improving Blood Compatibility. *Coatings* **2020**, *10*, 256. [CrossRef]
35. Rau, J.V.; De Bonis, A.; Curcio, M.; Schuhladen, K.; Barbaro, K.; De Bellis, G.; Teghil, R.; Boccaccini, A.R. Borate and Silicate Bioactive Glass Coatings Prepared by Nanosecond Pulsed Laser Deposition. *Coatings* **2020**, *10*, 1105. [CrossRef]
36. Bartmański, M.; Pawłowski, Ł.; Zieliński, A.; Mielewczyk-Gryń, A.; Strugała, G.; Cieślik, B. Electrophoretic Deposition and Characteristics of Chitosan–Nanosilver Composite Coatings on a Nanotubular TiO_2 Layer. *Coatings* **2020**, *10*, 245. [CrossRef]

coatings

MDPI

Review

Exploring the Role of Nanoparticles in Dental Materials: A Comprehensive Review

Rim Bourgi [1,2,*,†], Zahra Doumandji [3,†], Carlos Enrique Cuevas-Suárez [4], Teissir Ben Ammar [2], Chloé Laporte [2,3,5], Naji Kharouf [2,5] and Youssef Haikel [2,3,5]

1 Department of Restorative Dentistry, School of Dentistry, Saint-Joseph University, Beirut 1107 2180, Lebanon
2 Department of Biomaterials and Bioengineering, INSERM UMR_S 1121, University of Strasbourg, 67000 Strasbourg, France; tbenammar@unistra.fr (T.B.A.); chloelaporte9@gmail.com (C.L.); dentistenajikharouf@gmail.com (N.K.); youssef.haikel@unistra.fr (Y.H.)
3 Pôle de Médecine et Chirurgie Bucco-Dentaire, Hôpital Civil, Hôpitaux Universitaire de Strasbourg, 67000 Strasbourg, France; yasmine.doumandji@etu.unistra.fr
4 Dental Materials Laboratory, Academic Area of Dentistry, Autonomous University of Hidalgo State, San Agustín Tlaxiaca 42160, Mexico; cecuevas@uaeh.edu.mx
5 Department of Endodontics and Conservative Dentistry, Faculty of Dental Medicine, University of Strasbourg, 67000 Strasbourg, France
* Correspondence: rim.bourgi@net.usj.edu.lb or rim.bourgi@etu.unistra.fr or rim.bourgi@hotmail.com; Tel.: +96-171455529
† These authors contributed equally to this work.

Abstract: In recent decades, the integration of nanotechnology into dentistry has led to groundbreaking advancements in dental materials and applications. This article explores the role of nanoparticles (NPs) in modern dentistry, highlighting their definitions, unique properties, and various applications. The introduction establishes the significance of nanotechnology in dental health care, emphasizing the potential of NPs to transform traditional practices. The overview includes a discussion of the fundamental properties of NPs, which contribute to their effectiveness in dental applications. The article categorizes NPs into three main groups: antimicrobial, therapeutic, and material property-improving NPs, detailing their clinical uses and mechanisms of action. Furthermore, it addresses current innovations in dental products incorporating NPs and examines emerging trends in the field. The research for this review was conducted using high-quality, peer-reviewed scientific databases, including PubMed, Scopus, Web of Science, and Google Scholar, with no time restriction as an inclusion criterion. These databases were selected for their credibility and comprehensive collections of relevant studies. In conclusion, NPs represent a promising avenue for innovation in dental materials and therapeutics. Their unique properties enable the development of enhanced antimicrobial agents, effective drug delivery systems, and improved material performance. However, the risks associated with cytotoxicity and stability must be carefully managed to ensure safe and effective use. Ongoing research is essential to fully understand and optimize the applications of NPs in dentistry, balancing their benefits against potential health risks. As the field advances, the integration of NPs into clinical practice will likely revolutionize approaches to dental care and treatment.

Keywords: dentistry; multifunction nanoparticle; nanomaterials; nanoparticles; nanoscience; nanotechnology

1. Introduction

Nanotechnology has emerged as one of the most dynamic fields of research in recent decades, particularly within the realm of health sciences [1]. Maintaining oral health

 https://doi.org/10.3390/coatings15010033

presents significant challenges in dentistry. Various materials have been employed to address different dental conditions; however, the success of these treatments is often constrained by the properties of the biomaterials used. To address these limitations, the incorporation of nanoparticles (NPs) into dental applications offers promising solutions in fields such as endodontics, periodontics, tissue engineering, oral surgery, and imaging [2–5].

NPs, defined as ultrafine units with dimensions ranging from 1 to 100 nm, exhibit unique physical, chemical, mechanical, and biological properties that distinguish them from macro materials [6]. Their small size contributes to a high surface-to-volume ratio, enhancing their reactivity and allowing for novel applications across various fields, including medicine, engineering, and environmental remediation [7]. Within the dental sector, the integration of NPs has catalyzed significant advancements in dental materials and practices.

Prior research has highlighted the diverse applications of NPs in dentistry, encompassing areas such as tissue regeneration, antimicrobial strategies, and enhancement of material properties [8]. NPs can be categorized into several types based on their origin—natural or synthetic—and composition, including organic (e.g., dendrimers, liposomes) and inorganic (e.g., metal or metal oxide-based) [9]. While many NPs possess advantageous characteristics such as biocompatibility, chemical reactivity, and mechanical strength [10], not all NPs are inherently biocompatible. For instance, certain plate-like structures can pose risks to human health due to their cytotoxic or inflammatory potential. This underscores the need for careful evaluation and selection of NPs to ensure their safe use in dental applications.

In restorative dentistry, NPs are incorporated into composites to enhance mechanical strength, reduce shrinkage, and improve aesthetics [11]. Antimicrobial NPs, such as silver and zinc oxide (ZnO), are used to inhibit bacterial growth, thus reducing the risk of infections [12]. Additionally, NPs are instrumental in endodontics, where they are employed for tissue regeneration, drug delivery systems, and the effective elimination of biofilms and bacteria, ultimately aiming to improve overall oral health [13].

In preventive dentistry, NPs like hydroxyapatite and calcium phosphate are recognized for their ability to remineralize enamel and prevent caries [14]. Moreover, in the context of dental implants, NPs enhance osseointegration and surface adhesion, contributing to better treatment outcomes [15]. Recent advancements have also seen the use of NPs in developing adhesives that reduce biofilm accumulation and promote oral health [16].

Despite the numerous advantages that NPs present, their long-term effects and potential toxicity in biological systems necessitate careful consideration in both research and clinical applications [2,17]. Regulatory challenges and safety assessments remain critical for the integration of nanotechnology into dental practice. A thorough understanding of the physicochemical properties of NPs and their interactions within the oral environment is essential for optimizing their use in dentistry [2–4].

Specifically, the incorporation of NPs in dentistry not only enhances the efficacy of treatments but also holds the potential to revolutionize patient care by promoting personalized and targeted therapeutic strategies. For instance, with advancements in drug delivery systems utilizing NPs, there is the capability to deliver medications directly to the site of action, thus minimizing systemic exposure and side effects. This targeted approach could improve the effectiveness of treatments for conditions such as periodontal disease and oral cancer [18,19]. Furthermore, the multifunctional nature of NPs allows for the simultaneous delivery of therapeutic agents and diagnostic imaging, paving the way for novel applications in precision dentistry [2,20,21].

As research continues to uncover the versatile applications of NPs, their integration into clinical practice could lead to more effective, efficient, and patient-centered care in the field of dentistry [2,17,22]. Thus, the objective of this comprehensive review is to explore the role of NPs in dental materials by analyzing their properties, applications, and innovations. This review aims to provide a thorough understanding of the current state of knowledge regarding NPs in dentistry, identify emerging trends, and discuss potential future applications and challenges associated with their use.

2. Overview of Nanoparticles in Dentistry

2.1. Definition of Nanoparticles

NPs are typically defined as particles with at least one dimension ranging between 1 and 100 nm, which corresponds to one billionth of a meter (1 nm = 10^{-9} m) [23,24]. This definition is generally accepted across scientific disciplines, though specific regulations or organizations may have slight variations in their interpretation of this size range [25]. NPs occupy a unique size domain, larger than individual atoms yet smaller than cells, placing them in a category where they exhibit distinctive properties that bulk materials lack [26]. Their high surface area, quantum effects, and enhanced chemical reactivity provide them with unique physicochemical characteristics, such as increased mechanical strength, conductivity, and reactivity, making them highly suitable for various applications across industries [27].

NPs can occur naturally (e.g., volcanic ash, ocean spray, and fine dust) or be engineered synthetically to meet specific needs [28]. Synthetic NPs are created using two primary approaches: top-down or bottom-up methods [29]. In the top-down approach, larger bulk materials are broken down into NPs via processes like milling or lithography. In contrast, the bottom-up approach involves assembling NPs from atomic or molecular units through processes like chemical vapor deposition, sol-gel synthesis, or self-assembly [30]. As a result, NPs can exist in various forms, including powders, gels, or colloidal solutions, and are typically tailored to specific applications based on their size, shape, and surface properties [31–38].

Figure 1 represents the definition of NPs.

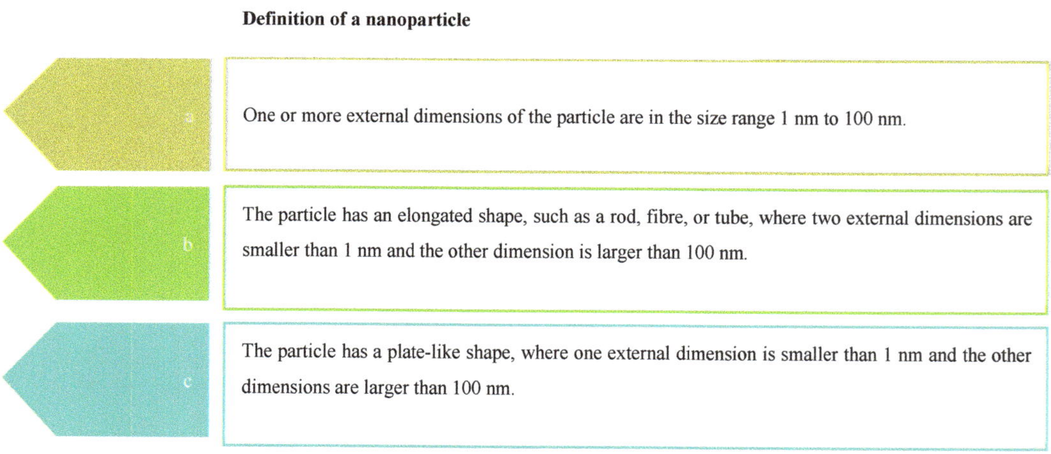

Figure 1. Nanoparticle means a natural, incidental or manufactured material consisting of solid particles that are present, either on their own or as identifiable constituent particles in aggregates or agglomerates, and where 50% or more of these particles in the number-based size distribution fulfil at least one of the following conditions presented in this figure [39].

Figure 2 illustrates the comparative scale of NPs.

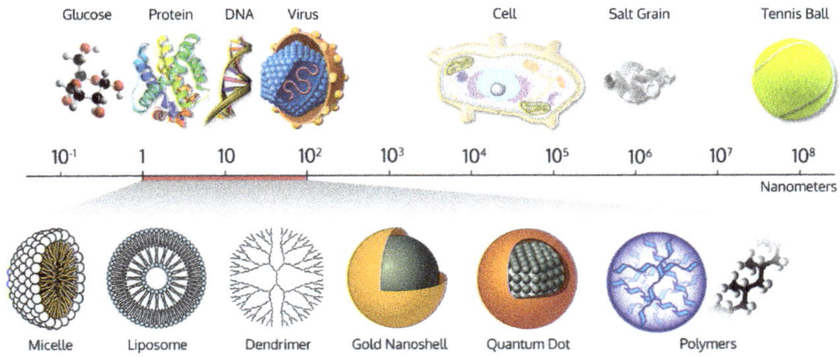

Figure 2. Comparative scale of nanoparticles (NPs)-("Size-comparison of Bio-NPs: nanometer scale comparison and nanotechnology chart ruler", 2017) [40].

NPs have a wide range of applications, spanning from medicine to aerospace and cosmetics [41]. Their unique size allows them to impart properties that are otherwise unattainable at larger scales. For instance, gold, which is chemically inert in its bulk form, acts as an efficient catalyst at the nanoscale due to the increased surface area and altered electronic properties [42]. However, despite these advantages, NPs also pose potential risks to both health and the environment. Their small size enables them to penetrate biological barriers, potentially leading to adverse effects such as cytotoxicity, oxidative stress, and inflammation in living organisms [43,44]. Additionally, the environmental impact of nanoparticle accumulation and their behavior in ecosystems remain critical concerns [30,45]. Ongoing research is crucial to balance the beneficial applications of NPs with the need to mitigate their potential risks [46].

2.2. *Nanotechnology*

In recent years, the primary focus of utilizing nanotechnologies in dental materials has been to enhance mechanical properties, improve abrasion resistance, reduce shrinkage, and optimize optical and aesthetic aspects [47]. Today, it is widely recognized that NPs possess a broad range of indispensable properties, including bioactive and antimicrobial characteristics [48]. Moreover, in the field of dentistry, specifically in endodontics, NPs are employed for tissue regeneration and drug delivery. Enhancing oral health is the primary objective, with the eradication of biofilms and bacteria using NPs at the core of ongoing research [49].

In this context, the incorporation of NPs into certain dental treatment materials has shown the potential to reduce biofilm accumulation, restore a safe level of oral pH, and promote remineralization while improving the durability of the material [50].

Biofilm refers to a complex community of microorganisms, primarily bacteria, that adhere to surfaces such as teeth or dental materials, embedded within a self-produced extracellular matrix. This protective layer makes biofilms highly resistant to conventional antimicrobial treatments, contributing to persistent infections, inflammation, and tooth decay. The dangers of biofilm formation in the oral cavity include an increased risk of periodontal disease, caries, and other oral infections. Additionally, biofilms can facilitate bacterial resistance to antibiotics, complicating treatment efforts. Furthermore, protein corona refers to the layer of proteins that adsorb onto the surface of NPs once they enter a biological environment. This interaction alters the NPs' surface properties and influences their behavior in the body. The protein corona can affect how NPs interact with cells, tissues, and the immune system, potentially influencing their effectiveness and safety. Understanding the protein corona is crucial for predicting the biological interactions of NPs, as it may impact their biocompatibility and therapeutic outcomes. With an in-depth understanding of the physical principles of nanomaterials, their chemical properties, strengths, and specific advantages, as well as knowledge of their limitations, nanotechnology research seeks to harness the potential of NPs to further enhance dental practices, allowing the practitioner to overcome some of the inadequacies found in currently available products [51].

Nanotechnology can be defined as a technology that deals with small structures or materials of very small size [52]. It is the science of designing, manufacturing, and applying products at the molecular level, continually developing more efficient materials through the unique properties permitted by the nanoscale. Nanoscience involves the exploration of NPs to understand their implications on a macroscopic scale. As such, nanotechnology builds upon the principles established by nanoscience.

Nanotechnology and nanoscience have proven highly effective for applications in the biological sciences, giving rise to the concept of nanobiotechnology. This emerging field represents a distinct and specialized scientific domain, integrating methods, techniques, and protocols from diverse disciplines such as nanotechnology, biology, and biochemistry. This convergence has led to the development of unique and innovative methodologies and materials [53].

Similarly, NPs' ability to penetrate cellular membranes and interact with cellular organelles enables them to induce specific effects, which may contribute to cytotoxicity and inflammation [53,54]. Consequently, nanostructures play a key role in the development of drug delivery systems, contrast agents, photothermal phenomena, and medical imaging techniques [55].

In dental surgery, the applications of nanotechnology are vast, ranging from diagnostics and prevention to treatment materials across different specialties, including endodontics, restorative care, periodontics, aesthetics, and even implantology [56].

However, although nanoscience and nanotechnology represent rapidly growing and intrinsically interesting scientific and technological fields, the addition of the "nano" prefix does not always guarantee better quality. Sometimes, this prefix is used to market dental products without sufficient clinical evidence proving that the nanometric version of a material is significantly superior to its conventional counterpart. Nonetheless, it seems that adding the "nano" prefix can boost sales, suggesting that the motive may sometimes be purely commercial [51].

2.3. Design of Nanoparticles

A variety of synthesis approaches for NPs have been developed, which can be categorized into either the top-down or bottom-up methods, as illustrated in Figure 3. These techniques allow for precise control over the size, shape, and surface properties of the NPs, contributing to their diverse applications in various fields [57–59].

2.3.1. Top-Down Approach

In the top-down approach, also known as the "descending method," nanoparticle design begins with bulk materials or larger structures that are gradually reduced to nanoscale dimensions. This method employs advanced miniaturization techniques to transform larger materials into NPs with tailored properties. Techniques such as ultrasonication, laser ablation, thermal decomposition, lithography, ion etching, and mechanical milling are commonly used to break down bulk materials into nanoscale particles [60–62]. These methods, particularly mechanical grinding and mechano-synthesis rely on the application of high energy or pressure [61,63].

For instance, mechanical milling and ball milling are effective techniques among the top-down approaches that can reduce particle dimensions to the nanoscale range (typically in the 1–5-micron range) through grinding or shearing [62,63]. While these techniques may not always produce NPs in the strictest sense (e.g., sub-100 nm sizes), they remain important for achieving controlled particle sizes and are commonly used for large-scale production, providing consistent and reproducible results, which are crucial for industrial and commercial applications, as noted by Ahmed et al. [62].

Furthermore, the top-down approach offers precise control over the size, shape, and surface properties of the NPs, which is beneficial for various fields, including biotechnology, materials science, and pharmaceuticals. However, one potential limitation of the top-down method, apart from the price, is the possible generation of defects during the milling or etching processes, which can impact the quality of the NPs produced [64].

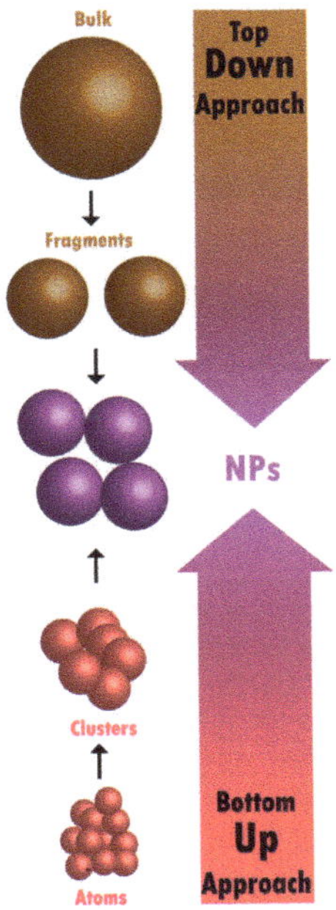

Figure 3. Diagram illustrating the two approaches for synthesizing nanoparticles: the top-down method and the bottom-up method [58].

2.3.2. Bottom-Up Approach

The bottom-up approach, widely used in research centers and nanoscience laboratories, involves synthesizing NPs by accumulating materials from atoms to aggregates, gradually forming nanoscale structures through chemical and biological processes [65]. Methods such as photochemical reduction, chemical precipitation, microemulsion, microbial reduction, and hydrothermal methods are commonly employed in the synthesis of NPs. Both approaches have distinct advantages and drawbacks. The bottom-up method offers significant flexibility in terms of control and diversity of nanoparticle structures, while the top-down approach is more suited for large-scale production, though controlling the particle size can be more challenging. The bottom-up approach predominantly relies on chemical and physical processes, whereas the top-down method favors mechanical processes [61].

3. Properties of Nanoparticles

The properties of NPs are summarized in Figure 4.

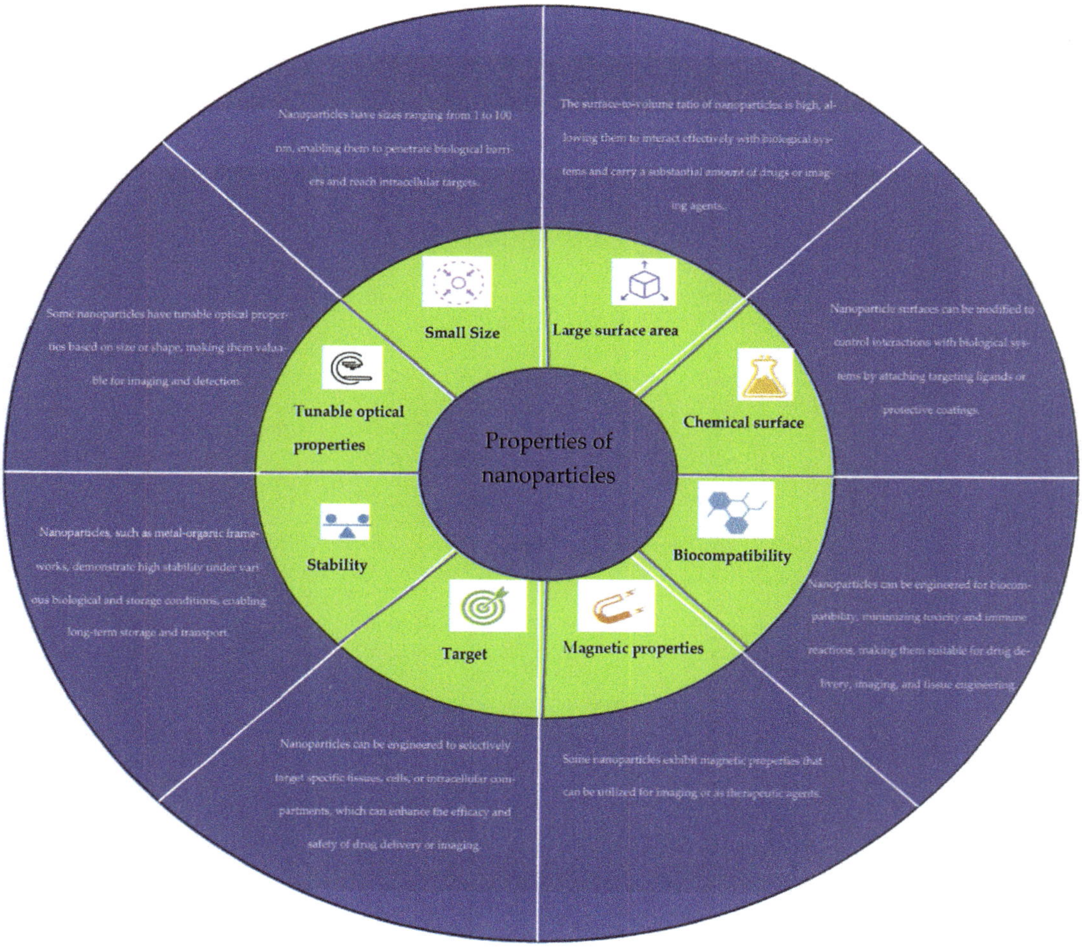

Figure 4. Schematic summarizing the properties of nanoparticles, adapted from Altammar et al. [58].

3.1. Morphological Properties

3.1.1. Size

The properties of NPs are often influenced by their morphological characteristics, which have garnered significant interest. Various techniques are available to characterize the size and morphology of NPs, but transmission electron microscopy (TEM) and scanning electron microscopy (SEM) are the most commonly used methods [66].

Several studies have evaluated the biodistribution of NPs of varying sizes in the human body following exposure [66,67]. For instance, Sonavane et al. [66] intravenously injected NPs ranging from 15 to 200 nm and observed that the accumulation of NPs in tissues such as the liver, lungs, spleen, and kidneys depended on their size. The smaller NPs were able to cross the blood–brain barrier [66,67].

It has also been demonstrated that the size of NPs affects their clearance from circulation. Particles with diameters smaller than 5–6 nm are rapidly cleared by the kidneys, while larger particles (over 200 nm) are more efficiently removed by the liver and spleen [68].

Larger particles, particularly those exceeding 100 nm, are cleared by the mononuclear phagocyte system (MPS), which is mainly facilitated by hepatic, splenic, and bone marrow cells [69,70].

The cytotoxicity of NPs is also closely linked to their size, with smaller NPs generally being more toxic. This was demonstrated by Gao et al. [71] who compared the cellular response to exposure to NPs with diameters of 8 nm and 37 nm. The results indicated that the toxicity was significantly higher after cellular exposure to the 8 nm NPs. The biodistribution of NPs can be clearly seen in Figure 5.

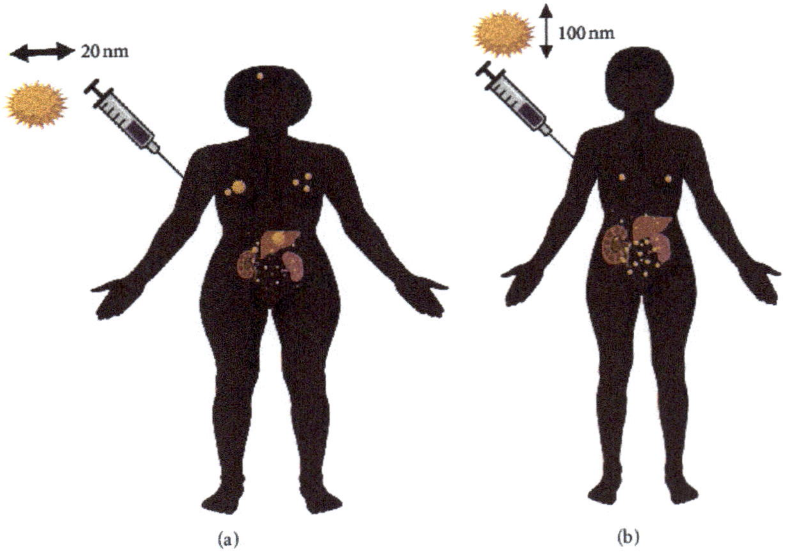

Figure 5. The diagram illustrates the biodistribution of nanoparticles in the human body based on size, specifically 20 nm (**a**) and 100 nm (**b**) [72].

3.1.2. Shapes of Nanoparticles

The size of NPs is the primary determinant of their absorption by macrophages; however, their shape also significantly contributes to regulating this absorption. NPs can have different forms:

One-dimensional (1D) NPs include nanotubes, nanowires, and nanofilaments.
Two-dimensional (2D) NPs are found as sheets or disks.
Three-dimensional (3D) NPs can take non-spherical shapes [73].

3.1.3. Interaction with Macrophages

When these non-spherical NPs interact with macrophages, the initial contact angle (CA) plays a crucial role in determining the internalization rate [73,74]. NPs that align with their major axis parallel to the cell membrane are internalized more slowly than those aligned with their minor axis. Furthermore, filamentous NPs are internalized more rapidly when they are perpendicular to the cellular axis ($\theta = 90°$). The internalization rate decreases when NPs are tangential to the macrophage membrane. For spherical NPs, the internalization rate is independent of the angle (θ) due to their symmetrical nature [72–75] (Figure 6).

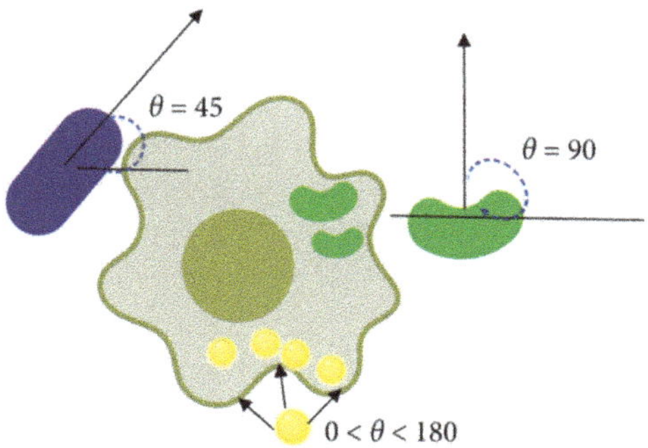

Figure 6. Effect of contact angle on internalization efficiency [72].

3.1.4. Structural Characterization of Nanoparticles

The structural characterization of NPs is crucial before any compositional study, as it provides essential information about the overall properties of the material being studied. Various methods exist for evaluating the shape of NPs, including the following:

- X-ray diffraction (XRD);
- Energy dispersive X-ray spectroscopy (EDX);
- X-ray photoelectron spectroscopy (XPS);
- Infrared spectroscopy (IR);
- Raman spectroscopy;
- Brunauer–Emmett–Teller (BET) surface analysis;
- Zetasizer analysis;
- TEM;
- SEM.

Certain types of NPs can induce toxicity, which, in some cases, is related to the shape of the particles. In this regard, Auclair and Gagné demonstrated that silver NPs (AgNPs) exhibit cytotoxicity depending on their morphology. Their study evaluated the toxicity of three distinct shapes (spherical, cubic, and prismatic) of AgNPs on *Hydra vulgaris*. To focus on morphology-related effects, the NPs were kept within the same size range and had identical surface coatings. Shape-dependent toxicity was observed, with spherical AgNPs being the most toxic, followed by prismatic AgNPs, while no significant toxicity was detected with cubic AgNPs [76].

Similarly, Champion and Mitragotri compared various shapes of polymeric NPs and concluded that elongated particles with a higher length-to-width ratio are less likely to be cleared by the immune system. Furthermore, elongated particles exhibited longer blood circulation times and avoided phagocytosis, depending on the CA when interacting with macrophages [77,78].

3.1.5. Specific Surface Area

Due to their minute size, NPs possess a large surface area, a characteristic that offers a wide range of applications. Consequently, measuring the surface area of nanomaterials would enhance the understanding of their properties, behaviors, and potential hazards. The BET method is considered the paramount technique for determining the surface area of particulate materials. This technique relies on the principles of adsorption and desorption, as well as the BET theorem.

It has been reported that, at equal unit mass, smaller NPs allow more particles and a larger surface area to participate in biological actions compared to larger particles, which could be the source of their more significant toxic effects. Lu et al. [79] injected silica particles of 30, 70, and 300 nm into mice intravenously at different doses. The results showed that when the surfaces of the injected particles were similar, despite the differing number of particles, the extent of hepatic lesions was also similar. This indicates that while the number of particles may exert some influence on in vivo toxicity, surface area could be a more critical factor for toxicity in both nanometric and micrometric particles. Supporting conclusions were drawn from the study by Nemmar et al. [80], which revealed that vascular homeostasis alteration was more pronounced in the group treated with 50 nm NPs than in the group treated with 500 nm particles after intraperitoneal administration. The authors attributed this to the high surface-to-volume ratio, which decreases inversely with size and favors biological interactions, leading to superior vascular and systemic toxicity [81–84].

3.2. Chemical Properties

3.2.1. Charge

The surface charge is often measured as the zeta potential (ξ), which is quantified using a Zetasizer. Several studies indicate that surface charge determines the fate of NPs. Surface charge plays a crucial role in the stability of particles and directly influences their level of toxicity. In fact, positively charged NPs exhibit a higher affinity for cell membranes compared to those that are negatively charged or neutral [85,86]. It appears that the surface characteristics of particles also play an important role regarding toxicity, as they determine the initial direct contact with biological materials and cell surfaces, along with their components.

It was demonstrated that gold NPs with a positive charge induce cell death through apoptosis, while neutral gold NPs cause necrosis in human adult low calcium temperature (HaCaT) cell lines [87]. It is important to note that surface charge influences the composition of biological proteins on the surface of NPs, thereby conditioning all subsequent chemical and biochemical interactions [88].

3.2.2. (Bio)Chemical Surface

The chemical or biochemical surface of NPs is formed upon their interaction with biological materials. As a result, NPs become coated with a diverse array of proteins, collectively known as the "protein corona". This corona can significantly modify the properties of NPs by obscuring their inherent surface characteristics [89,90]. Furthermore, it has been established that the duration of exposure in the bloodstream is a critical factor that influences the biomolecular composition of NPs. The novel properties imparted to NPs by the corona also become a primary determinant of their nanotoxicity and/or therapeutic efficacy within the body [91]. To date, considerable research has focused on elucidating the composition and biological implications of the protein corona [92]. A comprehensive understanding of NP–protein interactions is vital for the advancement of targeted delivery systems for nanomaterials in healthcare [93].

Extensive studies have been conducted to investigate the formation of protein coronas on various forms of NPs [94,95]. These investigations have revealed that multiple factors can significantly affect the thickness and composition of protein coronas. These factors include the physicochemical properties of NPs, such as surface chemistry, charge, size, shape, solubility, protein binding affinities, and exposure duration [89,96,97].

3.2.3. Biocompatibility

The concept of biocompatibility is based on the appropriate interaction between a material and its biological environment, characterized by the absence of toxic or immune responses from the treated biomaterial (cell, tissue, or organism) [98,99]. Biocompatibility is often described as the ability of a specific material or device to be compatible with living tissue or organisms. It is achieved when the interaction between the nanomaterial and the host does not lead to undesirable outcomes, such as oxidative stress, Deoxyribonucleic Acid (DNA) damage, mutagenesis, or apoptosis [100,101]. Cytotoxicity is generally linked to adverse effects on a specific cell line and is typically assessed first through specific in vitro tests, followed by in vivo evaluations. In practice, cytotoxicity and biocompatibility are influenced by several factors, including the inherent physicochemical properties of the NPs and their delivery methods within the body [102].

There are surface modification methods available for NPs aimed at optimizing their biocompatibility. The surfaces of most NPs can be functionalized with polymer linkers, hybridized DNA, proteins, cell membranes, or inorganic chemicals such as metals and ceramic coatings [103]. However, the role of surface coatings remains unclear to date [104]. Improving the specific interactions of different coatings and their delivery mechanisms is still a work in progress [72,105].

3.2.4. Stability

The term "stability of NPs" refers to the preservation of specific properties of a nanostructure, such as size, surface chemistry, aggregation, and shape. This stability is maintained only for a limited period, as all nanostructures are intrinsically thermodynamically and energetically unfavorable compared to larger structures with the same chemical composition [106].

The nature and concentration of reducing and stabilizing agents significantly influence the size distribution and shape of NPs during their synthesis, largely determining their functional properties. Widoniak et al. [107] described the preparation of colloidal silver solutions with average sizes ranging from 1 nm to 6 μm and various shapes (spheres, plates, needles, or sheets) through the reduction of silver ions using different reducing agents and various polymeric stabilizing agents. The control of nanoparticle shape and size distribution can also be achieved by carefully adjusting other experimental conditions, such as temperature, stirring speed, agitation, and reaction time [108]. Some studies have indicated that the stability of NPs could be preserved long-term when stored under standard conditions: at room temperature, protected from light, and shielded from humidity [109,110].

Surface modification of NPs can also effectively maintain their stability [111,112]. Encapsulation provides complex protection for the payload core and allows for modulation of cytotoxicity, as demonstrated by the work of Lv et al. [113]. The encapsulation of therapeutic gold NPs in hydrogel nanospheres led to increased stability and enhanced cellular uptake efficiency while significantly reducing oxidative stress levels in mesenchymal stem cells.

3.3. *Magnetic Properties*

3.3.1. Targeting

The majority of NPs exhibit magnetic properties, providing significant advantages. These properties allow for selective attachment to functional molecules, impart magnetic characteristics to the target, and facilitate manipulation and transportation to a desired location through the control of a magnetic field produced by an electromagnet or permanent magnet. Magnetic nanoparticle carriers consist of three functional components: a magnetic core, a surface coating, and a functionalized outer layer [114].

Targeting capability is a crucial aspect of various nanoparticle applications. Specifically, metallic NPs can direct therapeutic agents to specific sites within the body, thereby enhancing the efficacy of medical treatments [115].

3.3.2. Optical Properties

The optical properties of metallic NPs are largely influenced by the collective excitation of conduction electrons when interacting with electromagnetic radiation. This behavior is particularly evident in NPs made of gold, silver, and copper, owing to their conduction electron availability. When the electric field from the incident radiation interacts with these particles, it generates an electric dipole, leading to a force that counteracts this effect at a specific resonance frequency. Notably, the optical characteristics are influenced by the size, shape, and composition of the NPs [116–118].

4. Categories of Nanoparticles in Dental Applications

4.1. Antimicrobial Nanoparticles

4.1.1. Types (e.g., Silver, Zinc Oxide)

Antimicrobial NPs are increasingly utilized in dentistry to combat microbial infections. The most prominent types include the following:

- AgNPs: These are widely recognized for their broad-spectrum antimicrobial activity. They disrupt bacterial cell walls, interfere with metabolic processes, and can generate reactive oxygen species (ROS) that are harmful to microbial cells [119,120].
- ZnO NPs: ZnO NPs exhibit antibacterial properties by producing ROS and direct interaction with bacterial membranes, leading to cell death [121,122]. They also promote wound healing and have biocompatibility, making them suitable for dental applications.

4.1.2. Mechanisms of Action and Applications in Preventing Infections

The mechanisms of action for antimicrobial NPs typically involve the following:

- Membrane disruption: antimicrobial NPs can bind to the microbial cell membrane, leading to structural damage and eventual cell lysis.
- ROS: NPs like silver and ZnO can induce oxidative stress within microbial cells, resulting in DNA damage and apoptosis [123,124].

In dental applications, antimicrobial NPs are employed in various products:

- Dental cements: incorporation of AgNPs and ZnO NPs enhances the antimicrobial properties of dental cements, reducing the risk of secondary infections.
- Composite resins: these NPs are added to restorative materials to prevent biofilm formation on dental surfaces [125,126].

4.2. Therapeutic Nanoparticles

Drug Delivery Systems and Their Role in Pain Management and Healing

Therapeutic NPs play a crucial role in the targeted delivery of drugs, improving the effectiveness and reducing the side effects of treatments. Key features include the following:

- Enhanced bioavailability: NPs improve the solubility and stability of therapeutic agents, allowing for lower doses and minimizing systemic toxicity [127].
- Controlled release: NPs can be engineered to release drugs at a controlled rate, providing sustained therapeutic effects over time [128].
- Pain management: for dental procedures, therapeutic NPs can deliver analgesics directly to the site of pain, enhancing pain relief while minimizing the need for systemic medication [129].

4.3. Material Property-Improving Nanoparticles

Enhancements in Mechanical Strength, Wear Resistance, and Aesthetics of Dental Materials

NPs are incorporated into dental materials to improve their physical properties:

- Mechanical strength: NPs such as silica and alumina can significantly enhance the mechanical strength of composite resins and dental cements, making them more durable under occlusal forces [130].
- Wear resistance: the addition of ceramic NPs improves the wear resistance of restorative materials, prolonging their lifespan and maintaining their functionality [131].
- Aesthetics: NPs can improve the optical properties of dental materials, providing better color matching and translucency, which is critical for aesthetic restorations [132,133].

5. Current Applications and Innovations

5.1. Dental Products Incorporating Nanoparticles

5.1.1. Dental Composites

Modern dental composites integrate NPs like nanosilica to improve filler content, which enhances wear resistance, reduces polymerization shrinkage, and increases the composite's aesthetic quality and mechanical durability. These modifications result in more durable restorations, with reduced wear and cracking over time [134]. The incorporation of NPs also optimizes translucency, allowing composites to better mimic the appearance of natural teeth and improving their longevity and color stability [134,135]. In recent years, the integration of NPs into dental materials has significantly enhanced their mechanical properties. For instance, the incorporation of graphene platelets into epoxy nanocomposites has led to an increase in Young's modulus and an improvement in tensile strength compared to pristine epoxy. Similarly, the addition of nanosilica particles to cementitious composites reinforced with polyvinyl alcohol fibers has resulted in a 38% increase in tensile strength. These findings underscore the substantial impact of NPs on enhancing the mechanical performance of dental materials [136,137].

5.1.2. Antimicrobial Agents

AgNPs are commonly incorporated into dental materials like sealants, adhesives, and cements due to their antimicrobial capabilities, which help prevent biofilm formation, reduce bacterial colonization, and mitigate the risk of secondary infections [135,138]. These materials release Ag ions that interact with bacterial membranes, leading to cellular damage and limiting microbial growth [139]. ZnO NPs, another common antimicrobial nanoparticle, exhibit ROS generation, directly inhibiting bacterial activity and offering biocompatibility suitable for dental applications [140,141].

5.1.3. Endodontic Materials

In endodontics, NPs such as ZnO NPs and AgNPs are used in root canal sealers to enhance antibacterial effects, reduce leakage, and improve sealing properties essential for successful long-term outcomes. ZnO NPs, in particular, are known to target endodontic pathogens like *Enterococcus faecalis*, offering improved healing post procedure. The addition of ZnO NPs into root canal sealers has been shown to enhance antibacterial activity while maintaining biocompatibility with surrounding tissues, which helps in the prevention of reinfection and improves the overall success rate of endodontic treatments [142,143].

5.1.4. Dental Implants

Nanotechnology has significantly advanced dental implant surfaces, with nanocoatings applied to titanium and other metals to promote osseointegration and reduce infection risks. Nanostructured surfaces on implants enhance cell attachment, bone growth, and implant stability, thereby improving overall success rates and reducing healing times [144]. Titanium dioxide (TiO_2) NPs have shown promising results in creating antibacterial surfaces, reducing biofilm formation, and enhancing biocompatibility [145,146]. These modifications contribute to a faster healing process and lower rejection rates. The introduction of nanostructures to implant surfaces has been found to increase surface roughness at the nanoscale level, facilitating improved bone formation and promoting direct bone–implant contact, which is crucial for implant stability [2–4].

5.1.5. Amalgam

While dental amalgam is traditionally a blend of mercury with silver, tin, copper, and zinc, recent research has explored the addition of NPs like silver–copper and TiO_2 to improve the amalgam's properties. These enhancements have demonstrated improved antimicrobial activity, reduced cytotoxicity, and increased strength, making amalgam restorations safer and more effective. For example, the incorporation of TiO_2 NPs into dental amalgams has demonstrated the potential for reducing mercury content, improving the material's resistance to corrosion, and enhancing its overall durability [147,148].

5.1.6. Glass Ionomer Cements (GICs)

NPs such as chitosan, hydroxyapatite, TiO_2, silicone dioxide (SiO_2), and zirconia (ZrO_2) are added to GICs to enhance aesthetics, bond strength, and antimicrobial properties. In resin-modified GICs, fluoroaluminosilicate NPs are widely used, and silica nanofillers improve wear resistance and curing efficiency. Chitosan NPs increase fluoride release and material strength, while TiO_2 NPs inhibit biofilm formation and enhance physical properties, contributing to better clinical outcomes [149]. The addition of these NPs has shown to improve the overall longevity and functional performance of GICs, making them more versatile in clinical dentistry [2–4].

5.1.7. Dental Prosthetics

NPs are incorporated into dental prosthetics to enhance wear resistance, aesthetic qualities, and surface smoothness, which reduces bacterial adhesion. Nanosized fillers improve the flexural strength and elasticity of prosthetic materials, which is critical for enduring the stresses of mastication. ZrO_2 NPs, in particular, improve toughness and translucency, making them favorable for prosthetic frameworks and crowns [150].

In addition to their known roles, NPs are being explored for other innovative uses in dentistry. For example, NPs have been integrated into repair resins to enhance the mechanical properties of denture repairs, offering improved durability and resistance to fractures [151].

5.1.8. Periodontal Applications

NPs are used in periodontal treatments to facilitate drug delivery for targeted antibacterial action and tissue regeneration. Hydroxyapatite NPs, for example, support bone regrowth and are compatible with periodontal ligament cells, promoting the repair of periodontal defects. Additionally, silver and ZnO NPs serve as local antimicrobials, contributing to infection control and improving healing rates in periodontal surgeries [152]. Research is also exploring the potential of nano-hydroxyapatite in repairing damaged bone tissues and preventing periodontal disease progression [2–4].

5.1.9. Whitening Agents

NPs in whitening products are utilized for their effectiveness and reduced damage to enamel. Carbamide peroxide polymeric NPs enhance stability and efficacy in whitening agents, while nano-hydroxyapatite helps with remineralization and provides lasting whiteness in oral hygiene products. Nano-encapsulated sodium metabisulfite allows for safe, gradual whitening through liposomal enclosures, which reduces enamel erosion and sensitivity [153,154]. This application of NPs has been shown to significantly improve whitening effects while ensuring that enamel integrity is preserved during the treatment process.

5.1.10. Enamel Repair and Remineralization

NPs have shown promising potential in enamel repair and remineralization, offering innovative solutions to prevent and reverse early carious lesions. Among these, bioactive glass NPs have gained attention for their ability to release ions, such as calcium and phosphate, which are essential for remineralizing enamel surfaces. A recent study by Raszewski et al. [155] demonstrated the efficacy of bioactive glass-infused gels in restoring enamel mineral content in vitro, highlighting their potential as a non-invasive treatment option for enamel demineralization. These NPs not only promote the regeneration of lost mineral structure but also enhance the surface properties of enamel, providing added protection against future demineralization. Such advancements position bioactive glass NPs as a pivotal component of therapeutic strategies aimed at strengthening dental tissues and improving overall oral health.

5.2. Highlight Recent Advancements in Nanoparticle Technology

5.2.1. Targeted Drug Delivery

Advances in targeted drug delivery systems using NPs allow for the localized release of therapeutics in dental applications, enhancing treatment efficacy while minimizing systemic side effects [156].

5.2.2. Smart Biomaterials

The development of smart biomaterials that respond to environmental stimuli (e.g., pH, temperature) is a significant innovation. These materials can release drugs in a controlled manner when triggered by specific conditions in the oral environment [157].

5.2.3. Enhanced Imaging Techniques

NPs, particularly quantum dots and gold NPs, are being utilized to improve imaging techniques such as fluorescence imaging, enhancing the visualization of dental tissues and the detection of oral diseases [158].

6. Future Directions and Challenges

6.1. Potential Future Applications of Nanoparticles in Dentistry

6.1.1. Regenerative Dentistry

NPs may play a crucial role in regenerative approaches, such as stem cell therapy and tissue engineering, to promote the regeneration of dental tissues and structures [159].

6.1.2. Personalized Dental Care

The integration of NPs in diagnostic tools could lead to personalized treatment plans based on the specific needs of patients, enhancing the effectiveness of dental care [160].

6.1.3. Nanoparticle-Based Vaccines

Research into nanoparticle-based vaccines targeting oral diseases (e.g., periodontal disease, dental caries) is promising and could revolutionize preventive dental care [161].

The incorporation of NPs in dental applications presents a wealth of opportunities to enhance therapeutic effectiveness, improve material properties, and develop innovative solutions for oral health. However, ongoing research is essential to address potential challenges, including biocompatibility, safety, and regulatory hurdles [2–4].

6.2. Consideration of Risks and Regulatory Challenges in Their Use

The cytotoxic effects of nanomaterials are primarily observed when they exist in free form—meaning they are not integrated into the material's structure—or when they are released into the surrounding tissue. In dentistry, this cytotoxicity can be strategically utilized against bacterial and fungal pathogens that are resistant to traditional antibiotic treatments. By modifying dental sealers with NPs to impart antimicrobial properties, their effectiveness can be significantly enhanced through molecular interactions [2–4].

In a study evaluating three different commercial endodontic sealants modified with nanostructured silver vanadate ($AgVO_3$), cytotoxicity assessments revealed that two sealants, Sealer 26 and Endometasone N, exhibited cytotoxic effects both in their pure forms and when combined with NPs. Conversely, the AH Plus sealant demonstrated cytotoxicity only when used in conjunction with the nanomaterial, impacting the viability of human gingival fibroblasts (HGF) as the sole tested material [162]. The mechanisms of cytotoxicity associated with AgNPs on HGF have been further explored, illustrating the dual nature of their therapeutic and toxic potentials [107,163].

While NPs offer a plethora of therapeutic applications, they also pose risks related to cytotoxicity and potential immune responses. For example, magnetic NPs (MNPs) can enhance the translucency and abrasion resistance of dental composites without introducing significant risk. However, using MNPs in free form, especially in dental adhesives aimed at providing anti-biofilm properties or in drug delivery systems for caries and periodontal disease management, raises concerns about triggering undesirable immune responses [2,164].

Striking a balance between the benefits and risks associated with nanoparticle use presents a challenge for clinicians. Despite their cytotoxic potential, MNPs often exhibit cellular activity that is targeted, which helps to mitigate systemic toxicity. Consequently, in recent years, the application of NPs and magnetic forces in dentistry has gained traction for delivering drugs aimed at the prevention and treatment of dental diseases [165,166].

It is essential for future research to focus on overcoming these challenges by developing standardized testing protocols to evaluate the long-term performance and safety of nanoparticle-based materials in clinical settings. Additionally, innovations in manufacturing technologies will be crucial for producing NPs at scale while maintaining their unique properties [167]. Future studies should also explore the synergistic effects of combining different types of NPs (e.g., antimicrobial and material-enhancing) in a single product. Finally, the development of regulatory frameworks to ensure the safe incorporation of NPs in dental products will be vital for their successful commercialization. As these advancements occur, the research anticipates that NPs will play an increasingly significant role in revolutionizing dental care and treatment.

7. Conclusions

In conclusion, NPs represent a promising avenue for innovation in dental materials and therapeutics. Their unique properties enable the development of enhanced antimicrobial agents, effective drug delivery systems, and improved material performance. However, the risks associated with cytotoxicity and stability must be carefully managed to ensure safe and effective use. Ongoing research is essential to fully understand and optimize the applications of NPs in dentistry, balancing their benefits against potential health risks. As the field advances, the integration of NPs into clinical practice will likely revolutionize approaches to dental care and treatment.

Author Contributions: Conceptualization, R.B. and Y.H.; methodology, R.B.; software, Z.D., C.L. and N.K.; validation, Y.H., C.E.C.-S. and R.B.; formal analysis, Y.H., Z.D., R.B. and C.E.C.-S.; investigation, R.B., Y.H., C.L. and T.B.A.; resources, R.B.; data curation, R.B.; writing—original draft preparation, R.B.; writing—review and editing, Y.H., C.E.C.-S., N.K. and T.B.A.; visualization, T.B.A., R.B. and Z.D.; supervision, Y.H.; project administration, Y.H.; funding acquisition, R.B. and Y.H. All authors have read and agreed to the published version of the manuscript.

Funding: This research received no external funding.

Institutional Review Board Statement: Not applicable.

Informed Consent Statement: Not applicable.

Data Availability Statement: Data sharing is not applicable. No new data were created or analyzed in this study.

Conflicts of Interest: The authors declare no conflicts of interest.

References

1. Yakop, F.; Abd Ghafar, S.A.; Yong, Y.K.; Saiful Yazan, L.; Mohamad Hanafiah, R.; Lim, V.; Eshak, Z. Silver nanoparticles Clinacanthus Nutans leaves extract induced apoptosis towards oral squamous cell carcinoma cell lines. *Artif. Cells Nanomed. Biotechnol.* **2018**, *46*, 131–139. [CrossRef]
2. Gronwald, B.; Kozłowska, L.; Kijak, K.; Lietz-Kijak, D.; Skomro, P.; Gronwald, K.; Gronwald, H. Nanoparticles in Dentistry—Current Literature Review. *Coatings* **2023**, *13*, 102. [CrossRef]
3. Moraes, G.; Zambom, C.; Siqueira, W.L. Nanoparticles in Dentistry: A Comprehensive Review. *Pharmaceuticals* **2021**, *14*, 752. [CrossRef] [PubMed]
4. Bapat, R.A.; Joshi, C.P.; Bapat, P.; Chaubal, T.V.; Pandurangappa, R.; Jnanendrappa, N.; Gorain, B.; Khurana, S.; Kesharwani, P. The Use of Nanoparticles as Biomaterials in Dentistry. *Drug Discov. Today* **2019**, *24*, 85–98. [CrossRef] [PubMed]
5. Pecci-Lloret, M.P.; Gea-Alcocer, S.; Murcia-Flores, L.; Rodríguez-Lozano, F.J.; Oñate-Sánchez, R.E. Use of Nanoparticles in Regenerative Dentistry: A Systematic Review. *Biomimetics* **2024**, *9*, 243. [CrossRef] [PubMed]
6. Auffan, M.; Rose, J.; Bottero, J.-Y.; Lowry, G.V.; Jolivet, J.-P.; Wiesner, M.R. Towards a Definition of Inorganic Nanoparticles from an Environmental, Health and Safety Perspective. *Nat. Nanotechnol.* **2009**, *4*, 634–641. [CrossRef] [PubMed]
7. McNamara, K.; Tofail, S.A. Nanoparticles in Biomedical Applications. *Adv. Phys. X* **2017**, *2*, 54–88. [CrossRef]
8. Ghadiri, M.; Stokes, R. Nanotechnology in Dentistry: A Comprehensive Review. *Materials* **2020**, *13*, 2700. [CrossRef]
9. An, K.; Somorjai, G.A. Size and Shape Control of Metal Nanoparticles for Reaction Selectivity in Catalysis. *ChemCatChem* **2012**, *4*, 1512–1524. [CrossRef]
10. Arms, L.; Smith, D.W.; Flynn, J.; Palmer, W.; Martin, A.; Woldu, A.; Hua, S. Advantages and Limitations of Current Techniques for Analyzing the Biodistribution of Nanoparticles. *Front. Pharmacol.* **2018**, *9*, 802. [CrossRef]
11. ElSheikh, S.K.; Eid, E.G.; Abdelghany, A.M.; Abdelaziz, D. Physical/Mechanical and Antibacterial Properties of Composite Resin Modified with Selenium Nanoparticles. *BMC Oral Health* **2024**, *24*, 1245. [CrossRef] [PubMed]
12. Mercan, D.A.; Niculescu, A.G.; Grumezescu, A.M. Nanoparticles for Antimicrobial Agents Delivery—An Up-to-Date Review. *Int. J. Mol. Sci.* **2022**, *23*, 13862. [CrossRef] [PubMed]
13. Capuano, N.; Amato, A.; Dell'Annunziata, F.; Giordano, F.; Folliero, V.; Di Spirito, F.; More, P.R.; De Filippis, A.; Martina, S.; Amato, M.; et al. Nanoparticles and Their Antibacterial Application in Endodontics. *Antibiotics* **2023**, *12*, 1690. [CrossRef] [PubMed]
14. Bossù, M.; Saccucci, M.; Salucci, A.; Di Giorgio, G.; Bruni, E.; Uccelletti, D.; Sarto, M.S.; Familiari, G.; Relucenti, M.; Polimeni, A. Enamel Remineralization and Repair Results of Biomimetic Hydroxyapatite Toothpaste on Deciduous Teeth: An Effective Option to Fluoride Toothpaste. *J. Nanobiotechnol.* **2019**, *17*, 17. [CrossRef] [PubMed]
15. Yıldız, C.; Kılıç, E.; Kurt, K.; Özdemir, H.; Korkmaz, A. Nanoparticles for Dental Implant Applications: Enhancing Osseointegration. *Mater. Today Proc.* **2021**, *46*, 3469–3472. [CrossRef]
16. Gutiérrez, M.F.; Alegría-Acevedo, L.F.; Méndez-Bauer, L.; Bermudez, J.; Dávila-Sánchez, A.; Buvinic, S.; Hernández-Moya, N.; Reis, A.; Loguercio, A.D.; Farago, P.V.; et al. Biological, Mechanical, and Adhesive Properties of Universal Adhesives Containing Zinc and Copper Nanoparticles. *J. Dent.* **2019**, *82*, 45–55. [CrossRef]
17. Ekrikaya, S.; Yilmaz, E.; Arslan, S.; Karaaslan, R.; Ildiz, N.; Celik, C.; Ocsoy, I. Dentin Bond Strength and Antimicrobial Activities of Universal Adhesives Containing Silver Nanoparticles Synthesized with *Rosa canina* Extract. *Clin. Oral Investig.* **2023**, *27*, 6891–6902. [CrossRef] [PubMed]

18. Sahu, A.; Pramanik, K.; Mohapatra, A.; Dandapat, S.; Sinha, A.; Patra, S. Nanoparticle-based targeted drug delivery systems: Applications in cancer therapy. *Nanomaterials* **2020**, *10*, 883.
19. Chen, Z.; Zhang, H.; Yang, X.; Zhang, Y.; Zhu, G.; Yang, J.; Zhang, M. Multifunctional nanoparticles for cancer diagnosis and therapy. *Front. Chem.* **2020**, *8*, 212. [CrossRef]
20. Lu, H.; Li, J.; Huang, X.; Yan, H.; Gao, H.; Liu, Y.; Wu, Y.; Chen, Z.; Wang, Y. The application of nanoparticles in oral drug delivery systems. *Int. J. Nanomed.* **2021**, *16*, 1233–1248.
21. Sadeghi, A.; Dastjerdi, R.; Asgarian, A.; Shafiei, M.; Moudi, M.; Shahmoradi, K. Nanotechnology in dentistry: A review of the current literature. *Int. J. Dent.* **2020**, *2020*, 4568019.
22. Ramesh, R.; Babu, R.S.; Manickam, P.; Shyamaladevi, R.; Raja, V.S. Recent advancements in nanoparticle-based drug delivery systems for cancer therapy: A review. *Biotechnol. Rep.* **2020**, *25*, e00424.
23. Christian, P.; Von der Kammer, F.; Baalousha, M.; Hofmann, T. Nanoparticles: Structure, Properties, Preparation and Behaviour in Environmental Media. *Ecotoxicology* **2008**, *17*, 326–343. [CrossRef] [PubMed]
24. Kreuter, J. Nanoparticles—A Historical Perspective. *Int. J. Pharm.* **2007**, *331*, 1–10. [CrossRef] [PubMed]
25. Iavicoli, I.; Leso, V.; Fontana, L. Esposizione a Nanoparticelle nei Laboratori di Ricerca [Nanoparticle Exposure in Research Laboratories]. *G. Ital. Med. Lav. Ergon.* **2019**, *41*, 349–353.
26. Rahman, A.; Ghosh, M. Nanoparticles and Their Applications in Dental Materials: An Overview. *Nanomaterials* **2021**, *11*, 674. [CrossRef]
27. Niu, L.N.; Zhang, Y.; Tsoi, J.K.H.; Matinlinna, J.P. Nanotechnology in Dental Applications. *J. Dent. Res.* **2019**, *98*, 337–347.
28. Sharma, V.K.; Filip, J.; Zboril, R.; Varma, R.S. Natural Inorganic Nanoparticles—Formation, Fate, and Toxicity in the Environment. *Chem. Soc. Rev.* **2015**, *44*, 8410–8423. [CrossRef]
29. Ghosh, A.; Banerjee, S.; Paul, P. A Review on the Use of Nanoparticles for Dental Tissue Regeneration. *J. Biomater. Sci. Polym. Ed.* **2020**, *31*, 797–816.
30. Zhang, J.; Wang, Y.; Zhao, Y.; Chen, H. The Role of Nanoparticles in Restorative Dentistry. *Front. Mater.* **2021**, *8*, 583934.
31. Xu, X.; Liu, Y.; Li, J.; Wang, Y. Recent Advances in Protein-Repellent Adhesives Using Nanotechnology. *J. Adhes. Sci. Technol.* **2021**, *35*, 490–507.
32. Hasan, S. A Review on Nanoparticles: Their Synthesis and Types. *Res. J. Recent Sci.* **2015**, *2277*, 2502.
33. Tuncer, M.; Büyükyılmaz, T.; Tütüncü, M.; Korkmaz, Y. Nanoparticles in Endodontics: Current Trends and Future Directions. *J. Endod.* **2021**, *47*, 1204–1216. [CrossRef]
34. Daraee, H.; Eatemadi, A.; Abbasi, E.; Fekri Aval, S.; Kouhi, M.; Akbarzadeh, A. Application of Gold Nanoparticles in Biomedical and Drug Delivery. *Artif. Cells Nanomed. Biotechnol.* **2016**, *44*, 410–422. [CrossRef] [PubMed]
35. Zhang, Y.; Wang, Y.; Wang, Y.; Yang, X. Nanotechnology for Controlled Drug Delivery in Dental Treatments. *Dent. Mater. J.* **2019**, *38*, 80–86. [CrossRef]
36. Naguib, G.; Maghrabi, A.A.; Mira, A.I.; Mously, H.A.; Hajjaj, M.; Hamed, M.T. Influence of Inorganic Nanoparticles on Dental Materials' Mechanical Properties: A Narrative Review. *BMC Oral Health* **2023**, *23*, 897. [CrossRef] [PubMed]
37. Kim, D.; Shin, K.; Kwon, S.G.; Hyeon, T. Synthesis and Biomedical Applications of Multifunctional Nanoparticles. *Adv. Mater.* **2018**, *30*, e1802309. [CrossRef] [PubMed]
38. Gao, W.; Zhang, Y.; Zhang, Q.; Zhang, L. Nanoparticle-Hydrogel: A Hybrid Biomaterial System for Localized Drug Delivery. *Ann. Biomed. Eng.* **2016**, *44*, 2049–2061. [CrossRef]
39. European Commission. Commission Recommendation (EU) 2022/1089 of 14 June 2022 on Defining Nanomaterials. *Off. J. Eur. Union* **2022**, *L176*, 1–4. Available online: https://eur-lex.europa.eu/legal-content/EN/TXT/?uri=CELEX:32022H0614(01) (accessed on 23 December 2024).
40. Size-Comparison-Bio-Nanoparticles. Size Comparison of Bio-Nanoparticles: Nanometer Scale Comparison and Nanoparticle Size Comparison Nanotechnology Chart Ruler. Nanotechnology Chart Ruler. 2017. Available online: https://www.wichlab.com/nanometer-scale-comparison-nanoparticle-size-comparison-nanotechnology-chart-ruler-2/ (accessed on 23 December 2024).
41. Ghosh, A.; Banerjee, S. Nanoparticles in Medicine: Current Status and Future Directions. *Molecules* **2019**, *24*, 747. [CrossRef]
42. Baran, I.; Alavi, S.; Baghery, A.; Vatanpour, M. Potential Toxicity of Nanoparticles in Dental Applications: A Review. *Int. J. Mol. Sci.* **2020**, *21*, 5003. [CrossRef]
43. Grande, F.; Tucci, P. Titanium Dioxide Nanoparticles: A Risk for Human Health? *Mini Rev. Med. Chem.* **2016**, *16*, 762–769. [CrossRef]
44. Khan, Y.; Ali, S.; Zia, A.; Murtaza, G.; Ali, I.; Khan, M.N. Antimicrobial Nanoparticles in Dental Applications. *Crit. Rev. Microbiol.* **2020**, *46*, 421–438.
45. Bundschuh, M.; Filser, J.; Lüderwald, S.; McKee, M.S.; Metreveli, G.; Schaumann, G.E.; Schulz, R.; Wagner, S. Nanoparticles in the Environment: Where Do We Come from, Where Do We Go To? *Environ. Sci. Eur.* **2018**, *30*, 6. [CrossRef] [PubMed]
46. Meena, K.; Kumar, V.; Kumar, A.; Yadav, A.; Kumar, D. Hydroxyapatite Nanoparticles for Remineralization of Dental Enamel. *Mater. Sci. Eng. C* **2020**, *110*, 110704. [CrossRef]

47. Horikoshi, S.; Serpone, N. Introduction to Nanotechnology. In *Nanotechnology for Environmental Decontamination*; Springer: New York, NY, USA, 2013.
48. Takallu, M.; Mohammadi, M.; Hajikhani, M. Advances in Nanoparticles for Endodontic Regeneration. *J. Nanomater. Dent.* **2024**, *12*, 85–97.
49. Hayat, K.; Malik, A.; Khan, A. Role of nanoparticles in combating oral biofilms. *Int. J. Nanomed.* **2022**, *17*, 521–533.
50. Zhang, N.; Ma, J.; Li, Y. Remineralization of enamel with nanohydroxyapatite and its role in dentistry. *J. Dent. Sci.* **2018**, *13*, 170–180. [CrossRef]
51. Jandt, K.D.; Watts, D.C. Nanotechnology in dentistry: Present and future perspectives. *J. Dent. Res.* **2020**, *99*, 1242–1249.
52. Cao, G. *Nanostructures and Nanomaterials: Synthesis, Properties and Applications*; Imperial College Press: London, UK, 2004.
53. Jain, K.K. Nanobiotechnology and its applications. *Pharm. Nanotechnol.* **2012**, *4*, 215–229.
54. Banerjee, A.; Dutta, K.; Panda, A. Nanomaterials in dentistry: Applications and toxicological risks. *Mater. Sci. Eng. C* **2022**, *125*, 112086. [CrossRef]
55. Kaur, A.; Thombre, R. Nanoparticle-based drug delivery systems in dentistry. *J. Cont. Release* **2021**, *330*, 42–57. [CrossRef]
56. AlKahtani, R. Nanotechnology applications in dentistry: A review of recent advances. *Saudi Dent. J.* **2018**, *30*, 107–116. [CrossRef]
57. Bhushan, B. *Springer Handbook of Nanotechnology*, 4th ed.; Springer: Berlin, Germany, 2017. [CrossRef]
58. Altammar, K. A review on nanoparticles: Characteristics, synthesis, applications, and challenges. *Front Microbiol.* **2023**, *14*, 1155622. [CrossRef] [PubMed]
59. La, D.D.; Truong, T.N.; Pham, T.Q.; Vo, H.T.; Tran, N.T.; Nguyen, T.A.; Nguyen, T.B.; Pham, H.D.; Tran, V.T.; Nguyen, D.D. Scalable Fabrication of Modified Graphene Nanoplatelets as an Effective Additive for Engine Lubricant Oil. *Nanomaterials* **2020**, *10*, 877. [CrossRef]
60. Subhan, M.A.; Alharthi, A.I.; Kumar, M.; Bhowmik, S. Nanoparticle fabrication using various approaches. *Adv. Colloid Sci.* **2022**, *83*, 69–94.
61. Baig, U.; Kamal, S.; Gondal, M.A. Top-down and bottom-up approaches for the synthesis of nanomaterials: A review. *Nanomater. Sci. Eng.* **2021**, *10*, 224–230.
62. Ahmed, K.; Rashid, A.; Numan, A. Recent trends in nanoparticle synthesis via top-down approaches. *J. Nanotechnol.* **2021**, *22*, 331–345.
63. Kalaiselvan, S.; Malek, M.; Al-Abed, S. Advances in top-down nanoparticle fabrication techniques: Applications and challenges. *Int. J. Nanotechnol.* **2020**, *17*, 101–110.
64. Sinha, R.; Shukla, P.; Singh, A. Top-down nanofabrication techniques: Advancements and applications. *J. Appl. Nanotechnol.* **2022**, *13*, 58–75.
65. Vijayaram, T.R.; Sundaresan, R.; Shanmugam, K. Nanoparticle synthesis: Bottom-up approaches and their advantages in nanotechnology. *J. Nanomater. Res.* **2023**, *18*, 114–126.
66. Sonavane, G.; Tomoda, K.; Makino, K. Biodistribution of colloidal gold nanoparticles after intravenous administration: Effect of particle size. *Colloids Surf. B Biointerfaces* **2008**, *66*, 274–280. [CrossRef] [PubMed]
67. Sarin, H.; Kanevsky, A.S.; Wu, H.; Brimacombe, K.R.; Fung, S.H.; Sousa, A.A.; Auh, S.; Wilson, C.M.; Sharma, K.; Aronova, M.A.; et al. Physiologic upper limit of pore size in the blood-tumor barrier of malignant solid tumors. *J. Transl. Med.* **2009**, *7*, 51. [CrossRef] [PubMed]
68. Longmire, M.; Choyke, P.L.; Kobayashi, H. Clearance properties of nano-sized particles and molecules as imaging agents: Considerations and caveats. *Nanomedicine* **2008**, *3*, 703–717. [CrossRef] [PubMed]
69. Sm, P.; Qin, Y.; Jia, Z.; Wang, Y.; Tian, H. Clearance and biodistribution of nanoparticles in vivo. *Pharm. Biomed. Sci.* **2001**, *15*, 337–343.
70. Xu, M.; Zhao, X.; Huang, Y.; Yang, Y. Size-dependent biodistribution and clearance of nanoparticles. *Int. J. Nanomed.* **2023**, *18*, 2279–2290.
71. Gao, X.; Yin, L.; Zhang, Y.; Yuan, Y.; Wang, Y.; Li, Y. Toxicity of 8 nm and 37 nm silica nanoparticles in murine macrophages. *J. Hazard. Mater.* **2011**, *195*, 228–233.
72. Zein, R.; Sharrouf, W.; Selting, K. Physical Properties of Nanoparticles That Result in Improved Cancer Targeting. *J. Oncol.* **2020**, *2020*, 5194780. [CrossRef]
73. Zein, I.; Mosa, A.; Alkhazaleh, M. Nanoparticle shape and its influence on cellular uptake and internalization. *J. Nanotechnol.* **2020**, *15*, 125–134.
74. Zhang, L.; Chen, K.; Wang, Y. The role of nanoparticle morphology in cellular uptake. *Adv. Drug Deliv. Rev.* **2015**, *95*, 57–67. [CrossRef]
75. Jarai, B.M.; Fromen, C.A. Nanoparticle Internalization Promotes the Survival of Primary Macrophages. *Adv. Nanobiomed Res.* **2022**, *2*, 2100127. [CrossRef] [PubMed]
76. Auclair, K.; Gagné, F. Toxicity of silver nanoparticles: Influence of morphology. *Environ. Sci. Pollut. Res.* **2022**, *29*, 123–134.

77. Champion, J.A.; Mitragotri, S. Role of nanoparticle size, shape, and surface chemistry in oral drug delivery. *Adv. Drug Deliv. Rev.* **2009**, *61*, 1032–1045.
78. Shukla, R.; Cerniglia, G.; Wang, H. Impact of nanoparticle geometry on cellular uptake and cytotoxicity. *Int. J. Nanomed.* **2013**, *8*, 1897–1913.
79. Lu, J.; Liong, M.; Zink, J.I.; Tamanoi, F. Biocompatible silica nanoparticles for cancer therapy. *Small* **2010**, *6*, 1787–1790. [CrossRef] [PubMed]
80. Nemmar, A.; Albarwani, S.; Beegam, S.; Yuvaraju, P.; Yasin, J.; Attoub, S.; Ali, B.H. Amorphous silica nanoparticles impair vascular homeostasis and induce systemic inflammation. *Int. J. Nanomed.* **2014**, *9*, 2779–2789. [CrossRef]
81. Liu, Y.; Hardie, J.; Zhang, X.; Rotello, V.M. Effects of engineered nanoparticles on the innate immune system. *Semin. Immunol.* **2017**, *34*, 25–32. [CrossRef] [PubMed]
82. Greulich, C.; Kittler, S.; Epple, M.; Muhr, G.; Köller, M. Studies on the biocompatibility and the interaction of silver nanoparticles with human mesenchymal stem cells (hMSCs). *Langenbecks Arch. Surg.* **2009**, *394*, 495–502. [CrossRef]
83. Yazdimamaghani, M.; Moos, P.J.; Dobrovolskaia, M.A.; Ghandehari, H. Genotoxicity of amorphous silica nanoparticles: Status and prospects. *Nanomedicine* **2019**, *16*, 106–125. [CrossRef]
84. Nejati, K.; Dadashpour, M.; Gharibi, T.; Mellatyar, H.; Akbarzadeh, A. Biomedical applications of functionalized gold nanoparticles: A review. *J. Cluster Sci.* **2021**, 1–16. [CrossRef]
85. Fröhlich, E. The role of surface charge in the interactions of nanoparticles with biological systems. *Nanotoxicology* **2012**, *6*, 120–130.
86. Gwinn, M.R.; Vallyathan, V. Nanoparticles: Health effects—Pros and cons. *Environ. Health Perspect.* **2006**, *114*, 1818–1825. [CrossRef]
87. Sun, H.; Jia, J.; Jiang, C.; Zhai, S. Gold Nanoparticle-Induced Cell Death and Potential Applications in Nanomedicine. *Int. J. Mol. Sci.* **2018**, *19*, 754. [CrossRef] [PubMed]
88. Kopac, T. The influence of surface charge on the biological interactions of nanoparticles. *J. Nanobiotechnol.* **2021**, *19*, 43.
89. Lundqvist, M.; Stigler, J.; Elia, G.; Dawson, K. The evolution of the protein corona around nanoparticles. *Nat. Nanotechnol.* **2008**, *3*, 392–397.
90. Nel, A.E.; Madler, L.; Velegol, D.; Xia, T.; Hoek, E.M.V.; Somasundaran, P.; Klaessig, F.; Castranova, V. Understanding biophysicochemical interactions at the nano-bio interface. *Nat. Mater.* **2009**, *8*, 543–557. [CrossRef]
91. Yallapu, M.M.; Chauhan, N.; Othman, S.F.; Khalilzad-Sharghi, V.; Ebeling, M.C.; Khan, S.; Jaggi, M.; Chauhan, S.C. Implications of Protein Corona on Physico-Chemical and Biological Properties of Magnetic Nanoparticles. *Biomaterials* **2015**, *46*, 1–12. [CrossRef]
92. Nienhaus, K.; Nienhaus, G.U. Mechanistic Understanding of Protein Corona Formation Around Nanoparticles: Old Puzzles and New Insights. *Small* **2023**, *19*, 2301663. [CrossRef]
93. Ghosh, G.; Panicker, L. Protein–Nanoparticle Interactions and a New Insight. *Soft Matter* **2021**, *17*, 3855–3875. [CrossRef]
94. Del Pino, P.; Pelaz, B.; Zhang, Q.; Maffre, P.; Nienhaus, G.U.; Parak, W.J. Protein Corona Formation Around Nanoparticles–From the Past to the Future. *Mater. Horiz.* **2014**, *1*, 301–313. [CrossRef]
95. García-Álvarez, R.; Hadjidemetriou, M.; Sánchez-Iglesias, A.; Liz-Marzán, L.M.; Kostarelos, K. In Vivo Formation of Protein Corona on Gold Nanoparticles. The Effect of Their Size and Shape. *Nanoscale* **2018**, *10*, 1256–1264. [CrossRef]
96. Lynch, I.; Dawson, K.A. Protein–Nanoparticle Interactions. In *Nano-Enabled Medical Applications*; Elsevier: Amsterdam, The Netherlands, 2020; pp. 231–250.
97. Kurtz-Chalot, A.; Villiers, C.; Pourchez, J.; Boudard, D.; Martini, M.; Marche, P.N.; Cottier, M.; Forest, V. Impact of Silica Nanoparticle Surface Chemistry on Protein Corona Formation and Consequential Interactions with Biological Cells. *Mater. Sci. Eng. C* **2017**, *75*, 16–24. [CrossRef]
98. Black, J. *Biological Performance of Materials: Fundamentals of Biocompatibility*; CRC Press: Boca Raton, FL, USA, 2005.
99. Ratner, B.D. The Biocompatibility of Implant Materials. In *Host Response to Biomaterials*; Academic Press: Cambridge, MA, USA, 2015; pp. 37–51.
100. Kaur, J.; Tikoo, K. Evaluating Cell Specific Cytotoxicity of Differentially Charged Silver Nanoparticles. *Food Chem. Toxicol.* **2013**, *51*, 1–14. [CrossRef]
101. Suresh, A.K.; Pelletier, D.A.; Wang, W.; Morrell-Falvey, J.L.; Gu, B.; Doktycz, M.J. Cytotoxicity Induced by Engineered Silver Nanocrystallites Is Dependent on Surface Coatings and Cell Types. *Langmuir* **2012**, *28*, 2727–2735. [CrossRef] [PubMed]
102. Han, D.W.; Woo, Y.I.; Lee, M.H.; Lee, J.H.; Lee, J.; Park, J.C. In-Vivo and In-Vitro Biocompatibility Evaluations of Silver Nanoparticles with Antimicrobial Activity. *J. Nanosci. Nanotechnol.* **2012**, *12*, 5205–5209. [CrossRef] [PubMed]
103. Sanità, G.; Carrese, B.; Lamberti, A. Nanoparticle Surface Functionalization: How to Improve Biocompatibility and Cellular Internalization. *Front. Mol. Biosci.* **2020**, *7*, 587012. [CrossRef] [PubMed]
104. Malvindi, M.A.; Matteis, V.D.; Galeone, A.; Brunetti, V.; Anyfantis, G.C.; Athanassiou, A.; Cingolani, R.; Pompa, P.P. Toxicity Assessment of Silica Coated Iron Oxide Nanoparticles and Biocompatibility Improvement by Surface Engineering. *PLoS ONE.* **2014**, *9*, e85835. [CrossRef]

105. Kyriakides, T.R.; Raj, A.; Tseng, T.H.; Xiao, H.; Nguyen, R.; Mohammed, F.S.; Halder, S.; Xu, M.; Wu, M.J.; Bao, S.; et al. Biocompatibility of Nanomaterials and Their Immunological Properties. *Biomed. Mater.* **2021**, *16*, 1042005. [CrossRef] [PubMed]
106. Phan, T.T.; Haes, A.J. Stability of nanoparticles: Fundamentals and applications. *Nanomaterials* **2019**, *9*, 30.
107. Widoniak, J.; Eiden-Assmann, S.; Maret, G. Silver Particles Tailoring of Shapes and Sizes. *Colloids Surf. A Physicochem. Eng. Asp.* **2005**, *270*, 340–344. [CrossRef]
108. Pinto, V.V.; Ferreira, M.J.; Silva, R.; Santos, H.A.; Silva, F.; Pereira, C.M. Long-Time Effect on the Stability of Silver Nanoparticles in Aqueous Medium: Effect of the Synthesis and Storage Conditions. *Colloids Surf. A Physicochem. Eng. Asp.* **2010**, *364*, 19–25. [CrossRef]
109. Korshed, P.; Li, L.; Ngo, D.-T.; Wang, T. Effect of Storage Conditions on the Long-Term Stability of Bactericidal Effects for Laser Generated Silver Nanoparticles. *Nanomaterials* **2018**, *8*, 218. [CrossRef]
110. Popa, M.; Pradell, T.; Crespo, D.; Calderón-Moreno, J.M. Stable Silver Colloidal Dispersions Using Short Chain Polyethylene Glycol. *Colloids Surf. Physicochem. Eng. Asp.* **2007**, *303*, 184–190. [CrossRef]
111. Mahato, K.; Nagpal, S.; Shah, M.A.; Srivastava, A.; Maurya, P.K.; Roy, S.; Jaiswal, A.; Singh, R.; Chandra, P. Gold Nanoparticle Surface Engineering Strategies and Their Applications in Biomedicine and Diagnostics. *3 Biotech* **2019**, *9*, 57. [CrossRef]
112. Sani, A.; Cao, C.; Cui, D. Toxicity of Gold Nanoparticles (AuNPs): A Review. *Biochem. Biophys. Rep.* **2021**, *26*, 100991. [CrossRef]
113. Lv, Y.; Yu, C.; Li, X.; Bao, H.; Song, S.; Cao, X.; Lin, H.; Huang, J.; Zhang, Z. ROS-Activatable Nanocomposites for CT Imaging Tracking and Antioxidative Protection of Mesenchymal Stem Cells in Idiopathic Pulmonary Fibrosis Therapy. *J. Control. Release.* **2023**, *357*, 249–263. [CrossRef]
114. Vatta, L.L.; Sanderson, R.D.; Koch, K.R. Magnetic Nanoparticles: Properties and Potential Applications. *Pure Appl. Chem.* **2006**, *78*, 1793–1801. [CrossRef]
115. Mody, V.V.; Siwale, R.; Singh, A.; Mody, H.R. Introduction to Metallic Nanoparticles. *J. Pharm. Bioallied Sci.* **2010**, *2*, 282–289. [CrossRef] [PubMed]
116. Fu, H.B.; Yao, J.N. Size Effects on the Optical Properties of Organic Nanoparticles. *J. Am. Chem. Soc* **2001**, *123*, 1434–1439. [CrossRef]
117. Kelly, K.L.; Coronado, E.; Zhao, L.L.; Schatz, G.C. The Optical Properties of Metal Nanoparticles: The Influence of Size, Shape, and Dielectric Environment. *J. Phys. Chem. B* **2003**, *107*, 668–677. [CrossRef]
118. Scholes, G.D. Controlling the Optical Properties of Inorganic Nanoparticles. *Adv. Funct. Mater.* **2008**, *18*, 1157–1172. [CrossRef]
119. Khan, S.T.; Al-Khedhairy, A.A.; Musarrat, J. ZnO and TiO2 Nanoparticles as Novel Antimicrobial Agents for Oral Hygiene: A Review. *J. Nanopart. Res.* **2015**, *17*, 276. [CrossRef]
120. More, P.R.; Pandit, S.; Filippis, A.; Franci, G.; Mijakovic, I.; Galdiero, M. Silver Nanoparticles: Bactericidal and Mechanistic Approach against Drug Resistant Pathogens. *Microorganisms* **2023**, *11*, 369. [CrossRef] [PubMed]
121. Grenho, L.; Salgado, C.L.; Fernandes, M.H.; Monteiro, F.J.; Ferraz, M.P. Antibacterial Activity and Biocompatibility of Three-Dimensional Nanostructured Porous Granules of Hydroxyapatite and Zinc Oxide Nanoparticles—An In Vitro and In Vivo Study. *Nanotechnology* **2015**, *26*, 315101. [CrossRef] [PubMed]
122. Lallo da Silva, B.; Abuçafy, M.P.; Berbel Manaia, E.; Oshiro Junior, J.A.; Chiari-Andréo, B.G.; Pietro, R.C.R.; Chiavacci, L.A. Relationship between Structure and Antimicrobial Activity of Zinc Oxide Nanoparticles: An Overview. *Int. J. Nanomed.* **2019**, *14*, 9395–9410. [CrossRef]
123. Liu, X.; Lu, B.; Fu, J.; Zhu, X.; Song, E.; Song, Y. Amorphous Silica Nanoparticles Induce Inflammation via Activation of NLRP3 Inflammasome and HMGB1/TLR4/MYD88/NF-kB Signaling Pathway in HUVEC Cells. *J. Hazard. Mater.* **2021**, *404*, 124050. [CrossRef]
124. Gharpure, S.; Ankamwar, B. Synthesis and Antimicrobial Properties of Zinc Oxide Nanoparticles. *J. Nanosci. Nanotechnol.* **2020**, *20*, 5977–5996. [CrossRef] [PubMed]
125. Noronha, V.T.; Paula, A.J.; Durán, G.; Galembeck, A.; Cogo-Müller, K.; Franz-Montan, M.; Durán, N. Silver Nanoparticles in Dentistry. *Dent. Mater.* **2017**, *33*, 1110–1126. [CrossRef] [PubMed]
126. Mahamuni-Badiger, P.P.; Patil, P.M.; Badiger, M.V.; Patel, P.R.; Thorat-Gadgil, B.S.; Pandit, A.; Bohara, R.A. Biofilm Formation to Inhibition: Role of Zinc Oxide-Based Nanoparticles. *Mater. Sci. Eng. C.* **2020**, *108*, 110319. [CrossRef] [PubMed]
127. Wang, Y.; Pi, C.; Feng, X.; Hou, Y.; Zhao, L.; Wei, Y. The Influence of Nanoparticle Properties on Oral Bioavailability of Drugs. *Int. J. Nanomed.* **2020**, *15*, 6295–6310. [CrossRef]
128. Wang, J.J.; Sanderson, B.J.S.; Wang, H. Cytotoxicity and Genotoxicity of Ultrafine Crystalline SiO_2 Particulate in Cultured Human Lymphoblastoid Cells. *Environ. Mol. Mutagen.* **2007**, *48*, 151–157. [CrossRef] [PubMed]
129. Khan, I.; Saeed, K.; Khan, I. Nanoparticles: Properties, Applications and Toxicities. *Arab. J. Chem.* **2019**, *12*, 908–931. [CrossRef]
130. Priyadarsini, S.; Mukherjee, S.; Mishra, M. Nanoparticles Used in Dentistry: A Review. *J. Oral Biol. Craniofac. Res.* **2018**, *8*, 58–67. [CrossRef]
131. Yesil, Z.D.; Alapati, S.; Johnston, W.; Seghi, R.R. Evaluation of the Wear Resistance of New Nanocomposite Resin Restorative Materials. *J. Prosthet. Dent.* **2008**, *99*, 435–443. [CrossRef] [PubMed]

132. Silikas, N.; Masouras, K.; Satterthwaite, J.; Watts, D.C. Effect of Nanofillers on Adhesive and Aesthetic Properties of Dental Resin-Composites. *Int. J. Nano Biomater.* **2007**, *1*, 116–127. [CrossRef]
133. De Souza, G.M. Nanoparticles in Restorative Materials. In *Nanotechnology in Endodontics: Current and Potential Clinical Applications*; Springer: Berlin/Heidelberg, Germany, 2015; pp. 139–171.
134. Jongrungsomran, S.; Pissuwan, D.; Yavirach, A.; Rungsiyakull, C.; Rungsiyakull, P. The Integration of Gold Nanoparticles into Dental Biomaterials as a Novel Approach for Clinical Advancement: A Narrative Review. *J. Funct. Biomater.* **2024**, *15*, 291. [CrossRef] [PubMed]
135. Subramani, K.; Ahmed, W. *Nanobiomaterials in Clinical Dentistry*; Elsevier: Amsterdam, The Netherlands, 2012.
136. Yusuf, J.; Sapuan, S.M.; Rashid, U.; Ilyas, R.A.; Hassan, M.R. Thermal, Mechanical, Thermo-Mechanical, and Morphological Properties of Graphene Nanoplatelets Reinforced Green Epoxy Nanocomposites. *Polym. Compos.* **2024**, *45*, 1998–2011. [CrossRef]
137. Gkaliou, K. Developing Nanocomposites with Highly Aligned Nanoscale Reinforcement. Doctoral Dissertation, Cardiff University, Cardiff, Wales, 2021.
138. Mallineni, S.K.; Sakhamuri, S.; Kotha, S.L.; AlAsmari, A.R.G.M.; AlJefri, G.H.; Almotawah, F.N.; Mallineni, S.; Sajja, R. Silver Nanoparticles in Dental Applications: A Descriptive Review. *Bioengineering* **2023**, *10*, 327. [CrossRef] [PubMed]
139. Radzig, M.A.; Nadtochenko, V.A.; Koksharova, O.A.; Kiwi, J.; Lipasova, V.A.; Khmel, I.A. Antibacterial Effects of Silver Nanoparticles on Gram-Negative Bacteria: Influence on the Growth and Biofilm Formation, Mechanisms of Action. *Colloids Surf. B Biointerfaces* **2013**, *102*, 300–306. [CrossRef]
140. Mandal, A.K.; Katuwal, S.; Tettey, F.; Gupta, A.; Bhattarai, S.; Jaisi, S.; Parajuli, N. Current Research on Zinc Oxide Nanoparticles: Synthesis, Characterization, and Biomedical Applications. *Nanomaterials* **2022**, *12*, 3066. [CrossRef] [PubMed]
141. Mishra, P.K.; Mishra, H.; Ekielski, A.; Talegaonkar, S.; Vaidya, B. Zinc Oxide Nanoparticles: A Promising Nanomaterial for Biomedical Applications. *Drug Discovery Today* **2017**, *22*, 1825–1834. [CrossRef] [PubMed]
142. Samiei, M.; Farjami, A.; Dizaj, S.M.; Lotfipour, F. Nanoparticles for Antimicrobial Purposes in Endodontics: A Systematic Review of In Vitro Studies. *Mater. Sci. Eng. C.* **2016**, *58*, 1269–1278. [CrossRef] [PubMed]
143. Ibrahim, A.I.O.; Petrik, L.; Moodley, D.S.; Patel, N. Use of Antibacterial Nanoparticles in Endodontics. *S. Afr. Dent. J.* **2017**, *72*, 105–112.
144. Parnia, F.; Yazdani, J.; Javaherzadeh, V.; Dizaj, S.M. Overview of Nanoparticle Coating of Dental Implants for Enhanced Osseointegration and Antimicrobial Purposes. *J. Pharm. Pharm. Sci.* **2017**, *20*, 148–160. [CrossRef]
145. Tomsia, A.P.; Launey, M.E.; Lee, J.S.; Mankani, M.H.; Wegst, U.G.; Saiz, E. Nanotechnology Approaches for Better Dental Implants. *Int. J. Oral Maxillofac. Implants* **2011**, *26*, 25.
146. Thomas, B.; Ramesh, A. Nanotechnology in Dental Implantology. In *Nanomaterials in Dental Medicine*; Springer Nature Singapore: Singapore, 2023; pp. 159–175.
147. Tolou, N.B.; Fathi, M.H.; Monshi, A.; Mortazavi, V.S.; Shirani, F.; Mohammadi, M. The Effect of Adding TiO2 Nanoparticles on Dental Amalgam Properties. *Iranian J. Mater. Sci. Eng.* **2013**, *10*, 46–56.
148. Rajih, A.K.; Al-Sultani, K.F.; Al-Kinani, M.A. Mechanical Properties Improvement of Dental Amalgam Using TiO2 and ZnO. *Life Sci. J.* **2015**, *12*, 86–90.
149. Amin, F.; Rahman, S.; Khurshid, Z.; Zafar, M.S.; Sefat, F.; Kumar, N. Effect of Nanostructures on the Properties of Glass Ionomer Dental Restoratives/Cements: A Comprehensive Narrative Review. *Materials* **2021**, *14*, 6260. [CrossRef] [PubMed]
150. Nikkerdar, N.; Golshah, A.; Mobarakeh, M.S.; Fallahnia, N.; Azizie, B.; Shoohanizad, E. Recent Progress in Application of Zirconium Oxide in Dentistry. *J. Med. Pharm. Chem. Res.* **2024**, *6*, 1042–1071.
151. Basmacı, F.; Avukat, E.N.; Akay, C.; Aykent, F. Effect of Graphene Oxide Incorporation on the Strength of Denture Repair Resin. *ECS J. Solid State Sci. Technol.* **2024**, *13*, 061004. [CrossRef]
152. Kachoei, M.; Divband, B.; Tabriz, F.D.; Helali, Z.N.; Esmailzadeh, M. A Comparative Study of Antibacterial Effects of Mouthwashes Containing Ag/ZnO or ZnO Nanoparticles with Chlorhexidine and Investigation of Their Cytotoxicity. *Nanomed. J.* **2018**, *5*, 102–110.
153. Lima, F.V.; Mendes, C.; Zanetti-Ramos, B.G.; Nandi, J.K.; Cardoso, S.G.; Bernardon, J.K.; Silva, M.A.S. Carbamide Peroxide Nanoparticles for Dental Whitening Application: Characterization, Stability and In Vivo/In Situ Evaluation. *Colloids Surf. B Biointerfaces* **2019**, *179*, 326–333. [CrossRef] [PubMed]
154. Shang, R.; Kunzelmann, K.-H. Biomimetic Tooth-Whitening Effect of Hydroxyapatite-Containing Mouthrinses after Long-Term Simulated Oral Rinsing. *Am. J. Dent.* **2021**, *34*, 307–312.
155. Raszewski, Z.; Chojnacka, K.; Mikulewicz, M. Investigating Bioactive-Glass-Infused Gels for Enamel Remineralization: An In Vitro Study. *J. Funct. Biomater.* **2024**, *15*, 119. [CrossRef] [PubMed]
156. Javidi, M.; Zarei, M.; Naghavi, N.; Mortazavi, M.; Nejat, A.H. Zinc Oxide Nano-Particles as Sealer in Endodontics and Its Sealing Ability. *Contemp. Clin. Dent.* **2014**, *5*, 20–24. [PubMed]
157. Montoya, C.; Roldan, L.; Yu, M.; Valliani, S.; Ta, C.; Yang, M.; Orrego, S. Smart Dental Materials for Antimicrobial Applications. *Bioact. Mater.* **2023**, *24*, 1–19. [CrossRef]

158. Joseph, B. Nanotechnology in Oral and Dental Diagnosis. In *Nanomaterials in Dental Medicine*; Springer Nature Singapore: Singapore, 2023; pp. 33–49.
159. Makvandi, P.; Josic, U.; Delfi, M.; Pinelli, F.; Jahed, V.; Kaya, E.; Ashrafizadeh, M.; Zarepour, A.; Rossi, F.; Zarrabi, A.; et al. Drug Delivery (Nano) Platforms for Oral and Dental Applications: Tissue Regeneration, Infection Control, and Cancer Management. *Adv. Sci.* **2021**, *8*, 2004014. [CrossRef]
160. Alghamdi, M.A.; Fallica, A.N.; Virzì, N.; Kesharwani, P.; Pittalà, V.; Greish, K. The Promise of Nanotechnology in Personalized Medicine. *J. Pers. Med.* **2022**, *12*, 673. [CrossRef] [PubMed]
161. Elizabeth, P.S.; Néstor, M.M.; David, Q.G. Nanoparticles as Dental Drug-Delivery Systems. In *Nanobiomaterials in Clinical Dentistry*; Elsevier: Amsterdam, The Netherlands, 2019; pp. 567–593.
162. Bapat, R.A.; Parolia, A.; Chaubal, T.; Dharamadhikari, S.; Abdulla, A.M.; Sakkir, N.; Kesharwani, P. Recent Update on Potential Cytotoxicity, Biocompatibility, and Preventive Measures of Biomaterials Used in Dentistry. *Biomater. Sci.* **2021**, *9*, 3244–3283. [CrossRef]
163. Sun, X.; Wang, Z.; Zhai, S.; Cheng, Y.; Liu, J.; Liu, B. In Vitro Cytotoxicity of Silver Nanoparticles in Primary Rat Hepatic Stellate Cells. *Mol. Med. Rep.* **2013**, *8*, 1365–1372. [CrossRef] [PubMed]
164. Neagu, C.S.; Cojocariu, A.C.; Zaharia, C.; Romînu, M.; Negruţiu, M.L.; Duma, V.F.; Sinescu, C. The Evaluation of Dental Adhesives Augmented with Magnetic Nanoparticles. In *Advances in 3OM: Opto-Mechatronics, Opto-Mechanics, and Optical Metrology*; SPIE: Bellingham, WA, USA, 2022; Volume 12170, pp. 127–135.
165. Kim, J.E.; Shin, J.Y.; Cho, M.H. Magnetic Nanoparticles: An Update of Application for Drug Delivery and Possible Toxic Effects. *Arch. Toxicol.* **2012**, *86*, 685–700. [CrossRef] [PubMed]
166. Mathew, D.M.; Pushpalatha, C.; Anandakrishna, L. Magnetic Nanoparticles: A Novel Adjunct for Dentistry. *Mater. Today Proc.* **2022**, *50*, 173–180. [CrossRef]
167. Karunakaran, H.; Krithikadatta, J.; Doble, M. Local and Systemic Adverse Effects of Nanoparticles Incorporated in Dental Materials—A Critical Review. *Saudi Dent. J.* **2024**, *36*, 158–167. [CrossRef] [PubMed]

coatings

Article

Self-Etching Ceramic Primer Application Time Influences Durability of Bonding to Lithium Disilicate and Advanced Lithium Disilicate Ceramics

Ashwaq Alsolami [1] and Khaled M. Alzahrani [2,*]

[1] Graduate Prosthodontics, Faculty of Dentistry, Riyadh Elm University, Riyadh 12734, Saudi Arabia; dr.ashwaq.alsolami@gmail.com
[2] Department of Prosthetic Dental Sciences, College of Dentistry, Prince Sattam Bin Abdulaziz University, Al-Kharj 11942, Saudi Arabia
* Correspondence: dr_kmq@hotmail.com or k.alzahrani@psau.edu.sa

Abstract: Background: Monobond Etch & Prime (MEP; Ivoclar) has a material-dependent effect on dental ceramics. This study evaluated the impact of MEP application time on the resin–ceramic shear bond strength (SBS) of lithium disilicate (LD) and advanced lithium disilicate (ALD) ceramics. Methods: LD and ALD discs were distributed into four groups (n = 20): no surface treatment; 5% HF etching followed by the application of a silane-containing primer (S); MEP applied for 40 s and maintained for 20 s; MEP applied for 2 min and maintained for 40 s. Resin–ceramic SBS was evaluated at 24 h and after thermocycling. The statistical analysis of SBS was performed by using the three-way ANOVA and Tukey's multiple comparisons tests. Additional ceramic samples were prepared to assess the surface topography. Results: There was no statistically significant difference in the SBS of the MEP groups (20 s and 2 min) and HF + S. However, only the HF + S and MEP (2 min) groups showed durable adhesion after thermocycling. MEP resulted in less distinct surface alterations compared with HF etching. Conclusion: Increasing the application time of MEP to 2 min resulted in more durable resin–ceramic adhesion compared with the application time (40 s) recommended by the manufacturer. The adhesion of resin–ceramic to LD and ALD did not significantly differ.

Keywords: ceramic; adhesion; etching; priming

1. Introduction

Several ceramic materials satisfy patients' and dentists' increasing expectations for esthetic restorations [1]. With the development of computer-assisted design (CAD)/computer-assisted manufacturing (CAM) technology, which has improved in strength and marginal fit, dental ceramics have become more popular as restorative materials [2]. Currently, zirconia is one of the most common ceramic restorative materials, along with lithium disilicate ceramic (LD), which has higher translucency and lower mechanical strength than zirconia. LD is a glass-matrix ceramic that is composed of a matrix of silicon dioxide, generally silica or quartz, in which additional crystals are incorporated. The glassy matrix alone does not have sufficient strength to resist defects; therefore, to increase the strength and improve the mechanical properties, dispersion strengthening is usually required [3]. Dentsply Sirona has introduced the advanced lithium disilicate (ALD) CAD/CAM blocks, a new material for anterior and posterior CAD/CAM-fabricated restorations. ALD ceramic is characterized by its strength and highly esthetic nature. This type of ceramic has a better wear behavior than lithium disilicate [4]. ALD is composed of lithium disilicate crystals incorporated into a glassy matrix with lithium aluminosilicate [4]. LD is an etchable ceramic because of its glass-matrix content. Hydrofluoric acid (HF) etching is recommended as a surface treatment that can alter the surface morphology, increasing the surface roughness of LD and rendering the LD substrate more favorable for resin adhesion [5]. However, the toxicity of HF is well documented and known [6,7]. In addition, considering the aggressiveness of HF, prolonged etching might deteriorate the mechanical properties of LD restorations. HF etching is followed by a deactivation step and/or a rinsing step. The inadequate cleaning of the etching residue can impair resin–ceramic adhesion. Thus, several alternative ceramic surface treatments have been attempted [8–10]. However, none of these have yielded comparable results to HF etching. Despite the altered surface topography of HF-etched LD, a priming step using a silane-containing primer (S) is essential to promoting bonding between the resin cement and LD surface. Silane is a bifunctional monomer that bonds to the ceramic surface via its silanol group and to resin cements via its methacrylate group [11]. Silane primers contain large amounts of organic solvents; thus, an efficient air-drying step is required to evaporate such solvents [12]. Otherwise, a loose silane layer can be formed on the LD surface, resulting in inferior bonding [13]. This drying step can be controversial for some clinicians, particularly for restorations of complex shape and geometry [14]. In an attempt to avoid clinical errors associated with HF etching and silane priming steps, Monobond Etch & Prime, Ivoclar (MEP), a self-etching ceramic primer, has been introduced as a simplified approach that combines the etching effect and silane priming capacity in one step [13,15]. MEP is composed of a tetrabutylammonium dihydrogen trifluoride (TADF) as an etchant, silane methacrylate, methacrylate phosphate monomer, alcohol and water as solvents, and colorant as a pigment to enhance its visibility [16]. Several studies have assessed the influence of MEP on adhesion to dental ceramics. It was found that the etching (application) duration of MEP might influence the obtained resin–ceramic adhesion. Up until now, the effect of MEP on the surface topography and adhesion of resin to advanced LD has not been investigated yet. This study involved the evaluation of the resin–ceramic shear bond strength (SBS) of LD and advanced LD following MEP surface treatment. The hypothesis were as follows: (1) there would be no significant difference in SBS between lithium disilicate and advanced lithium disilicate materials; (2) there would be no significant differences in SBS between the use of 5% HF followed by the application of S and that of MEP; (3) there would be no effect of increasing the application time of the self-etching ceramic primer (MEP) on SBS; (4) there would be no significant difference in SBS before thermocycling at 24 h and after 10 k cycles of thermocycling.

2. Materials and Methods

2.1. Specimen Preparation

Lithium disilicate and advanced lithium disilicate ceramic blocks (Table 1) were sectioned under running water by using a low-speed prevision cutting saw (Isomet Low Speed, Buehler, Lake Bluff, IL, USA) into a total of 168 smaller blocks (Figure 1). The dimensions of each prepared block were approximately 6 mm × 6 mm × 3 mm. This was followed by an ultrasonic cleaning step for 5 min in distilled water (Sonicer, Yoshida Dental Manufacturing, Tokyo, Japan) and an air-drying step. The IPS e.max CAD and CEREC Tessera polished blocks were sintered and glazed and then embedded in self-curing orthodontic resin. Figure 1 shows a flowchart of the study design. The samples were cleaned, polished under running water by using 600-grit silicon carbide polishing paper and a grinder–polisher machine (MetaServ® 250 grinder-polisher machine; Buehler, Lake Bluff, IL, USA) at 150 rpm, ultrasonically cleaned as described above, and air-dried.

Table 1. The materials utilized and their composition.

Description	Material	Manufacturer	Composition
Lithium disilicate ceramic (LD)	IPS e.max CAD	Ivoclar, Schaan, Liechtenstein	SiO_2: 57–80%; Li_2O: 11–19%; K_2O: 0–13%; P_2O_5: 0–11%; ZrO_2: 0–8%; ZnO: 0–8%; coloring oxides: 0–8%
Advanced lithium disilicate ceramic (ALD)	CEREC Tessera	Dentsply Sirona, Charlotte, NC, USA	$Li_2Si_2O_5$: 90%; Li_3PO_4: 5%; $Li_{0.5}Al_{0.5}Si_{2.5}O_6$ (virgilite): 5%
Ceramic etchant (HF)	IPS ceramic etching gel	Ivoclar, Schaan, Liechtenstein	5% hydrofluoric acid
Self-etching ceramic primer (MEP)	Monobond, Etch & Prime, Ivoclar, Schaan, Liechtenstein.	Ivoclar, Schaan, Liechtenstein	Tetrabutyl ammonium dihydrogen trifluoride, methacrylated phosphoric acid ester, trimethoxysilylpropyl methacrylate, alcohol, and water
Silane-containing universal primer (S)	Monobond N	Ivoclar, Schaan, Liechtenstein	Ethanol, 3-trimethoxysilylpropyl methacrylate,10-MDP, and disulfide acrylate
Dual-cure resin cement	Variolink esthetic DC	Ivoclar, Schaan, Liechtenstein	Urethane dimethacrylate, methacrylate monomers, inorganic fillers (ytterbium trifluoride and spheroid mixed oxide), initiators, stabilizers, and pigments

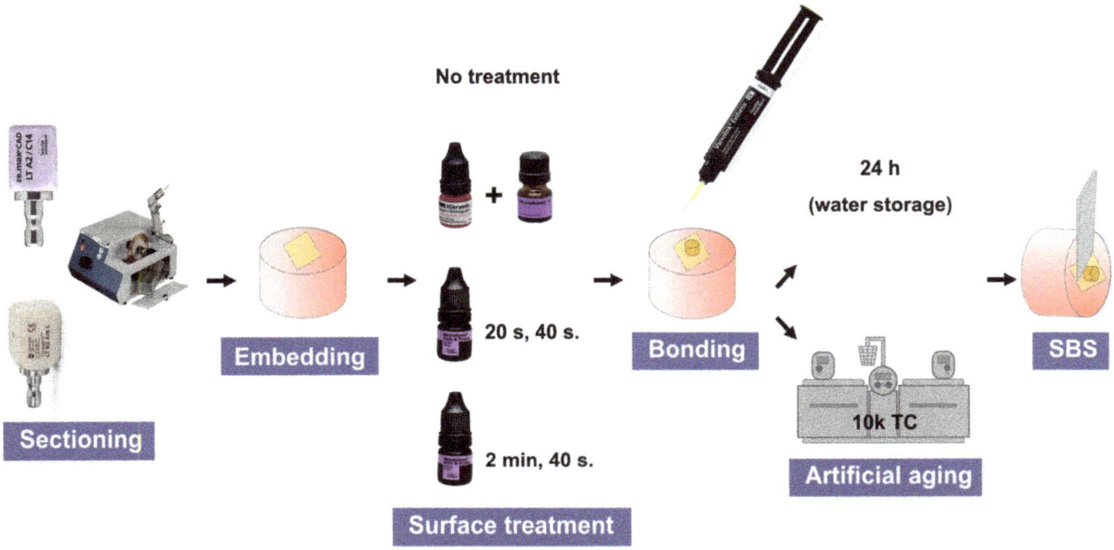

Figure 1. Study design.

2.2. Surface Treatment

The blocks of each ceramic material were allocated to a total of four groups (n = 20 per material) based on the following protocol:

Group 1 (NT): No surface treatment was performed.

Group 2 (HF + S): Surface treatment with 5% HF etching following the manufacturer's instructions. HF was applied for 20 s for LD and 30 s for ALD and rinsed thoroughly. One coat of a silane-containing universal primer (S) was applied to the etched ceramic surface, left to react for 1 min according to the manufacturer's instructions, and then thoroughly air-dried.

Group 3 (MEP 20 s + 40 s): MEP was applied following the manufacturer 's instructions (rubbed for 20 s and left on the surface for 40 s), rinsed thoroughly with water, and then air-dried for 10 s.

Group 4 (MEP 2 min + 40 s): MEP was applied with modifications with respect to the manufacturer's instructions (rubbed for 2 min and left on the surface for 40 s), rinsed thoroughly with water, and then air-dried for 10 s.

A silicon mold was used for building cylinders (3 mm in diameter and 2 mm in height) of a dual-cure resin cement (Variolink esthetic DC, Ivoclar). While the mold was placed, the light curing of the resin cement was performed for 1 min by using a light-curing device (Elipar^{TM} S10; 3M ESPE, St. Paul, MN, USA) operated at an intensity of approximately 1100 mW/cm^2. Additional light curing for 1 min was performed at approximately 1 mm after the silicone mold was carefully removed. A total of 10 specimens of each group were tested for resin–ceramic SBS at 24 h. The other 10 samples of each group were artificially aged through 10k thermal cycles by using a thermocycler machine (THE-1100; SD Mechatronik GmbH, Feldkirchen-Westerham, Germany). Each cycle involved fully dipping the specimens in a distilled water bath at a temperature of 5 $\pm$ 1 °C for 30 s followed by fully dipping them in another water bath at a temperature of 55 $\pm$ 1 °C for 30 s each, with a 5 s transfer time between the two baths.

2.3. Evaluation of Resin–Ceramic SBS

The cross-sectional area of the resin–ceramic interface was determined by using a digital caliper with a surface area range of 3 $\pm$ 0.03 mm. A stainless steel chisel connected to a universal testing machine (Instron 5965; Instron Corporation, Canton, MA, USA) equipped with a load cell of 1 kN was utilized to shear the specimens at the resin–ceramic interface until failure. The cross-speed was 0.5 mm/min. SBS was obtained by diving the load reported at specimen failure by the premeasured resin cement cross-sectional surface area.

2.4. Failure Mode Analysis

All the tested specimens were investigated by using a digital microscope (Hirox Co., Ltd., Tokyo, Japan) at $\times$50 magnification to determine the failure mode patterns that had occurred. The failure modes were classified as adhesive (type 1), with specimen failure (debonding) at the resin–ceramic interface; cohesive in the ceramic (type 2), in which the failure (fracture) had occurred within the ceramic material (substrate) with no adhesive failure; cohesive in the resin cement (type 3), with whole specimen failure (fractured) within the resin cement build-up with no adhesive failure; and mixed failure (type 4), which combined both adhesive and cohesive failure. Samples that presented failure before testing (pre-test failures) were recorded and accounted as 0 MPa in the statistical analysis.

2.5. Statistical Analysis

The SBS values were checked for data normality by using the Shapiro–Wilk statistical test. The effect of the three independent variables identified (ceramic material, surface treatment, and thermocycling), as well as their interactions, on resin–ceramic SBS was statistically analyzed by using three-way analysis of variance (ANOVA) test. Further, Tukey's multiple comparisons test was used for the inter-group comparisons. The distribution of failure modes among the groups was determined by using a chi-squared test.

2.6. Assessment of Surface Topography

Additional ceramic samples were prepared and treated as described for SBS evaluation except for the silane priming step. Ceramic samples were cleaned and then dehydrated in ascending concentrations of ethanol. Dried ceramic samples were fixed on the sample stub with the treated surface facing upwards and then sputter-coated with gold by using a sputter coater (Q150RS; Quorum Technologies, Laughton, East Sussex, UK). The ceramic samples were observed by using a scanning electron microscope (SEM) (EVO LS10; Carl Zeiss SMT, Cambridge, UK) at 10,000$\times$ magnification operated at 15 kV EHT voltage.

3. Results

3.1. SBS

The SBS data showed a normal distribution ($p > 0.05$). The SBS results are illustrated in Figure 2. The results of the three-way ANOVA are illustrated in Table 2.

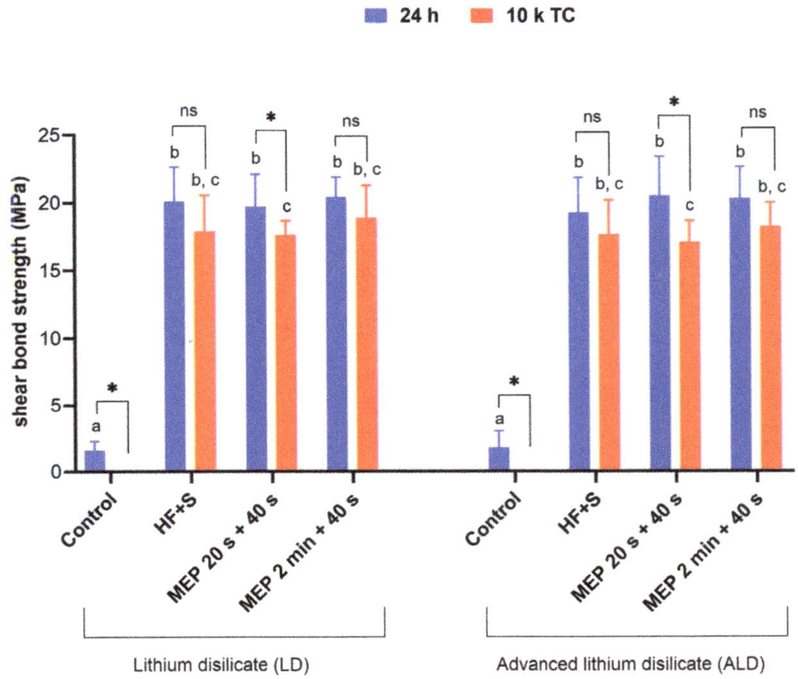

Figure 2. Bar chart of mean and standard deviation of resin–ceramic shear bond strength (SBS) following different surface treatments of lithium disilicate (LD) and advanced lithium disilicate (ALD) ceramics. Different letters indicate statistically significant differences between the groups. * indicates a statistically significant difference between immediate (24 h) and aged (TC) bond strength.

Table 2. Results of three-way analysis of variance (ANOVA) statistical analysis.

Variable	*p*-Value
Ceramic material	0.587
Surface treatment	0.000 *
Thermocycling	0.000 *
Ceramic material × surface treatment	0.818
Ceramic material × thermocycling	0.554
Surface treatment × thermocycling	0.583
Ceramic material × surface treatment × thermocycling	0.720

* indicates a statistically significant difference.

There was no significant difference in SBS between lithium disilicate and advanced lithium disilicate materials in all surface treatment groups at 24 h and after thermocycling (p-value = 0.587). However, there was a statistical difference among different surface treatments and different storage conditions (24 h vs. 10k thermocycling) (p-value = 0.000). The interactions among study factors (material, surface treatment, and thermocycling) were tested, which showed that there were no interactions among the test factors. When comparing surface treatments, both MEP and HF + S (groups 2, 3, and 4) presented significantly increased SBS compared with group 1 (no surface treatment) at 24 h and after thermocycling (p-value = 0.000). However, HF + S showed comparable SBS to MEP in both materials (p-value > 0.05). Group 4 (MEP 2 min + 40 s) showed relatively higher SBS than group 3 (MEP 20 s + 40 s) in both materials, regardless of the evaluation time; however, the difference was not statically significant (p-value > 0.05). Regarding the storage condition, thermocycling significantly decreased SBS for most of the surface treatment groups ($p < 0.05$), but in group 2 and group 4, the difference was not statistically significant ($p > 0.05$) for both materials.

3.2. Failure Mode Analysis

The distribution of failure modes in terms of frequency and percentage is shown in Table 3. Cohesive failure within resin (Figure 3B) and mixed failures (Figure 3C) were the most common among surface treatment groups, whereas the adhesive failure type (Figure 3A) and pre-test failure were most common in the control group (group 1). However, the failure mode cohesive in ceramic was not detected in any of the study samples. Pre-test failures occurred in the control group.

Table 3. The distribution of failure modes among study groups.

Material	Artificial Aging	Surface Treatment	Adhesive (Type 1)	Cohesive in Ceramic (Type 2)	Cohesive in Resin (Type 3)	Mixed (Type 4)	Pre-Test Failure
LD	24 h	No treatment	9 (90%)	0 (0%)	0 (0%)	0 (0%)	1 (10%)
		Hydrofluoric acid	0 (0%)	0 (0%)	8 (80%)	2 (20%)	0 (0%)
		MEP	1 (10%)	0 (0%)	7 (70%)	2 (20%)	0 (0%)
		MEP extended	0 (0%)	0 (0%)	7 (70%)	3 (30%)	0 (0%)
	10k TC	No treatment	0 (0%)	0 (0%)	0 (0%)	0 (0%)	10 (100%)
		Hydrofluoric acid	3 (30%)	0 (0%)	2 (20%)	5 (50%)	0 (0%)
		MEP	1 (10%)	0 (0%)	2 (20%)	7 (70%)	0 (0%)
		MEP extended	0 (0%)	0 (0%)	3 (30%)	7 (70%)	0 (0%)
ALD	24 h	No treatment	7 (70%)	0 (0%)	0 (0%)	0 (0%)	3 (30%)
		Hydrofluoric acid	1 (10%)	0 (0%)	6 (60%)	3 (30%)	0 (0%)
		MEP	0 (0%)	0 (0%)	7 (70%)	3 (30%)	0 (0%)
		MEP extended	0 (0%)	0 (0%)	5 (50%)	5 (50%)	0 (0%)
	10k TC	No treatment	0 (0%)	0 (0%)	0 (0%)	0 (0%)	10 (100%)
		Hydrofluoric acid	0 (0%)	0 (0%)	3 (30%)	7 (70%)	0 (0%)
		MEP	0 (0%)	0 (0%)	3 (30%)	7 (70%)	0 (0%)
		MEP extended	0 (0%)	0 (0%)	6 (60%)	4 (40%)	0 (0%)

LD: lithium disilicate ceramic; ALD: advanced lithium disilicate ceramic; TC: thermocycling.

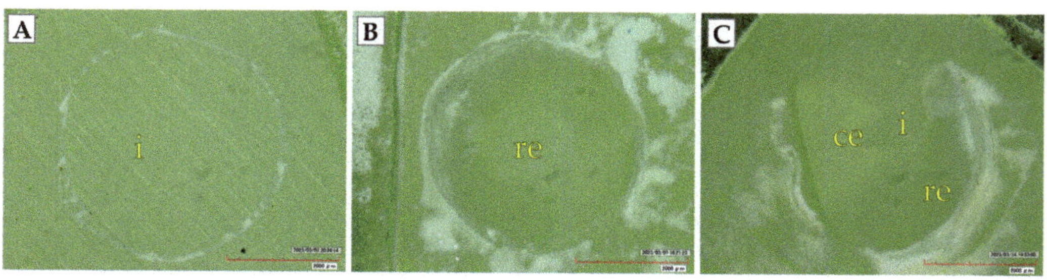

Figure 3. Representative images of different failure modes examined under digital microscope (**A**–**C**): (**A**) adhesive failure, (**B**) cohesive failure within resin, and (**C**) mixed failure. (i) Resin–ceramic interface, (re) resin cement, and (ce) ceramic surface.

3.3. Surface Topography

Application of either HF or MEP yielded distinct surface topographic changes in both the ceramic materials compared with no treatment, which showed a smooth, more homogenous surface texture without any craters or pits in all ceramic materials (Figure 4). A more prominent etching pattern with micro-porosities, grooves, and striations was observed with HF application due to the dissolution of the slightly larger vitreous matrix in both materials compared with the MEP surface treatment in both groups 3 and 4, which showed a milder etching pattern and limited change in roughness. Increasing the application time to 2 min in group 4 (MEP 2 min +40 s) led to an increase in micro-porosities on the ceramic surface and more topographic changes, resulting in a slightly more prominent etching pattern compared with group 3 (MEP 20 s+ 40 s).

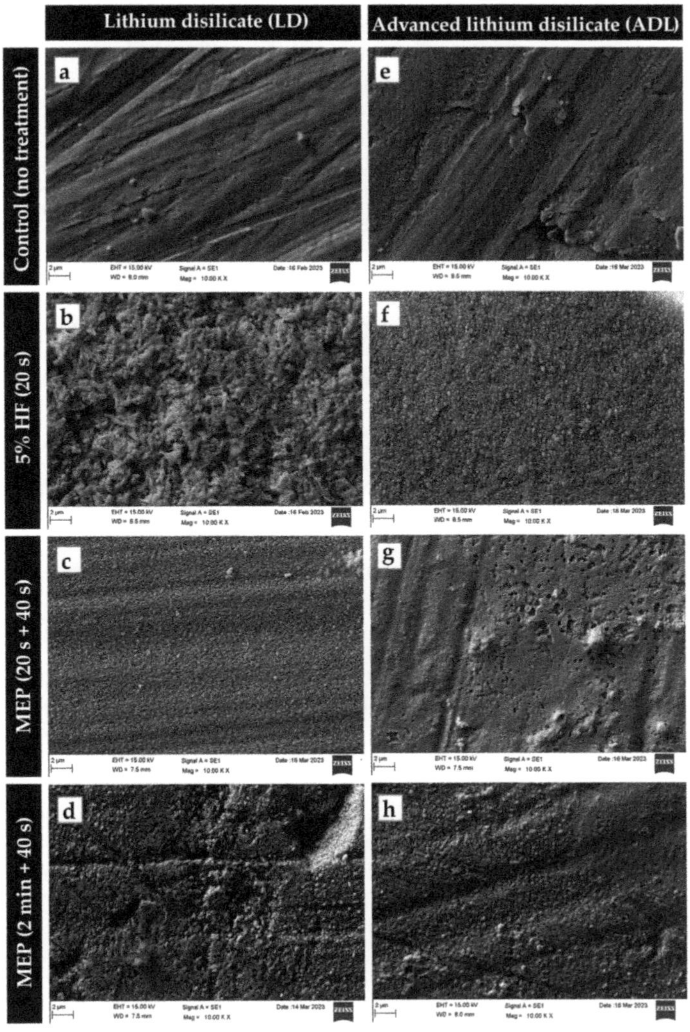

Figure 4. Representative SEM micrographs (10,000×) of lithium disilicate and advanced lithium disilicate: (**a**,**e**) control (no treatment); (**b**,**f**) hydrofluoric acid etching; (**c**,**g**) Monobond Etch & Prime (MEP) applied following the manufacturer's instructions (rubbed for 20 s and left on ceramic surface for 40 s); (**d**,**h**) MEP extended application time (rubbed for 2 min and left on ceramic surface for 40 s).

4. Discussion

This laboratory study tested the impact of MEP and HF on the surface topography and resin–ceramic adhesion of two ceramic materials. Although the correlation between adhesion strength studies and clinical performance of ceramic restorations is still debatable, in vitro studies can provide a reliable evaluation of new adhesive systems and/or techniques using standardized testing conditions that cannot be provided by clinical studies [17]. The SBS test utilized in this study has been extensively used due to uncomplicated specimen preparation. In addition, bonded resin–ceramic specimens do not necessitate additional preparation, i.e., micro-sectioning, which might influence the resin–ceramic interface. However, this method might be associated with inhomogeneous stress distribution at the resin–ceramic interface [18]. A dual-cure resin cement was applied to the ceramic materials in all the study groups. Although constructing the resin cement cylinders might not be the main clinical application of the resin cement used, this study followed the protocol applied in previous similar studies [19,20]. The SBS results of the current study showed no statistically significant differences between the two ceramic materials tested. Thus, the first null hypothesis of this study was accepted. This can be explained by the similar amount of glassy matrix in both ceramic materials which reacts with the ceramic etchant (HF) or MEP. Different surface treatments, including MEP and HF, followed by S priming were evaluated in this study. The second null hypothesis was accepted, since there was no statistically significant difference between the experimental groups (HF +S and MEP). Resin–ceramic adhesion relies on two main aspects. First, micromechanical retention of resin-based cement or inside micro irregularities created by the dissolution of silica matrix. Second, a chemical bond (adhesion) between the ceramic surface and methacrylate-based resin cement must be achieved through a silane priming step. The lack of both aspects well explains the reduced SBS means of the control group [21]. Interestingly, there was no difference between the HF and the MEP groups (groups 2, 3, and 4). Such findings come in accordance with those of other studies that showed comparable resin–ceramic adhesion following MEP application and 5% HF. The application and reaction times of MEP reported in the literature varies between 20 and 60 s for application time and between 40 s and 10 min for reaction time [13]. In the present study, increasing the application time of MEP to 2 min followed by 40 s of reaction time (group 4) did not significantly affect SBS. Therefore, the third null hypothesis was accepted. However, increasing the application time enhanced the durability of the resin–ceramic bond strength compared with that achieved by the 20 s application time, which significantly deteriorated after artificial aging by TC. A previous study reported similar results to our study, with no significant difference in SBS when MEP was applied either following the manufacturer's instructions (20 s of application and 40 s of reaction time) or by increasing the application and reaction times to 40 s and 80 s, respectively [22]. Another study by Alshihri et al. reported no significant differences in SBS between MEP-etched samples with standardized application time (20 s) and various reaction times (40, 80, and 120 s) [23]. However, this contradicts with the results of a previous study in which higher bond strength was associated with MEP ceramic surface treatment compared with HF when the application time of MEP was increased to 60 s and 40 s of reaction time. The variation in the results could relate to the different bond strength tests used, i.e., the macro-shear bond test used in the present study as opposed to the micro-shear test used in the study by Cardenas et al. [24].

Various techniques have been utilized in in vitro studies to simulate the physiological aging of adhesively bonded material, including water storage and thermocycling. Thermocycling involves subjecting samples to alternating temperatures (usually 5 and 55 °C) to simulate thermal changes that occur in the oral cavity. The interfacial stresses generated by temperature alterations can negatively impact resin–ceramic adhesion [25]. A total of 10 k thermal cycles were applied in this study. This might be equal to 1 year of clinical function [26]. Thermocycling caused a significant decrease in SBS in most of the treatment groups, except for HF for both materials and MEP 2 min + 40 s for lithium disilicate. Thus, the fourth null hypothesis was partially rejected. This coincides with the results of a recent study which indicated stable adhesion to MEP-treated glass ceramic after thermocycling [27]. Similarly, Azevedo et al. reported stable SBS with glass-ceramic samples treated with 5% HF for 20 s over 16 months of water storage [28].

The SEM results indicate almost no surface alterations in group 1 (no treatment) (Figure 4) compared with the other groups. This could account for the low mean bond strength and frequent pre-test failure of this group. Group 2 produced the highest dissolution of the matrix, showing a porous and rough surface (Figure 4, HF). This could explain the high SBS and stable bond observed after thermocycling. The MEP surface treatment, according to the SEM analysis, resulted in a milder etching pattern and less distinct topographical changes compared with HF etching. This could be a result of the partial dissolution of the glass-ceramic matrix, which could have affected the durability and resulted in a significant decrease in bond strength after thermocycling [24]. The self-etching ceramic primer uses ammonium polyfluoride and trimethoxypropyl methacrylate for etching and priming, respectively. Ammonium polyfluoride is frequently used to etch glass-based ceramics, and the lower acidity of this etchant could explain the milder etching pattern and limited changes in surface topography observed (Figure 4) [29]. However, more distinct surface topographic changes were noticed with extending the MEP application time to 2 min. This provides a plausible explanation for the durable adhesion of the MEP 2 min group compared with the 20 s group. On the other hand, the phosphate ester monomers of the universal primer or MEP utilized might make a minimal contribution to the adhesion of resin to glass ceramics [30].

Failure mode analyses can offer important information for interpreting bond strength results and identifying weaknesses in testing procedures. Although the incidence of frequent cohesive failures within resin cement might be an indicator of inhomogeneous stress distribution during testing and can be considered a limitation of the testing procedure, omitting the bond strength values associated with cohesive failures can lead to biased conclusions [31]. This information can then be used to increase the reliability of these methodologies to ensure that the results accurately reflect the strengths of these bonds [18]. The present study showed an association between the failure mode and surface treatment, where the highest percentage of adhesive failure was observed in the no-surface-treatment group, while a high percentage of both cohesive failure and mixed failure was observed in the MEP and HF groups. The high frequency of cohesive and mixed failures in the surface treatment groups could be explained by the robust resin–ceramic adhesion following ceramic surface treatment using MEP or HF followed by silane priming. However, the absence of cohesive failure within ceramics could be due to the mechanical properties of the ceramic materials used [32]. Despite the promising results of the MEP-treated groups, the outcome of this study should be interpreted considering the limitations of this study, which include the lack of multi-factorial artificial aging utilizing mechanical loading and exposure to oral biofilm and different beverages. One more limitation of this study is that one resin cement was used for build-up on resin–ceramic specimens. Further in-depth chemical characterizations of MEP-primed ceramic surface are required to verify the possible effect of water rinsing on the hydrolysis of the silane molecules.

5. Conclusions

Within the confines of this in vitro study, it can be concluded that although the self-etching ceramic primer (MEP) resulted in the partial dissolution of the ceramic glass content and a less aggressive ceramic etching pattern compared with HF etching, MEP can be considered an alternative to the protocol including hydrofluoric acid etching and silane priming. Increasing the application time of MEP to 2 min resulted in more durable resin–ceramic adhesion compared with the application time (20 s) recommended by the manufacturer. The strength of the adhesion of resin–ceramic to lithium disilicate and advanced lithium disilicate ceramic did not significantly differ.

Author Contributions: Conceptualization, K.M.A.; methodology, A.A. and K.M.A.; software, A.A.; validation, A.A. and K.M.A.; formal analysis, A.A.; investigation, A.A.; resources, K.M.A.; data curation, A.A.; writing—original draft preparation, A.A.; writing—review and editing, K.M.A.; visualization, A.A.; supervision, K.M.A.; project administration, K.M.A.; funding acquisition, K.M.A. All authors have read and agreed to the published version of the manuscript.

Funding: This study was supported via funding from Prince Sattam bin Abdulaziz University, project number PSAU/2024/R/1446.

Institutional Review Board Statement: Not applicable.

Informed Consent Statement: Not applicable.

Data Availability Statement: The data presented in this study are available upon request from the corresponding author.

Acknowledgments: This study was supported via funding from Prince Sattam bin Abdulaziz University, project number PSAU/2024/R/1446.

Conflicts of Interest: The authors declare no conflicts of interest.

References

1. Denry, I.; Holloway, J.A. Ceramics for Dental Applications: A Review. *Materials* **2010**, *3*, 351–368. [CrossRef]
2. Christensen, G.J. The all-ceramic restoration dilemma: Where are we? *J. Am. Dent. Assoc.* **2011**, *142*, 668–671. [CrossRef] [PubMed]
3. Gracis, S.; Thompson, V.P.; Ferencz, J.L.; Silva, N.R.; Bonfante, E.A. A new classification system for all-ceramic and ceramic-like restorative materials. *Int. J. Prosthodont.* **2015**, *28*, 227–235. [CrossRef]
4. Rosentritt, M.; Schmid, A.; Huber, C.; Strasser, T. In Vitro Mastication Simulation and Wear Test of Virgilite and Advanced Lithium Disilicate Ceramics. *Int. J. Prosthodont.* **2022**, *35*, 770–776. [CrossRef] [PubMed]
5. Awad, M.M.; Alqahtani, H.; Al-Mudahi, A.; Murayshed, M.S.; Alrahlah, A.; Bhandi, S.H. Adhesive Bonding to Computer-aided Design/ Computer-aided Manufacturing Esthetic Dental Materials: An Overview. *J. Contemp. Dent. Pract.* **2017**, *18*, 622–626. [PubMed]
6. Meldrum, M. Toxicology of hydrogen fluoride in relation to major accident hazards. *Regul. Toxicol. Pharmacol.* **1999**, *30*, 110–116. [CrossRef] [PubMed]
7. Lund, K.; Refsnes, M.; Ramis, I.; Dunster, C.; Boe, J.; Schwarze, P.; Skovlund, E.; Kelly, F.; Kongerud, J. Human exposure to hydrogen fluoride induces acute neutrophilic, eicosanoid, and antioxidant changes in nasal lavage fluid. *Inhal. Toxicol.* **2002**, *14*, 119–132. [CrossRef]
8. Klosa, K.; Boesch, I.; Kem, M. Long-term bond of glass ceramic and resin cement: Evaluation of titanium tetrafluoride as an alternative etching agent for lithium disilicate ceramics. *J. Adhes. Dent.* **2013**, *15*, 377.
9. Çömlekoğlu, M.E.; Dündar, M.; Güngör, M.A.; Şen, B.H.; Artunç, C. Preliminary evaluation of titanium tetrafluoride as an alternative ceramic etchant to hydrofluoric acid. *J. Adhes. Dent.* **2009**, *11*, 447.
10. Kukiattrakoon, B.; Thammasitboon, K. Optimal acidulated phosphate fluoride gel etching time for surface treatment of feldspathic porcelain: On shear bond strength to resin composite. *Eur. J. Dent.* **2012**, *6*, 63. [CrossRef]
11. Tian, T.; Tsoi, J.K.-H.; Matinlinna, J.P.; Burrow, M.F. Aspects of bonding between resin luting cements and glass ceramic materials. *Dent. Mater.* **2014**, *30*, e147–e162. [CrossRef] [PubMed]
12. Matinlinna, J.P.; Lassila, L.V.; Özcan, M.; Yli-Urpo, A.; Vallittu, P.K. An introduction to silanes and their clinical applications in dentistry. *Int. J. Prosthodont.* **2004**, *17*, 155. [PubMed]
13. Awad, M.M.; Alhalabi, F.; Alotaibi, N.; Alzamil, F.; Binalrimal, S.; Alrahlah, A.; Ahmed, M.H. A systematic review and meta-analysis of bond strength studies associated with self-etching primer and HF acid etching of dental glass-ceramics. *Int. J. Adhes. Adhes.* **2022**, *118*, 103216. [CrossRef]
14. Awad, M.M.; Alrahlah, A.; Matinlinna, J.P.; Hamama, H.H. Effect of adhesive air-drying time on bond strength to dentin: A systematic review and meta-analysis. *Int. J. Adhes. Adhes.* **2019**, *90*, 154–162. [CrossRef]
15. Alrahlah, A.; Awad, M.M.; Vohra, F.; Al-Mudahi, A.; Al Jeaidi, Z.A.; Elsharawy, M. Effect of self etching ceramic primer and universal adhesive on bond strength of lithium disilicate ceramic. *J. Adhes. Sci. Technol.* **2017**, *31*, 2611–2619. [CrossRef]
16. Siqueira, F.S.; Alessi, R.S.; Cardenas, A.F.; Kose, C.; Souza Pinto, S.C.; Bandeca, M.C.; Loguercio, A.D.; Gomes, J.C. New Single-bottle Ceramic Primer: 6-month Case Report and Laboratory Performance. *J. Contemp. Dent. Pract.* **2016**, *17*, 1033–1039. [CrossRef]
17. Sirisha, K.; Rambabu, T.; Ravishankar, Y.; Ravikumar, P. Validity of bond strength tests: A critical review-Part II. *J. Conserv. Dent.* **2014**, *17*, 420–426. [CrossRef]
18. Ismail, A.M.; Bourauel, C.; ElBanna, A.; Salah Eldin, T. Micro versus Macro Shear Bond Strength Testing of Dentin-Composite Interface Using Chisel and Wireloop Loading Techniques. *Dent. J.* **2021**, *9*, 140. [CrossRef]
19. El-Damanhoury, H.M.; Gaintantzopoulou, M.D. Self-etching ceramic primer versus hydrofluoric acid etching: Etching efficacy and bonding performance. *J. Prosthodont. Res.* **2018**, *62*, 75–83. [CrossRef]

20. Grasel, R.; Santos, M.J.; Rêgo, H.C.; Rippe, M.P.; Valandro, L.F. Effect of Resin Luting Systems and Alumina Particle Air Abrasion on Bond Strength to Zirconia. *Oper Dent.* **2018**, *43*, 282–290. [CrossRef]
21. Güler, A.U.; Yilmaz, F.; Yenisey, M.; Güler, E.; Ural, C. Effect of acid etching time and a self-etching adhesive on the shear bond strength of composite resin to porcelain. *J. Adhes. Dent.* **2006**, *8*, 21–25. [PubMed]
22. Donmez, M.B.; Okutan, Y.; Yucel, M.T. Effect of prolonged application of single-step self-etching primer and hydrofluoric acid on the surface roughness and shear bond strength of CAD/CAM materials. *Eur. J. Oral Sci.* **2020**, *128*, 542–549. [CrossRef] [PubMed]
23. Alshihri, A. Etching efficacy and bonding performance of resin to lithium disilicate ceramic using self-etching primer with different reaction times. *J. Adhes. Sci. Technol.* **2019**, *33*, 1215–1225. [CrossRef]
24. Cardenas, A.F.M.; Quintero-Calderon, A.S.; Siqueira, F.S.F.; Campos, V.S.; Wendlinger, M.; Pulido-Mora, C.A.; Masson-Palacios, M.J.; Sarmiento-Delgado, M.L.; Loguercio, A.D. Do Different Application Modes Improve the Bonding Performance of Self-etching Ceramic Primer to Lithium Disilicate and Feldspathic Ceramics? *J. Adhes. Dent.* **2019**, *21*, 319–327.
25. Ozcan, M.; Barbosa, S.H.; Melo, R.M.; Galhano, G.A.; Bottino, M.A. Effect of surface conditioning methods on the microtensile bond strength of resin composite to composite after aging conditions. *Dent. Mater.* **2007**, *23*, 1276–1282. [CrossRef]
26. Gale, M.S.; Darvell, B.W. Thermal cycling procedures for laboratory testing of dental restorations. *J. Dent.* **1999**, *27*, 89–99. [CrossRef] [PubMed]
27. Dapieve, K.S.; Aragonez, G.C.; Prochnow, C.; Burgo, T.A.L.; Rippe, M.P.; Pereira, G.K.R.; Venturini, A.B.; Valandro, L.F. Different Etching Times of a One-step Ceramic Primer: Effect on the Resin Bond Strength Durability to a CAD/CAM Lithium-Disilicate Glass-Ceramic. *J. Adhes. Dent.* **2021**, *23*, 133–143.
28. Azevedo, V.L.B.; de Castro, E.F.; Bonvent, J.J.; de Andrade, O.S.; Nascimento, F.D.; Giannini, M.; Cavalli, V. Surface treatments on CAD/CAM glass-ceramics: Influence on roughness, topography, and bond strength. *J. Esthet. Restor. Dent.* **2021**, *33*, 739–749. [CrossRef]
29. Scherer, M.M.; Prochnow, C.; Venturini, A.B.; Pereira, G.K.R.; Burgo, T.A.L.; Rippe, M.P.; Valandro, L.F. Fatigue failure load of an adhesively-cemented lithium disilicate glass-ceramic: Conventional ceramic etching vs etch & prime one-step primer. *Dent. Mater.* **2018**, *34*, 1134–1143.
30. Yao, C.; Zhou, L.; Yang, H.; Wang, Y.; Sun, H.; Guo, J.; Huang, C. Effect of silane pretreatment on the immediate bonding of universal adhesives to computer-aided design/computer-aided manufacturing lithium disilicate glass ceramics. *Eur. J. Oral Sci.* **2017**, *125*, 173–180. [CrossRef]
31. Franz, A.; Lettner, S.; Watts, D.C.; Schedle, A. Should statistical analysis of bond-strength data include or exclude cohesive failures? *Dent. Mater.* **2022**, *38*, 1942–1946. [CrossRef] [PubMed]
32. Awad, M.M.; Al Jeaidi, Z.A.; Almutairi, N.; Vohra, F.; Özcan, M.; Alrahlah, A. Effect of self-etching ceramic primer on bond strength of zirconia-reinforced lithium silicate ceramics. *J. Adhes. Sci. Technol.* **2020**, *34*, 91–101. [CrossRef]

coatings

MDPI

Article

Silane-Containing Universal Adhesives Influence Resin-Ceramic Microtensile Bond Strength

Mohamed M. Awad [1], Feras Alhalabi [1,*], Abdullah Alshehri [1], Mohammed Ali Salem [1], Ali Robaian [1], Shahad Alghannam [2], Abdullah S. Alayad [3], Basil Almutairi [3] and Ali Alrahlah [3,4]

[1] Department of Conservative Dental Sciences, College of Dentistry, Prince Sattam Bin Abdulaziz University, Al-Kharj 11942, Saudi Arabia
[2] College of Dentistry, Prince Sattam Bin Abdulaziz University, Al-Kharj 11942, Saudi Arabia
[3] Department of Restorative Dentistry, King Saud University College of Dentistry, Riyadh 11545, Saudi Arabia
[4] Engr Abdullah Bugshan Research Chair for Dental and Oral Rehabilitation, King Saud University, Riyadh 11545, Saudi Arabia
* Correspondence: f.alhalabi@psau.edu.sa

Abstract: Background: Silane-containing universal adhesives (UAs) are marketed as adhesion promotors for glass-ceramics. Objectives: This study aimed to evaluate the priming capacity of γ-methacryloxypropyltrimethoxysilane (γMPTS)-containing and γ-methacryloxypropyltriethoxysilane (γMPTES)/3-(aminopropyl)triethoxysilane (APTES)-containing universal adhesives (UAs) for lithium disilicate ceramic (LDC). Materials and Methods: Etched LDC discs were distributed into four groups according to the priming material used: (control), no priming; (MBN), LDC was primed with a universal primer (Monobond N); (SBU), γMPTS-containing UA (Single Bond Universal Adhesive) was used as a primer; and (SBP), γMPTES/APTES-containing UA (Scotchbond Universal Plus Adhesive) was used as a primer. LDC discs were cemented using a dual-cure resin cement, then sectioned into microbeams for microtensile bond strength (μTBS) evaluation. Failure modes were assessed. Results: MBN application showed the highest μTBS among all groups. γMPTES/APTES-containing UA (SBP) resulted in considerably higher μTBS compared with γMPTS-containing UA (SBU) or the control group. The mixed failures were the most predominant among all groups. Conclusions: The effect of silane-containing UAs on resin-ceramic μTBS is material dependent. Although γMPTES/APTES-containing UA improved bonding to LDC, the priming of LDC with either of the UAs tested cannot be considered as an alternative to a separate silanization (priming) step using a universal primer.

Keywords: adhesives; ceramics; microtensile bond strength; silanes

1. Introduction

Ceramic restorations are usually cemented using adhesive or self-adhesive resin cements which can effectively bond to both tooth structure and ceramic restoration while having enhanced esthetic and mechanical properties compared with conventional cements [1]. Robust resin-ceramic bond strength (adhesion) is crucial for the clinical outcome of indirect restorations made of lithium disilicate ceramic (LDC) [2,3]. Multiple steps are required to promote adequate bonding between LDC and methacrylate-based materials, such as adhesives or resin-based cements [4,5]. Ceramic surface treatments, such as priming in addition to using a type of luting cement, can influence resin-ceramic bond strength (adhesion) [6,7]. LDC is one of the most used glass-ceramic materials, which can be supplied in a fully or partly crystallized form, which requires crystallization (firing) or heat-pressing. Despite the wide range of chemical and microstructural differences among LDC materials, such materials are generally composed of a glass matrix supported by crystals [8]; hence, etching with hydrofluoric acid (HF) is required to dissolve the glass matrix [9] and increase the surface energy and wettability [10]. This step is usually followed by a priming procedure using a silane-containing or universal primer to promote chemical bonding (adhesion) between silane-treated LDC and methacrylate-based materials, such as adhesives or resin-based cements [9,11]. Dental silane primers usually contain organofunctional trialkoxysilanes, such as γ-methacryloxypropyltrimethoxysilane (γMPTS) and γ-methacryloxypropyltriethoxysilane (γMPTES), which are diluted and dissolved in ethanol and water at a specific pH [12]. Silane molecules chemically bond to methacrylate-based materials via the organofunctional group (methacrylate) and to HF-etched glass-ceramic via the silanol group resulting from silane activation (hydrolysis) [9,11,12]. Clinically, upon application of silane primer to a glass-ceramic restoration for a specific time (usually 60 s), an effective air-drying step is required to disperse any remaining excess of the primer solution and to evaporate the ethanol (solvent) and water content to form a homogenous silane layer onto the ceramic surface; otherwise, the resin-ceramic bond strength would be impaired [13], which would negatively affect the clinical performance of glass-ceramic restorations.

To simplify dental adhesive procedures, some universal adhesives (UAs) have been marketed as effective adhesion promotors (primers) to indirect materials owing to their silane and 10-methacryloyloxydecyl dihydrogen phosphate (10-MDP) content [14]. Such adhesives are claimed to be alternatives to 10-MDP-containing primers used for polycrystalline ceramics or silane-containing primers used for glass-ceramics. Recent studies indicated that although UAs can effectively promote bonding to zirconia ceramic or indirect composite [15], their ability to promote adequate bond strength (adhesion) to glass-ceramics is limited [15–19]. Similarly, the application of such silane-containing UAs as an adhesion promoter to hybrid ceramic failed to replace the use of commercially available silane-based primers [20]. Experimentally, the fresh mixing of silane compounds such as γMPTS with dental adhesives is more effective than adding silane compounds to an adhesive composition followed by long-term storage due to the acidity (pH) of the adhesive solution [21] and complexity of the contemporary UAs' composition [20,22]. Such adhesives contained only organofunctional trialkoxysilane compounds. Recently, Scotchbond Universal Plus Adhesive (SBP; 3M Deutschland GmbH, Neuss, Germany) was introduced to the dental market as the first UA containing both organofunctional trialkoxysilane and amino–silane compounds, such as 3-(aminopropyl)triethoxysilane (APTES). Such optimized silane content is expected to enhance the priming efficacy of SBP adhesive to glass-ceramics due to its optimized silane content (γMPTES and APTES) [23]. Nevertheless, it is essential to compare the priming capacity of the new γMPTES/APTES-containing UA with commercially available silane-based or universal primers used clinically. The use of SBP as an alternative to silane-based or universal primer is assumed to promote adequate resin-ceramic bond strength (adhesion), which can save more clinical time and the additional financial cost of silane-containing or universal primers. In addition, γMPTES/APTES-containing UA can be applied in more versatile clinical situations, such as in the intra-oral repair of LDC restorations. This study aimed to assess whether silane-containing UAs can influence the resin-ceramic microtensile bond strength (μTBS). Therefore, the effect of two silane-containing UAs on the resin-ceramic μTBS was evaluated in comparison with silane-containing universal primer. The null hypotheses were that: (1) Ceramic surface treatments (primer and UAs) would have no significant effect on the resin-ceramic μTBS; (2) There would be no significant difference between the resin-ceramic μTBS obtained after the use of either of the two silane-containing UAs tested.

2. Materials and Methods

The materials used in the study and their compositions are described in Table 1.

Table 1. Materials used in study.

Material (LOT Number)	Code	Composition
Initial LiSi Block, GC, GC corporation, Tokyo, Japan. (LOT: 2201201C)	LDC	Silicon dioxide: 81%; phosphorus oxide 8.1%; potassium oxide 5.9%; aluminum oxide 3.8%; titanium oxide 0.5%; and cerium oxide 0.6%.
Single Bond Universal Adhesive, 3M Deutschland GmbH, Neuss, Germany. (LOT: 10608B)	SBU	10-MDP phosphate monomer, Vitrebond, copolymer, HEMA, Bis-GMA, dimethacrylate resin, camphorquinone, silane (γMPTS), ethanol, and water.
Scotchbond Universal Plus Adhesive, 3M Deutschland GmbH, Neuss, Germany. (LOT: 7676509)	SBP	10-MDP phosphate monomer, Vitrebond, copolymer, HEMA, dimethacrylate resin, crosslinking radiopaque monomer, dual-cure accelerator, camphorquinone, optimized silane (γMPTES/APTES), ethanol, and water.
Monobond N, Ivoclar Vivadent, Schaan, Liechtenstein. (LOT: Z02S7Z)	MBN	Alcohol, silane methacrylate, phosphoric acid methacrylate, disulfide methacrylate.
RelyX Universal Resin Cement, 3M Oral Care, St. Paul, USA. (LOT: 7756479)	-	Dimethacrylate monomers, phosphorylated dimethacrylate adhesion monomers, photoinitiator system, novel amphiphilic redox initiator system, radiopaque fillers, and rheological additives and pigments.

Bis-GMA, bisphenol A glycidyl methacrylate; 10-MDP, 10-methacryloyloxydecyl dihydrogen thiophosphate; HEMA, 2-hydroxyethyl methacrylate; γMPTS, γ methacryloxypropyltrimethoxysilane; γMPTES, γ-methacryloxypropyltriethoxysilane; APTES, 3-(aminopropyl)triethoxysilane.

2.1. Specimens' Preparation

LDC (GC InitialLiSi Block, GC Corporation, Tokyo, Japan) blocks were cut into a total of 40 (approximately 5.5 mm × 6.5 mm) 5 mm thick discs using a water-cooled diamond cutting disc mounted on a computer-controlled precision cutter (IsoMet 5000 Linear Precision Saw, Buehler, Lake Bluff, IL, USA). The top surface of each ceramic disc was wet-polished for 1 min using 600-grit silicon carbide papers mounted on a grinding machine (Automata, Jean Wirtz, Germany) performing 200 revolutions per minute (rpm), and then the discs were subjected to ultrasonic cleaning in distilled water for 5 min using an ultrasonicator (Sonicer, Yoshida Dental Manufacturing. Co., Ltd., Tokyo, Japan). The top surface of each ceramic disc was acid-etched with 5% HF (IPS Ceramic Etching Gel, Ivoclar Vivadent, Schaan, Liechtenstein) for 20 s, then thoroughly washed with water, before being air-dried and subjected to additional ultrasonic cleaning in distilled water for 5 min, and then air-dried and randomly distributed into four groups according to the priming material utilized. In Group 1 (control), the LDC discs were not primed. In Group 2, a universal primer (MBN) was applied to the top surface using a microbrush according to the manufacturer's instructions, where it was left for 60 s; subsequently, any remaining excess was dispersed with a strong stream of air for approximately 5 s. In Group 3, γMPTS-containing UA (SBU) was mixed with one drop (1-to-1 ratio) of Scotchbond Universal DCA (3M Deutschland GmbH, Neuss, Germany) and immediately applied onto the ceramic surface with a microbrush for 20 s, before being air-dried for approximately 5 s without being light-cured. In Group 4, γMPTES/APTES-containing UA (SBP) was applied onto the ceramic surface with a microbrush for 20 s, before being air-dried for approximately 5 s without being light-cured. For each group, the two discs that received the same surface treatment were cemented using a dual-cure resin cement (RelyX Universal Resin Cement, 3M Oral Care, St. Paul, MN, USA). Equal amounts of resin cement were applied onto the HF-etched or surface-treated (primed) sides before both discs were vertically aligned using a custom-made metal cementation device; then, a cementation force of 1 kg was applied to the discs for 1 min (Figure 1) [24]. The excess cement was immediately and carefully removed with a small-size cotton pellet. The cemented blocks (n = 5/group) were light-cured from each side for 40 s. Then, additional light curing was performed from the top and bottom sides for an additional 40 s each using Elipar™ S10 (3M ESPE, St. Paul, MN, USA) operated at 1000 mW/cm^2, as verified by a hand-held radiometer (Bluephase Meter, Ivoclar Vivadent, Austria). The light-curing tip was kept at approximately 1 mm and at 0 angle to the cemented block. Then, the cemented blocks were kept in distilled water for 24 h at 37 °C before sectioning.

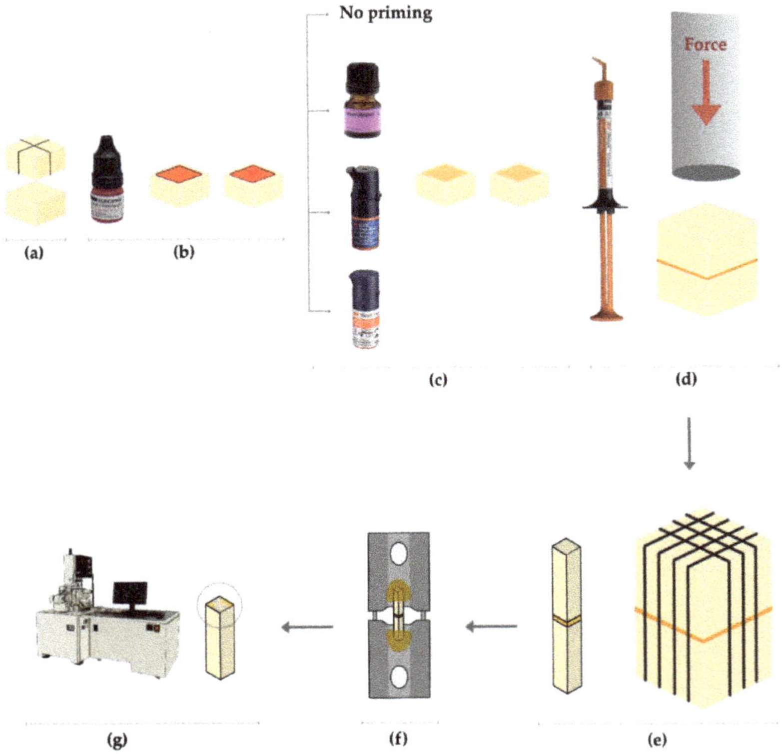

Figure 1. Schematic illustration of specimens' preparation and μTBS test. (**a**) Sectioning of ceramic blocks into discs; (**b**) acid etching with 5% HF; (**c**) priming of ceramic discs, control (no priming), MBN, SBU, SBP; (**d**) cementation of ceramic discs using RelyX Universal Resin Cement under 1 kg force; (**e**) sectioning of cemented discs into microbeams; (**f**) fixation of microbeams onto μTBS metal jig; (**g**) using a scanning electron microscope (SEM) to evaluate the failure/fracture mode of microbeams with failure-mode patterns unidentifiable with a light microscope.

2.2. μTBS

The cemented ceramic blocks were sectioned into ceramic–resin-ceramic microbeams with a cross-sectional area of approximately 1 mm^2 (Figure 1e) using a water-cooled diamond cutting disc mounted on a low-speed precision cutter (IsoMet 1000 Precision Cutter, Lake Bluff, IL, USA) for μTBS evaluation. Thirty microbeams were generated within each group. The generated microbeams were examined with a light microscope (Hirox Co., Ltd., Tokyo, Japan) at 15× magnification before μTBS testing. Defective microbeams were excluded. A digital micrometer (Mitutoyo, Kanagawa, Japan) was used to measure the microbeam cross-sectional surface area at the interface. Then, the microbeams were fixed (glued) to a custom-made metal jig using a cyanoacrylate glue (Loctite 416, Henkel, Düsseldorf, Germany) and attached to a universal testing machine (Instron 5965, Instron Corporation, Norwood, MN, USA) supplied with a 30 kN load cell. Before testing, it was ensured that the alignment of each microbeam was parallel to the direction of the tensile force. The microbeams were stressed under tension at a crosshead speed of 0.5 mm/min until failure or debonding. The μTBS (in megaPascal (MPa)) was calculated for each microbeam by dividing the maximum force (in Newton (N)) at fracture or debonding by the cross-sectional surface area (in mm^2). The μTBS test was performed in accordance with the protocol described by the Academy of Dental Materials [25].

2.3. Failure/Fracture Mode Assessment

Failure/fracture modes were assessed at 15× magnification using a light microscope (Hirox Co., Ltd., Tokyo, Japan). Failed/fractured microbeams with unidentifiable failure-mode patterns were ultrasonically cleaned in distilled water for 5 min and dehydrated in ascending concentrations of ethanol, before being gold sputtered for 180 s at 40 mA using a JFC-1100 Fine Coat Ion Sputter (JEOL Ltd., Tokyo, Japan). Gold-sputtered microbeams were then assessed for the failure/fracture mode at low (75×, 80×) and high magnifications (500×) using a scanning electron microscope (JSM-6610LV; JEOL Ltd., Tokyo, Japan) operated at 20 kV. Failure/fracture mode patterns were classified as cohesive failure/fracture in the ceramic; adhesive failure/fracture at the resin-ceramic interface; cohesive failure/fracture the resin cement; or mixed failure, including failure/fracture at the ceramic and the resin cement, cohesive failure in the resin cement, and adhesive failure/fracture at the resin-ceramic interface. Defective microbeams or those failed/fractured away (>2 mm) from the resin-ceramic interface were excluded.

2.4. Statistical Analysis

The sample size was calculated using G*Power software, version 3.1.9.7 (Heinrich-Heine-Universität Düsseldorf, Düsseldorf, Germany). A total of 20 cemented blocks are required to provide 5 blocks per group. This design achieves 98% power and a 5% significance level. The μTBS values of microbeams generated from the same cemented ceramic block were averaged, thus the cemented ceramic block was considered as the statistical (experimental) unit [26]. Upon the evaluation of the normality of μTBS data using the Kolmogorov–Smirnov test, the one-way analysis of variance (ANOVA) and Tukey's multiple comparison tests were applied in the statistical analysis considering the effect of the priming material on the resin-ceramic μTBS, with $p < 0.05$ considered as being statistically significant. The statistical analysis was performed using R software version 4.1.2 (R Foundation for Statistical Computing, Vienna, Austria).

3. Results

3.1. μTBS

The means ± standard deviations (SDs) μTBS for the tested groups are presented graphically and numerically in Figure 2. Universal primer (MBN) application showed the highest μTBS (32.21 ± 4.49 MPa) among all groups. γMPTES/APTES-containing UA (SBP) resulted in significantly higher μTBS (20.41 ± 6.26 MPa) compared with γMPTS-containing UA (SBU) (10.65 ± 3.5 MPa) or the control group (8.47 ± 3.11 MPa). There was no statistically significant difference between γMPTS-containing UA (SBU) and the control group. The results of Tukey's multiple comparisons statistical analysis are illustrated in Table 2.

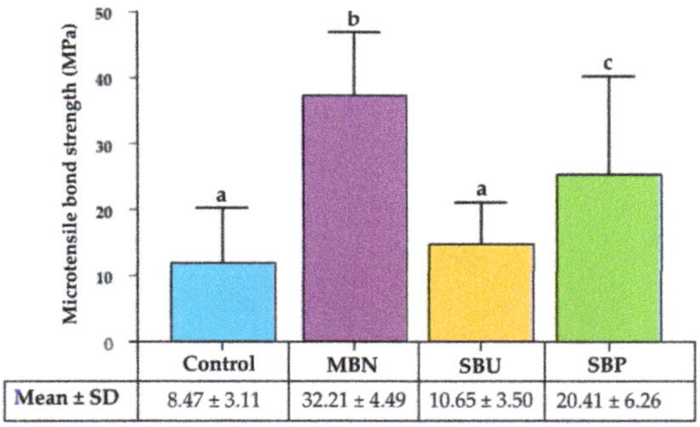

Figure 2. The means ± standard deviations (SDs) of resin-ceramic microtensile bond strength (μTBS) (expressed in MPa) of Group 1 (control), no priming; Group 2, primed with MBN; Group 3, primed with SBU; and Group 4, primed with SBP. Different superscript letters indicate statistically significant difference ($p < 0.05$) between groups.

Table 2. Results of Tukey's multiple comparisons statistical analysis of μTBS data.

Pair-Wise Groups Comparisons	Mean Difference	Lower Bound	Upper Bound	*p*-Value	Significance
Control-MBN	23.739	15.581	31.897	<0.001	*
Control-SBU	2.181	−5.977	10.339	0.869	
Control-SBP	11.942	3.784	20.100	0.003	*
MBN-SBU	21.557	13.399	29.715	<0.001	*
MBN-SBP	11.797	3.639	19.955	0.004	*
SBU-SBP	−9.761	−17.919	−1.603	0.017	*

*: indicates statistically significant difference.

3.2. Failure Mode Evaluation

The frequencies of failure modes (expressed as percentages) recorded for the tested μTBS microbeams are presented in Figure 3. The mixed failures were the most predominant among all groups at 60% (15 microbeams) for the control group, 80% (20 microbeams) for MBN, 68% (17 microbeams) for SBU, and 72% (18 microbeams) for SBP. Adhesive failures occurred next to mixed failures, while cohesive failures were the least among all groups. No cohesive failures in ceramic, or pretest failures, were recorded for any group. The representative SEM photomicrographs of the recorded failure modes are presented in Figure 4. Most of the mixed failure-mode patterns involved an adhesive interfacial failure at the ceramic side and a cohesive failure within the resin cement (Figure 4c,d).

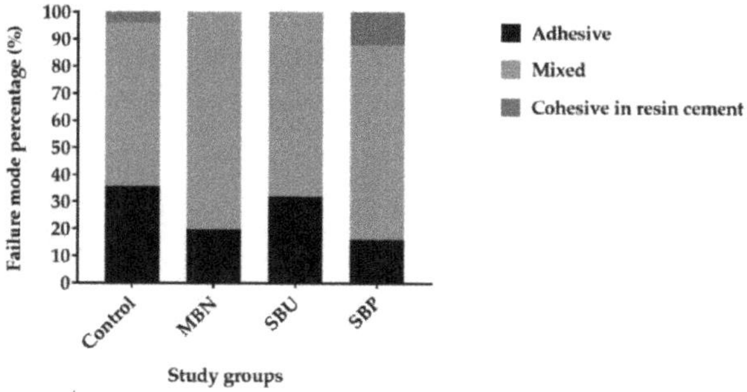

Figure 3. Frequencies of failure modes (expressed as percentages) recorded for tested μTBS microbeams of following groups: control (no priming), MBN, SBU, and SBP.

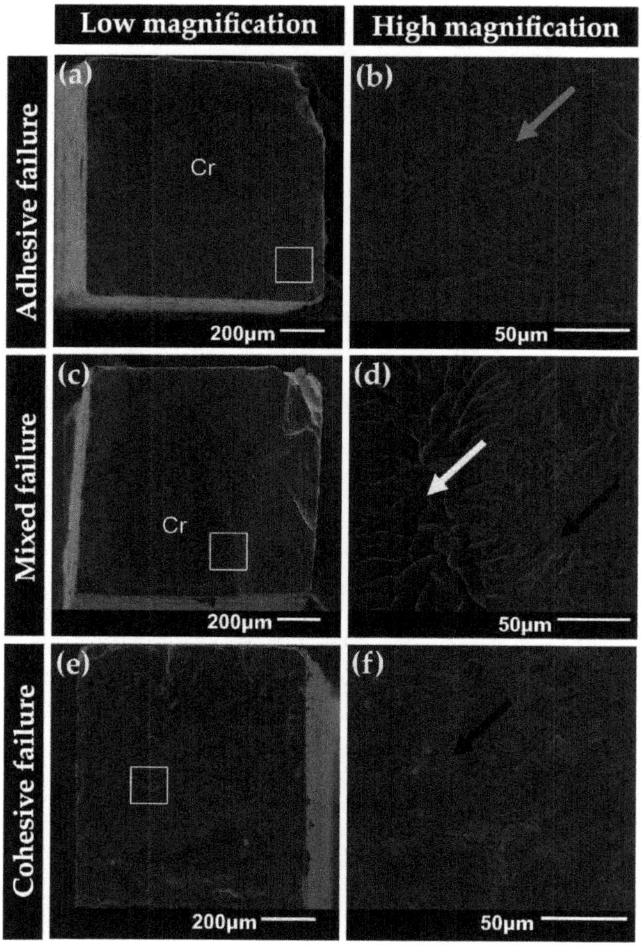

Figure 4. Representative SEM photomicrographs at low and high magnifications of tested microbeams. (**a**), at 75× and (**b**), at 500×: adhesive failure; (**c**), at 75× and (**d**), at 500×: mixed failure; (**e**), at 80× and (**f**), at 500×: cohesive failure in resin cement. Cr: ceramic surface; gray arrow: ceramic surface covered by adhesive layer; white arrow: adhesive layer, and black arrow: resin cement.

4. Discussion

Bond strength evaluations can be utilized to assess the effectiveness of adhesive procedures and materials [27] before clinical studies can be conducted. This study was designed to evaluate the effect of priming LDC with silane-containing UAs on resin-ceramic bond strength. In contrast to a previous study [23] in which a shear (macro) bond strength test was used to evaluate the bonding of light-cured (polymerized) silane-containing UAs, the current study utilized the μTBS test due to its better discriminative ability, more even stress distribution during testing, and less cohesive substrate failures compared with macro bond strength tests [28,29]. In addition, the ceramic discs were primed with UAs before they were cemented (under force) using a dual-cure resin-based cement to simulate the intra-oral cementation procedure.

The first null hypothesis was that ceramic surface treatments (universal primer or UAs) would have no significant effect on the resin-ceramic μTBS. The results of this study indicated that despite γMPTS-containing UA (SBU) having no effect on the resin-ceramic μTBS, both the universal primer (MBN) and γMPTES/APTES-containing UA (SBP) considerably improved the resin-ceramic μTBS. Thus, the first hypothesis was rejected.

Silane-containing universal primers, such as MBN, would not promote adequate adhesion to glass-ceramics without a prior HF etching step [30]. That is why the effect of MBN or silane-containing UAs should not be evaluated without prior HF etching. The application of HF on glass-ceramics results in the formation of tetrafluorosilane, which reacts with HF to form a soluble hydrofluorosilicic acid [12]. Then, a considerable amount of the glass matrix at the etched ceramic surface is dissolved and can be easily rinsed away with water. This not only results in the creation of surface microirregularities but also results in the formation of hydroxyl groups on the etched ceramic surface [9]. Upon the surface treatment of HF-etched glass-ceramic with silane-containing primers, a condensation reaction between such hydroxyl groups and the silanol groups of activated silane occurs, representing one aspect of ceramic bonding [9,31]. The other aspect is the reaction between organofunctional groups in the silane molecule and functional (methacrylate) groups of the methacrylate-based materials, such as resin-based cements or adhesives [9,31]. The universal primers utilized in this study, such as MBN, contain large percentages of organic solvents, such as ethanol or acetone [11,32]. Upon air-drying the primer layer applied on the ceramic surface, much of the solvent will be evaporated, resulting in the formation of a silane layer on the ceramic surface, which is a crucial factor affecting the resin-ceramic bond strength [31]. In contrast, compared with MBN, the composition of silane-containing UAs is much more complex because they are primarily manufactured to bond resin-based composite materials to the tooth structure [33]. To specify, other components within the UA composition, such as 10-MDP, Bis-GMA, HEMA, dimethacrylate resin, crosslinker monomers, and other components, cannot be eliminated after the application of UA. Thus, it is assumed that despite the ability of UAs to flow effectively into the microirregularities created by HF etching [21], the formation of a silane layer deposited onto the ceramic surface seems less likely, even with the effective air-drying of such UAs [20]. This can partly explain the inferior priming capacity of UAs compared with MBN. Meanwhile, UAs are acidic in nature [14,34], which might have affected the chemical stability of silane molecules within the UAs' formulations, impairing their priming potential [21] and resulting in a low resin-ceramic bond strength. γMPTS-containing UA (SBU) did not have any impact on resin-ceramic bonding. This is in accordance with a recent meta-analysis that indicated that γMPTS-containing UAs do not have a priming capacity similar to silane-containing or universal primers [35]. In contrast, γMPTES/APTES-containing UA (SBP) not only significantly improved the resin-ceramic μTBS compared with the control group but also resulted in significantly higher μTBS than the γMPTS-containing UA (SBU). Therefore, the second null hypothesis, that there would be no significant difference between the resin-ceramic μTBS obtained after the use of either of the two silane-containing UAs tested, was also rejected. The different bond strengths obtained can be explained—in part—by the better priming capacity of SBP due to its optimized (γMPTES/APTES) silane technology.

Combining the effects of the silanol-groups of both γMPTES and APTES molecules, as well as the amino groups of the APTES molecule, means they can simultaneously interact with the hydroxyl groups on the glass-ceramic surface [36], increasing UA's priming capacity; thus, the resin-ceramic μTBS is enhanced. The APTES molecule within SBP forms hydrogen bonds at the HF-etched glass-ceramic surface and reacts with silanol groups, resulting in the formation of amino-silanol groups [37]. In addition, the APTES molecule can act as a siloxane intermolecular crosslinker [36], which can increase the density of the silane interaction layer and enhance the bonding to the glass-ceramic material [23]. Moreover, the Si–O–C_2H_5 group within the γMPTES molecule in SBP hydrolyzes slower than the Si–O–CH_3 group within the γMPTS molecule in SBU [38], which may have decreased the dehydration condensation of silanol groups [23]. Another plausible explanation for the superior resin-ceramic μTBS obtained when SBP is used as a primer may be its improved dual-cure compatibility with the luting resin cement used (RelyX Universal Resin Cement). However, further in-depth investigations are required to prove this speculation.

Despite the precision of the μTBS test employed in this study, it is a technique-sensitive method associated with the generation of interfacial stress during specimens' sectioning. Such stress can result in failure of low-performance adhesive interfaces or brittle substrates [29]. Multiple factors, such as specimens' alignment in relation to the direction of tensile force, shape and dimensions, gripping jig, and testing machine, can also influence the outcome of the μTBS evaluation [25,39]. The outcome (μTBS data) should be interpreted in combination with the corresponding failure modes because a statistical evaluation of μTBS data can be affected by the inclusion or censoring microbeams with specific failure modes [40,41]. Both a light microscope and SEM were utilized to assess the failure mode patterns of the tested microbeams. The fractographic evaluation using SEM is more illustrative and can be performed in higher magnifications compared with using a light microscope [25]. We found that mixed failure patterns (Figure 4c,d) were the most predominant in all the groups. The control (no priming) group presented a higher incidence of adhesive failures compared with the other groups, which may be due to the relatively low μTBS. The failure mode patterns of the tested groups were affected by the setup of the μTBS test. Although sectioning the cemented ceramic blocks into microbeams is laboratory-intensive and technique-sensitive work, testing the ceramic-resin-ceramic microbeams employed in this study not only had the advantage of solely evaluating the resin-ceramic interface [24] but also eliminated the incidence of cohesive failures within the ceramic material, which can be explained by the high mechanical properties of LDC [42] and the homogenous stress distribution at the interface. Overall, the γMPTES/APTES-containing UA (SBP) showed promising performance as an LDC primer before cementation with a dual cure resin cement. However, this study has some limitations, including the lack of long-term artificial aging (water storage and thermocycling) of the microbeams and the use of one ceramic material and one resin cement. Moreover, mechanical loading before bond strength evaluation cannot be performed because of the geometry of the cemented blocks. Thus, the results of this study should be considered with much caution. In fact, in-depth chemical analyses of γMPTES/APTES ratio applied in SBP in addition to the reaction between γMPTES/APTES-containing UA (SBP) and HF-etched glass-ceramic are required to confirm its positive impact on resin-ceramic bond strength.

5. Conclusions

The effect of silane-containing UAs on resin-ceramic μTBS is material-dependent. γMPTS-containing UA (SBU) did not affect the resin-ceramic μTBS. Although γMPTES/APTES-containing UA (SBP) improved bonding to LDC compared with γMPTS-containing UA (SBU) or the control group (no priming), it resulted in considerably lower μTBS than a universal silane-containing primer (MBN). The priming of LDC with either of the UAs tested cannot be considered as an alternative to a separate silanization (priming) step using a universal silane-containing primer.

Author Contributions: Conceptualization, M.M.A.; methodology, M.M.A.; software, A.A. (Ali Alrahlah); validation, A.S.A.; formal analysis, M.M.A.; investigation, B.A. and S.A.; resources, A.A. (Abdullah Alshehri); data curation, M.M.A.; writing—original draft preparation, M.M.A. and F.A.; writing—review and editing, M.A.S.; visualization, M.M.A. and A.R.; supervision, A.A. (Ali Alrahlah); project administration, F.A.; funding acquisition, A.A. (Abdullah Alshehri). All authors have read and agreed to the published version of the manuscript.

Funding: The authors are grateful to the Deanship of Scientific Research, King Saud University, for funding this study through the Vice Deanship of Scientific Research Chairs and the Research Chair for Dental and Oral Rehabilitation, Engineer Abdullah Bugshan.

Institutional Review Board Statement: Not applicable.

Informed Consent Statement: Not applicable.

Data Availability Statement: The data presented in this study are available upon request from the corresponding author.

Conflicts of Interest: The authors declare no conflict of interest.

References

1. van den Breemer, C.R.; Gresnigt, M.M.; Cune, M.S. Cementation of Glass-Ceramic Posterior Restorations: A Systematic Review. *BioMed Res. Int.* **2015**, *2015*, 148954. [CrossRef] [PubMed]
2. Politano, G.; Van Meerbeek, B.; Peumans, M. Nonretentive Bonded Ceramic Partial Crowns: Concept and Simplified Protocol for Long-lasting Dental Restorations. *J. Adhes. Dent.* **2018**, *20*, 495–510. [CrossRef] [PubMed]
3. Phark, J.-H.; Duarte, S., Jr. Microstructural considerations for novel lithium disilicate glass ceramics: A review. *J. Esthet. Restor. Dent.* **2022**, *34*, 92–103. [CrossRef] [PubMed]
4. Awad, M.M.; Alqahtani, H.; Al-Mudahi, A.; Murayshed, M.S.; Alrahlah, A.; Bhandi, S.H. Adhesive Bonding to Computer-aided Design/ Computer-aided Manufacturing Esthetic Dental Materials: An Overview. *J. Contemp. Dent. Pract.* **2017**, *18*, 622–626. [CrossRef] [PubMed]
5. Malysa, A.; Wezgowiec, J.; Grzebieluch, W.; Danel, D.P.; Wieckiewicz, M. Effect of Thermocycling on the Bond Strength of Self-Adhesive Resin Cements Used for Luting CAD/CAM Ceramics to Human Dentin. *Int. J. Mol. Sci.* **2022**, *23*, 745. [CrossRef]
6. Malysa, A.; Wezgowiec, J.; Orzeszek, S.; Florjanski, W.; Zietek, M.; Wieckiewicz, M. Effect of Different Surface Treatment Methods on Bond Strength of Dental Ceramics to Dental Hard Tissues: A Systematic Review. *Molecules* **2021**, *26*, 1223. [CrossRef]
7. Akar, T.; Dündar, A.; Kırmalı, Ö.; Üstün, Ö.; Kapdan, A.; Er, H.; Kuştarcı, A.; Er, K.; Yılmaz, B. Evaluation of the shear bond strength of zirconia to a self-adhesive resin cement after different surface treatment. *Dent. Med. Probl.* **2021**, *58*, 463–472. [CrossRef]
8. Lubauer, J.; Belli, R.; Peterlik, H.; Hurle, K.; Lohbauer, U. Grasping the Lithium hype: Insights into modern dental Lithium Silicate glass-ceramics. *Dent. Mater.* **2022**, *38*, 318–332. [CrossRef]
9. Tian, T.; Tsoi, J.K.-H.; Matinlinna, J.P.; Burrow, M.F. Aspects of bonding between resin luting cements and glass ceramic materials. *Dent. Mater.* **2014**, *30*, e147–e162. [CrossRef]
10. Della Bona, A.; Anusavice, K.J.; Mecholsky, J.J., Jr. Failure analysis of resin composite bonded to ceramic. *Dent. Mater.* **2003**, *19*, 693–699. [CrossRef]
11. Matinlinna, J.P.; Lassila, L.V.; Özcan, M.; Yli-Urpo, A.; Vallittu, P.K. An introduction to silanes and their clinical applications in dentistry. *Int. J. Prosthodont.* **2004**, *17*, 155–164. [PubMed]
12. Lung, C.Y.; Matinlinna, J.P. Aspects of silane coupling agents and surface conditioning in dentistry: An overview. *Dent. Mater.* **2012**, *28*, 467–477. [CrossRef] [PubMed]
13. Shen, C.; Oh, W.S.; Williams, J.R. Effect of post-silanization drying on the bond strength of composite to ceramic. *J. Prosthet. Dent.* **2004**, *91*, 453–458. [CrossRef] [PubMed]
14. Nagarkar, S.; Theis-Mahon, N.; Perdigão, J. Universal dental adhesives: Current status, laboratory testing, and clinical performance. *J. Biomed. Mater. Res. B Appl. Biomater.* **2019**, *107*, 2121–2131. [CrossRef] [PubMed]
15. Cuevas-Suárez, C.E.; de Oliveira da Rosa, W.L.; Vitti, R.P.; da Silva, A.F.; Piva, E. Bonding Strength of Universal Adhesives to Indirect Substrates: A Meta-Analysis of in Vitro Studies. *J. Prosthodont.* **2020**, *29*, 298–308. [CrossRef]
16. Kalavacharla, V.K.; Lawson, N.C.; Ramp, L.C.; Burgess, J.O. Influence of Etching Protocol and Silane Treatment with a Universal Adhesive on Lithium Disilicate Bond Strength. *Oper. Dent.* **2015**, *40*, 372–378. [CrossRef]
17. Makishi, P.; Andre, C.B.; Silva, J.L.; Bacelar-Sa, R.; Correr-Sobrinho, L.; Giannini, M. Effect of Storage Time on Bond Strength Performance of Multimode Adhesives to Indirect Resin Composite and Lithium Disilicate Glass Ceramic. *Oper. Dent.* **2016**, *41*, 541–551. [CrossRef]
18. Yao, C.; Yang, H.; Yu, J.; Zhang, L.; Zhu, Y.; Huang, C. High Bond Durability of Universal Adhesives on Glass Ceramics Facilitated by Silane Pretreatment. *Oper. Dent.* **2018**, *43*, 602–612. [CrossRef]

19. Yao, C.; Zhou, L.; Yang, H.; Wang, Y.; Sun, H.; Guo, J.; Huang, C. Effect of silane pretreatment on the immediate bonding of universal adhesives to computer-aided design/computer-aided manufacturing lithium disilicate glass ceramics. *Eur. J. Oral Sci.* **2017**, *125*, 173–180. [CrossRef]
20. Awad, M.M.; Albedaiwi, L.; Almahdy, A.; Khan, R.; Silikas, N.; Hatamleh, M.M.; Alkhtani, F.M.; Alrahlah, A. Effect of universal adhesives on microtensile bond strength to hybrid ceramic. *BMC Oral Health* **2019**, *19*, 178. [CrossRef]
21. Yoshihara, K.; Nagaoka, N.; Sonoda, A.; Maruo, Y.; Makita, Y.; Okihara, T.; Irie, M.; Yoshida, Y.; Van Meerbeek, B. Effectiveness and stability of silane coupling agent incorporated in 'universal' adhesives. *Dent. Mater.* **2016**, *32*, 1218–1225. [CrossRef] [PubMed]
22. Alrahlah, A.; Awad, M.M.; Vohra, F.; Al-Mudahi, A.; Al Jeaidi, Z.A.; Elsharawy, M. Effect of self etching ceramic primer and universal adhesive on bond strength of lithium disilicate ceramic. *J. Adhes. Sci. Technol.* **2017**, *31*, 2611–2619. [CrossRef]
23. Yao, C.; Ahmed, M.H.; De Grave, L.; Yoshihara, K.; Mercelis, B.; Okazaki, Y.; Van Landuyt, K.L.; Huang, C.; Van Meerbeek, B. Optimizing glass-ceramic bonding incorporating new silane technology in an experimental universal adhesive formulation. *Dent. Mater.* **2021**, *37*, 894–904. [CrossRef] [PubMed]
24. Südbeck, J.M.; Jacker-Guhr, S.; Lührs, A.K. Do Different Pretreatment Methods Influence the Microtensile Bond Strength of Composite Cements to Silicate Ceramic? *J. Adhes. Dent.* **2021**, *23*, 335–345. [CrossRef] [PubMed]
25. Armstrong, S.; Breschi, L.; Özcan, M.; Pfefferkorn, F.; Ferrari, M.; Van Meerbeek, B. Academy of Dental Materials guidance on in vitro testing of dental composite bonding effectiveness to dentin/enamel using micro-tensile bond strength (μTBS) approach. *Dent. Mater.* **2017**, *33*, 133–143. [CrossRef]
26. Eckert, G.J.; Platt, J.A. A statistical evaluation of microtensile bond strength methodology for dental adhesives. *Dent. Mater.* **2007**, *23*, 385–391. [CrossRef]
27. El Zohairy, A.A.; De Gee, A.J.; Mohsen, M.M.; Feilzer, A.J. Microtensile bond strength testing of luting cements to prefabricated CAD/CAM ceramic and composite blocks. *Dent. Mater.* **2003**, *19*, 575–583. [CrossRef]
28. De Munck, J.; Mine, A.; Poitevin, A.; Van Ende, A.; Cardoso, M.V.; Van Landuyt, K.L.; Peumans, M.; Van Meerbeek, B. Meta-analytical review of parameters involved in dentin bonding. *J. Dent. Res.* **2012**, *91*, 351–357. [CrossRef]
29. Salz, U.; Bock, T. Testing adhesion of direct restoratives to dental hard tissue—A review. *J. Adhes. Dent.* **2010**, *12*, 343–371. [CrossRef]
30. Huang, T.H.; Kao, C.T. The shear bond strength of composite brackets on porcelain teeth. *Eur. J. Orthod.* **2001**, *23*, 433–439. [CrossRef]
31. Matinlinna, J.P.; Lung, C.Y.K.; Tsoi, J.K.H. Silane adhesion mechanism in dental applications and surface treatments: A review. *Dent. Mater.* **2018**, *34*, 13–28. [CrossRef] [PubMed]
32. Zakir, M.; Ashraf, U.; Tian, T.; Han, A.; Qiao, W.; Jin, X.; Zhang, M.; Tsoi, J.K.-H.; Matinlinna, J.P. The Role of Silane Coupling Agents and Universal Primers in Durable Adhesion to Dental Restorative Materials—A Review. *Curr. Oral Health Rep.* **2016**, *3*, 244–253. [CrossRef]
33. Chen, C.; Niu, L.N.; Xie, H.; Zhang, Z.Y.; Zhou, L.Q.; Jiao, K.; Chen, J.H.; Pashley, D.H.; Tay, F.R. Bonding of universal adhesives to dentine–Old wine in new bottles? *J. Dent.* **2015**, *43*, 525–536. [CrossRef] [PubMed]
34. Papadogiannis, D.; Dimitriadi, M.; Zafiropoulou, M.; Gaintantzopoulou, M.D.; Eliades, G. Universal Adhesives: Setting Characteristics and Reactivity with Dentin. *Materials* **2019**, *12*, 1720. [CrossRef]
35. Lima, R.B.W.; Muniz, I.A.F.; Campos, D.E.S.; Murillo-Gómez, F.; Andrade, A.K.M.; Duarte, R.M.; de Souza, G.M. Effect of universal adhesives and self-etch ceramic primers on bond strength to glass ceramics: A systematic review and meta-analysis of in vitro studies. *J. Prosthet. Dent.* **2022**. [CrossRef]
36. Acres, R.G.; Ellis, A.V.; Alvino, J.; Lenahan, C.E.; Khodakov, D.A.; Metha, G.F.; Andersson, G.G. Molecular Structure of 3-Aminopropyltriethoxysilane Layers Formed on Silanol-Terminated Silicon Surfaces. *J. Phys. Chem. C* **2012**, *116*, 6289–6297. [CrossRef]
37. Park, S.-J.; Jin, J.-S.; Lee, J.-R. Influence of silane coupling agents on the surface energetics of glass fibers and mechanical interfacial properties of glass fiber-reinforced composites. *J. Adhes. Sci. Technol.* **2000**, *14*, 1677–1689. [CrossRef]
38. Xiong, M.; Zhang, K.; Chen, Y. ATRP of 3-(triethoxysilyl)propyl methacrylate and preparation of "stable" gelable block copolymers. *Eur. Polym. J.* **2008**, *44*, 3835–3841. [CrossRef]
39. Roeder, L.; Pereira, P.N.; Yamamoto, T.; Ilie, N.; Armstrong, S.; Ferracane, J. Spotlight on bond strength testing–unraveling the complexities. *Dent. Mater.* **2011**, *27*, 1197–1203. [CrossRef]
40. Comba, L.; Tichy, A.; Bradna, P. The Effect of Censoring on the Statistical Evaluation of Composite-to-Composite Bond Strength. *J. Adhes. Dent.* **2020**, *22*, 183–193. [CrossRef]
41. Franz, A.; Lettner, S.; Watts, D.C.; Schedle, A. Should statistical analysis of bond-strength data include or exclude cohesive failures? *Dent. Mater.* **2022**, *38*, 1942–1946. [CrossRef] [PubMed]
42. da Silva, S.E.G.; de Araújo, G.M.; Souza, K.B.; Moura, D.M.D.; Aurélio, I.L.; May, L.G.; Vila-Nova, T.E.L.; Zhang, Y.; de Assunção, E.S.R.O. Biaxial flexure strength and physicochemical characterization of a CAD/CAM lithium disilicate ceramic: Effect of etching time, silane, and adhesive applications. *Clin. Oral Investig.* **2022**, *26*, 6753–6763. [CrossRef] [PubMed]

coatings

MDPI

Review

Degradation and Stabilization of Resin-Dentine Interfaces in Polymeric Dental Adhesives: An Updated Review

Faiza Amin [1], Muhammad Amber Fareed [2], Muhammad Sohail Zafar [3,4], Zohaib Khurshid [5,*], Paulo J. Palma [6,7,*] and Naresh Kumar [8]

1 Science of Dental Materials Department, Dow Dental College, Dow University of Health Sciences, Karachi 74200, Pakistan; faiza.ameen@duhs.edu.pk
2 Department of Adult Restorative Dentistry, College of Dentistry, Gulf Medical University, Ajman 4184, United Arab Emirates; prof.mafareed@gmu.ac.ae
3 Department of Restorative Dentistry, College of Dentistry, Taibah University, Al Madina, Al Munawwarra 41311, Saudi Arabia; mzafar@taibahu.edu.sa
4 Department of Dental Materials, Islamic International Dental College, Riphah International University, Islamabad 44000, Pakistan
5 Department of Prosthodontics and Dental Implantology, College of Dentistry, King Faisal University, Al-Ahsa 31982, Saudi Arabia
6 Center for Innovation and Research in Oral Sciences (CIROS), Faculty of Medicine, University of Coimbra, 3000-075 Coimbra, Portugal
7 Institute of Endodontics, Faculty of Medicine, University of Coimbra, 3000-075 Coimbra, Portugal
8 Science of Dental Materials Department, Dr. Ishrat Ul Ebad Khan Institute of Oral Health Sciences, Dow University of Health Sciences, Karachi 74200, Pakistan; kumar.naresh@duhs.edu.pk
* Correspondence: zsultan@kfu.edu.sa (Z.K.); ppalma@uc.pt (P.J.P.); Tel.: +966-558-420410 (Z.K.); +351-239-249-151 (P.J.P.)

Abstract: Instability of the dentine-resin interface is owed to the partial/incomplete penetration of the resin adhesives in the collagen fibrils. However, interfacial hydrolysis of the resin-matrix hybrid layer complex activates the collagenolytic and esterase enzymes that cause the degradation of the hybrid layer. Adequate hybridization is often prevented due to the water trapped between the interfibrillar spaces of the collagen network. Cyclic fatigue rupture and denaturation of the exposed collagen fibrils have been observed on repeated application of masticatory forces. To prevent interfacial microstructure, various approaches have been explored. Techniques that stabilize the resin–dentine bond have utilized endogenous proteases inhibitors, cross linking agents' incorporation in the exposed collagen fibrils, an adhesive system free of water, and methods to increase the monomer penetration into the adhesives interface. Therefore, it is important to discover and analyze the causes of interfacial degradation and discover methods to stabilize the hybrid layer to execute new technique and materials. To achieve a predictable and durable adhesive resin, restoration is a solution to the many clinical problems arising due to microleakage, loss of integrity of the restoration, secondary caries, and postoperative sensitivity. To enhance the longevity of the resin-dentine bond strength, several experimental strategies have been carried out to improve the resistance to enzymatic degradation by inhibiting intrinsic collagenolytic activity. In addition, biomimetic remineralization research has advanced considerably to contemporary approaches of both intrafibrillar and extrafibrillar remineralization of dental hard tissues. Thus, in the presence of biomimetic analog complete remineralization of collagen, fibers are identified.

Keywords: biomimetic; bond degradation; dental adhesive; dentine-resin interface; resin composites

1. Introduction

Restorative and adhesive dentistry has witnessed extraordinary improvements after the innovations in contemporary adhesive materials. These new adhesive systems do not require any mechanical retention through features such as dovetails, grooves, sharp internal angles, and undercuts [1]. For the success of modern restorative dentistry, these adhesive systems play a critical role, as sound tooth structure would be preserved using these newer systems. In addition, by using these contemporary and advanced adhesive systems, secondary caries due to microleakage may be reduced or eliminated [1].

Buonocore, in 1955, reported that enamel and dentine surfaces could be made more receptive to adhesion by altering it through acid pretreatment. Moreover, he revealed that by conditioning the human enamel with 85% phosphoric acid, the acrylic resin could be bonded with enamel. Simultaneously, this technique was used for pit and fissure sealants and class III and class V restorations. Until the 1950s, developments in dentine adhesives were sluggish. '*Surface active comonomer*', synthesized by Bowen, demonstrated that resins formed a chemical bond to dentinal calcium, but the commercial products developed on the basis of this co-monomer resulted in very poor clinical performance [2]. In the dentinal bonding systems over the past 45 years, a lot of variations have been discovered in terms of the chemistry of the bonding agent, the effectiveness of the bonding agent, the technique, and the mechanism. Due to the continuously increased demand of esthetic bonded restorations, an increase in the evolution of bonding agents occurred accordingly [3]. The paradigm shift of esthetic dentistry has transformed adhesive dentistry in the past few years and has gained much consideration due to extensive research on dental adhesives and continuously changing concepts. Prompt progress in adhesive dentistry is due to the increased demand for minimally invasive tooth restorations and esthetics [4]. Different types of dental adhesives have emerged due to this research and development. These adhesive systems are prone to degradation of the resins. Therefore, the longer-term success of clinical stability and durability of resin-based polymers depends upon many factors, primarily on the degradation of the dentinal collagen. Although there have still been some uncertain hitches in the adhesive system in the past 50–60 years, it is incredible to witness adhesive bonding. The performance of the adhesive polymers cannot be comprehended accurately, as they contain a complex mixture of various constituents. For the correct clinical use of these adhesives, sound and intense knowledge is required. The literature shows variable clinical data in which low bonding strength was observed for some materials but higher bond strength in other materials [5]. Bond failure of these resin polymers over time might be due to the elution of unreacted monomers, water sorption, and polymer swelling [6].

Thus, this review consulted several studies and the factors important for the degradation of adhesive resins and the strategies to eliminate these factors. The aim of this review was to present the factors that are responsible for the resin-bond interface degradation, and the possible strategies to minimize and prevent this degradation. In addition, the role of biomimetic remineralization and associated factors in preventing the degradation of the resin adhesive interface was explored.

2. Composition of Dental Adhesives

By composition, dental adhesives consist of resin monomer solutions [7]. These monomers consist of both hydrophobic and hydrophilic groups. The interaction and co-polymerization of the adhesive system with the restorative material are by the hydrophobic component, whereas the hydrophilic component improves the wettability of the material with the hard dental tissues [8]. Other components in the adhesive system are comprised of initiators, inhibitors, stabilizers, organic solvents, and inorganic fillers [8]. However, to achieve durable bonding with these adhesive techniques, the composition and structure of enamel and dentine need to be examined. Besides water and organic material, enamel is composed of crystalline hydroxyapatite, having a high-energy surface with strong intermolecular forces. On the contrary, dentine is a biological composite of hydroxyapatite, having a collagen network with low surface energy and intermolecular forces. Structurally, dentine is entirely different from enamel, and it is humid and less stiff than enamel. Dentine contains a smear layer, organic contents, and dentinal tubules [9]. In contrast to the enamel, dentine undergoes an aging physiological process, due to which dentine permeability decreases and the thickness of dentine increases [10]. The bonding technique for enamel is different from dentine bonding due to the fact that the enamel becomes dried easily. The loss of dentine bond strength is due to the degradation of hydrophilic resin constituents and due to the degradation of collagen fibrils. This degradation will damage the hybrid layer. To avoid collagenolysis in the resin-dentine interface, extensive research activity was conducted to gain further understanding of the role of enzymes in the hybrid layer. This review emphasized various factors accountable for the degradation of collagen fibrils and the hybrid layer, in addition to it discussing the strategies to prevent and control the hydrolytic enzyme-related loss of the bond strength of the adhesives and the damage of the hybrid layer (HL).

3. Classifications of Dental Adhesives

Due to the complex nature of adhesive agents, the concept of generations has been used by the dental industry and academia. Dental adhesive systems have evolved from no-etch to total-etch to Self-Etch (4th, 5th, 6th, 7th, and 8th generation) techniques, and the details of these are presented in Tables 1 and 2. While improving the chemistry of the adhesive systems, each subsequent generation has focused on minimizing the number of steps to simplify and reduce the procedural time, and they achieved a faster application technique, which is required during clinical applications.

Table 1. A description of the components of bonding agents.

Components	Ingredients	References
Resin components	HEMA, Bis GMA, TEGDMA, is the main component of adhesives systems Different monomers, cross-linkers and functional polymer group, methacrylamides and MDPB, methacryloxylethylcetyl ammonium chloride, PENTA	[11–13]
Functional monomer	4 META and 10 MDP and GPDM	[12]
Photo-initiators	Camphorquinone, 1-phenyl-1,2 propanedione (PPD), MAPO and BAPO	[14,15]
Chemical initiators	Benzoyl peroxide and tertiary amine	[16,17]
Inhibitors	Butylated hydroxytoluene and others	[18,19]
Solvents	The most common solvents are water, acetone, Solvent effect adhesion via wettability, collagen expansion and monomer ionization	[20,21]
Fillers	Bioactive fillers, nanofillers, fluoride releasing fillers Fillers to improve radio-opacity, montmorillonite nano-clay	[22,23]
Recent modification	Glutaraldehyde as the denaturation of collagen in dentine and the occlusion of the dentinal tubules MMPs, chlorhexidine, bioactive ingredients, and antimicrobial and remineralizing agents	[24–26]

HEMA: 2-Hydroxyethyl methacrylate, BIS-GMA: Bisphenol A-glycidyl-methacrylate, TEGDMA: Triethylene glycol dimethacrylate, MDBP: 12-Methacryloloxydodecyl pyridinium bromide, PENTA: Dipentaerythritol penta-acrylate phosphate, 4 META: 4-Methacryloxyethyl trimellitate anhydride, MDP:10-Methacryloloxydecyl dihydrogen phosphate, GDPM: Glycerol-Phosphate dimethacrylate, MAPO: (4-maleimidophenyl)oxirane, BAPO: (2,4,6 trimethyl benzoyl)-phosphine oxide, MMPs: Matrix Metalloproteinases.

Table 2. Summary of the different classification systems used for categories of dental adhesives.

Classification	Description and Characteristics of Several Adhesive Systems	References
Current adhesive system	**Etch and Rinse (total etch)**, Etchant removed the smear layer to form demineralized dentine Complete infiltration of monomers is not achieved leaving incompletely infiltrated zones.	[27–30]
	Self Etch, Separate acid-etch step was not needed Stability is dependent on the coupling between the collagen fibril substrate and the adhesive resin Reduced porosities, homogenous resin infiltration, and better collagen fibrils protection Immunohistochemical labeling with anti-type I collagen antibodies presents a weak, uniform hybrid layer The efficacy of bonding to enamel without the need for separate acid etching is questionable	[30–34]
Number of steps	**Three step (4th Generation)**, Involves etch, prime, and bond (three bottles) Highest in bond strength and greatest durability	[35]
	Two step (5th generation), Etch, prime, and bond in a single coating (two bottles) Simplified method Efficient and stable bonding less predictable and more difficult dentine bonding	[35]
The historical concept of dental adhesive generations **Historical concept of dental adhesive generations** *(continued)*	**First generation adhesives (One-step)** Contains glycerophosphoric acid dimethacrylate (NPG-GMA) Ionic bond with hydroxyapatite and covalent bond to collagen Smear layer not removed, polymerization shrinkage occurs, low bond strength (2–6 MPa)	[36–38]
	Second generation adhesives (One-step), Polymerizable phosphates incorporated to bis-GMA resins to enhance bonding Formation of ionic bond in calcium and chlorophosphate groups The smear layer was not removed Debonding, microleakage, and low bond strength 4–6 MPa are disadvantages	[2,37–39]
	Third generation adhesives (One-step), Acid-etch enamel and dentine to partially eliminate the smear layer A primer application after the acid rinsed away with water Greater bond strength then first and second	[39–42]
	Fourth generation adhesives (Three-step) Golden standard in dentine bonding Complete removal of smear layer Total-etch technique and concept of hybridization introduced were introduced Technique sensitive due to complexity of multiple bottles and steps, and it was time-consuming Bond strength is in 10–20 MPa range and reduced margin leakage	[39,43–47]

Table 2. *Cont.*

Classification	Description and Characteristics of Several Adhesive Systems	References
	Fifth generation adhesives (Two-step), Etch and Rinse concept, combining the primer and adhesive resin into one application Hybrid layer formation, more prone to water degradation than 4th generation adhesive Bond strength 3–25 MPa	[45,48,49]
	Sixth generation adhesives (Two Steps), Self-Etching primers Does not involve a separate etching step Efficacy is less dependent on the dentine hydration than the total-etch systems Sufficient bond strength to conditioned dentin while the bonding with enamel was less effective	[39,45,50,51]
	Seventh generation (One-step), One-bottle Self-Etching system Prone to hydrolysis, waster sorption, and chemical breakdown Limited resin infiltration and creates some porosities Lowest initial and long-term bond strengths	[49,52–55]
	Eight Generation (One Step), Nano-filler (12 nm) present Self Etch generations An acidic hydrophilic monomer Improved enamel and dentin bond strength and stress absorption and produced a longer shelf life Increased resin monomer penetration and improved hybrid layer thickness Due to nano-fillers clusters, it can form cracks and reduce the bond strength	[56–59]

Dental adhesives are generally characterized in historical generations that reflect the amended handling performance for improvements in novel preparations rather than new adhesion concepts or mechanisms. There are two major adhesive concepts based on chemistry and the mechanism of adhesion to the tooth structure:

a. Superficial demineralization of dentine and enamel, which depend on the complete removal of the smear layer.
b. Partial or superficial dissolution of the smear layer in the adhesive interface to create a hybrid layer [27].

Both models encourage resin adhesion by micro-mechanical retention to enamel and dentine. However, a supplementary chemical bond to the substrate is also present in both concepts [28]. Van Meerbeek et al. [29] proposed an adhesive classification into two categories, Etch and Rinse, and the selective etch technique. A distinct acid-etching step was not required in Self Etch systems since the adhesive resins simultaneously infiltrate and demineralize the tooth structure and create a more homogenous infiltration of adhesive resins in the demineralized collagen fibrils [30]. The bond strength of the Self Etch adhesive technique, therefore, depends on the collagen fibril and the resin adhesives coupling [31,32]. However, the effectiveness of Self Etch bonding to enamel without an acid-etch step is still questionable [30,31].

4. Degradation of Adhesive Interface/Hybrid Layer

The adhesion of direct restoratives to the tooth structures is due to the ongoing development and recent research in adhesive systems. In spite of the developments in chemistry, composition, and classification, the stability and strength of the resin-dentine interfacial bond remain questionable [60]. The consequences of resin degradation at the adhesive interface results in post-operative sensitivity, marginal staining, and secondary caries, due to which the durability of the restoration will be compromised [40]. This might be due to the fact that degradation will produce marginal deterioration and weaken the adhesion procedure [61]. It is a commonly established fact that dentine-resin adhesive interface degradation occurs after the use of current dental adhesives and has been an interesting research topic most recently. The following factors are proposed to influence the degradation of the hybrid layer and resin-dentine interface.

4.1. Degradation of Adhesive Resins

The main factor involved in the chronic degradation of the adhesive resins is the hydrolysis and adhesive resin leaching from the resin-dentine matrix [62,63]. Water diffusion into the hydrophilic adhesive initiates the leaching process, and, due to the adhesive phase separation, there is a limited degree of polymerization in the hydrophilic domains [64]. In the aqueous environment, the poorly polymerized hydrophilic phase undergoes degradation more quickly. The primary factor involved in the reduction of the bond strength of the resin adhesive interface is the hydrolysis within the hybrid layer, which contributes to poor adhesion after some time [64]. Water begins to penetrate the resin restoration interface after prolonged exposure of the resin restorations to oral fluids. The water acts as a plasticizer between the polymer chains of the adhesives and as a molecular lubricant. This molecular lubricant will cause mechanical wear of the exposed adhesives [64]. This allows for greater transport of both water and enzymes, along with increasing the surface area and abrading the resin dentine interface surface, leading to the acceleration of matrix degradation [38].

The other factor that enhances the degradation of the resin interface is chemical hydrolysis due to the water transport or salivary fluids in methacrylate materials, which results in damage to the ester bonds [38]. An immediate increase in the bond strength was observed after the infiltration of the exposed collagen fibrils by the hydrophilic 2-hydroxyethyl methacrylate (HEMA) monomer [63,65]. The limitation of these adhesive systems is that they compromise the longevity of the dentine-resin bond [66]. One of the most important factors besides the presence of water, that contributes to the degradation of adhesive resins, is the incomplete polymerization of variable degrees that can be associated to the extent of fluid movement in the adhesive hybrid layer [5,62,63,67]. Contemporary dental adhesive systems comprise both hydrophilic and hydrophobic components. Hydrophobic monomers continue to stay on the surface, whereas the hydrophilic components infiltrate the interior of the hybrid layer [60]. These systems produce heterogeneous resin layers due to the nanophase separation ration phenomenon [68].

One of the most significant causes involved in the degradation of the resins is the hydrophobic camphorquinone (CQ) photo-initiator [69]. The hydrophobic CQ initiator may potentially cause a suboptimal degree of conversion of hydrophilic monomers [69], resulting in deficient polymerization in the hybrid layer deep zone [69]. In that instance, to improve the degree of conversion of adhesives, it is suggested to use camphorquinone in addition to water-compatible photo-initiators, such as TPO (ethyl 4-dimethylaminobenzoate and diphenyl (2,4,6-trimethylbenzoyl)-phosphine oxide) [69]. This photo-initiator possibly reduces the damaging effect of the nanophase separation by increasing the degree of conversion of the hydrophilic and the hydrophobic components of the resin polymer [68,70,71]. Other minor causes of resin interface degradation include the expansions and the contraction in resin restorative materials due to temperature changes and occlusal forces. These factors compromised the dentine resin bond stability by allowing the penetration of oral fluids and water into the resin interface [72]. Moreover, hydrophobic and hydrophilic cytotoxic by-products, such as ethylene glycol and methacrylic acid, are released because of the breakage of ester bonds present in the HEMA [73]. This ester bond breakage also occurs when saliva, pulp, and bacteria release esterase enzymes [74].

4.2. Degradation of Collagen

Dayan et al. [75] and Tjäderhane et al. [76] have reported the collagenolytic activity in dentine. In aseptic conditions, collagen can degrade over time, as confirmed by Pashley et al. [77], because of the intrinsic matrix proteases. It was reported that treated specimens with enzyme inhibitors remarkably found a decrease in the intrinsic dentine gelatinolytic and collagenolytic activity [77]. Thenceforth, scientists put their efforts towards exploring the role of these enzymatic activities in the degradation of the HL, enzymes involved in the degradation, and their localization within the dentine, and they also investigated the approaches to reduce or stop this enzymatic activity [77]. The matrix metalloproteinase and cysteine cathepsins are the most noticeable collections of endogenous enzymes within dentine, and they are discussed below.

4.2.1. Matrix Metalloproteinase (MMP)

MMPs are proenzymes common to both bone and dentine. They are Zn^{2+} and Ca^{2+}-dependent endogenous proteases that consist of a bridge between the Zn^{2+} ions and the cysteine residue [78]. The tertiary structure of MMPs is preserved by their Ca^{2+} part, whereas the Zn^{2+} ions are responsible for the enzyme activation. In the intact form, these MMPs prevent the binding of the Zn^{2+} ions with the water molecules, thereby preventing enzyme activation (Figure 1) [78]. Their classification is based on the substrate on which they act similar to collagenases (MMP-1, -8, -13, and MMP-18), gelatinases (MMP-2 and MMP-9), stromelysins (MMP-3 and MMP-10), matrilysins (MMP-7 and MMP-26), and membrane-type MMPs (MMP-14, -15, -16, and MMP-24). MMPs that are present in human dentine are MMP-2, -3, -8, -9, and MMP-20 [79].

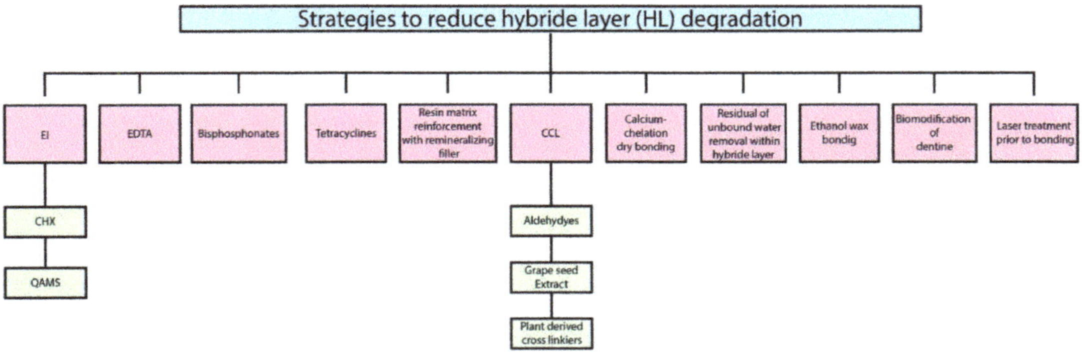

Figure 1. Summary of the strategies used to reduce hybrid layer degradation. EI → Enzymes Inhibitors; CHX → Chlorhexidine; QAMS → Quaternary ammonium methacrylate compounds; EDTA → Ethylene diamine tetra acetic acid (EDTA); CCL → Collagen Cross Linkers.

During dentine maturation, MMPs play an important role, but they become inactive after the collagen matrix mineralization is completed [80]. In addition, during growth, organogenesis, and normal tissue turnover, MMPs are responsible for the degradation of extracellular matrix proteins [81]. Nevertheless, the comparative influence of such enzymatic activity on bond degradation is still indistinguishable. The most relevant pathway of bond degradation is thought to be due to hydrolytic water sorption [81]. Although there is inconsistent literature available on the different kinds of adhesives about enzyme exposure and activation, the data regarding retarding/arresting bond degradation effects upon use of MMP inhibitors vary among studies [81]. The most plentiful MMPs found in dentine are MMP-2 and MMP-9 [81]. Mazzoni et al. [82] studied MMP-2 and MMP-9 molecular forms in the demineralized dentine by immunoassay, gelatin zymography, and western blotting [82] and reported that the organic matrix of the fibrillar network of human dentine consists of both MMPs as intrinsic constituents [82]. It was also found that both mineralized and un-demineralized dentine have distinctive distribution patterns and concentrations of MMPs, and they concluded that the bioavailability and activation of MMPs are affected by the demineralization [61]. The roles of MMPs in the Etch and Rinse technique and the Self Etch approach were compared in a study after mixing adhesives with human dentine powder, and it was reported that, when such adhesives were applied to acid-etched dentine, there was an increase in the quantity of activated, non-denatured enzymes and proteolytic activities in the tested adhesives [83,84]. However, the denaturation of MMPs can occur during the conditioning of dentine at a lower pH (0.7–1) of phosphoric acid [85]. Latent forms of the enzymes activated by low pH trigger the cysteine switch and employ their effect on the catalytic action of dentine. The inhibitory activity of tissue inhibitors of MMP also decreases through this process. However, other studies showed the disparity and demonstrated that, after mixing demineralized dentine powder with Self Etch adhesives, the low pH value was neutralized rapidly in the Self Etch approach, resulting in momentarily preventing the MMPs activity [86]. Many in vivo and in vitro techniques can be used to activate the MMPs. In vitro techniques can be physical (heat and low pH) and chemical (chaotropic agents, sulfhydryl groups, and reactive oxygen). In vivo activation can be by proteases and other MMPs, and they are associated with MMP activation, which was induced by the dentinal adhesive application [87]. Lehmann et al. [88] found that, in human odontoblasts, the synthesis of MMP-2 was increased after the application of the adhesive, due to which movement through the hybrid layer to dentinal tubules also increased [89].

4.2.2. Cysteine Cathepsins

In dentine matrices, the prominent endogenous proteolytic enzymes that are involved in the degradation of dentine are cysteine cathepsins (CTs). The presence of CTs in dentine was reported to compromise 98% of the cathepsin activity against dentine collagen [90,91]. These proteases are expressed by mature human odontoblasts cells and pulpal tissues. In human dentine, there are 11 CTs that have been found involving the CT-K, CT-L, and CT-B [92]. They can generate multiple collagen fragments due to the presence of glycosaminoglycan (GAG) side chains, as they are able to cleave helical collagen at multiple sites, whereas other CTs and MMPs can only cleave the non-helical telopeptide part of the collagen cleave [93]. The association of cysteine cathepsins with caries progression and hybrid layer degradation was demonstrated by a few researchers [90,91,94]. MMPs and cysteine cathepsins are located near the target substrate and are close to each other. Therefore, there is synergistic activity between them that was found in sound and carious dentine. In this way, the two different classes of these proteases worked as a cascade network [90,91,94]. Dentine-bounded MMPs were further activated by the acidic activation of cysteine cathepsins [95]. In addition, the spectrofluorometric use of fluorogenic substrates was carried out for these proteases (MMP and CT activities), and it was found that, in matrix degradation, two types of proteases have diverse coordinated roles. Interestingly, in dentine CT-MMP interactions, it was found and demonstrated that MMPs and CTs regulate the activities of each other [91].

4.3. Incomplete Infiltration of the Resin Adhesives

Incomplete hybridization of the adhesive in the collagen complex in the Etch and Rinse technique is a result of the difference between penetration of the adhesive and action of the conditioning acidic agents [96]. Nanopercolation occurs because of the incomplete hybridization of collagen networks, as they become more susceptible to hydrolytic degradation [96]. A stable and complete hybrid layer cannot be achieved such as in the inter- and intrafibrillar compartments; the monomers are not able to replace the free and collagen-bound water [97,98]. In addition, large monomers such as BIS-GMA were entrapped in inter-fibrillary spaces due to highly hydrated proteoglycan hydrogels. These proteoglycan hydrogels only allow small monomers such as HEMA to penetrate toward the base of the hybrid layer [99]. As HEMA creates weak linear chains, they will cause cyclic fatigue failure of the collagen chains when subjected to stress [100].

5. Strategies to Reduce Hybrid Layer (HL) Degradation

Several strategies have been reported in the literature to improve the longevity and decrease the degradation of the hybrid layer by using protease inhibitors, enzyme inhibitors, and collagen cross-linkers as primary agents [101–103]. Some of these agents that influence the endogenous dentinal enzymes are discussed in the section below.

5.1. Enzymes Inhibitors

5.1.1. Chlorhexidine (CHX)

Chlorhexidine is the most extensively investigated enzymatic inhibitor and an antimicrobial agent and is characterized by the excellent inhibition of MMP activity in both dentine and resin [104]. Due to chelating properties at low concentrations (0.2%), it showed excellent inhibition of protease enzymes [104–109]. The literature regarding the mechanism of action of CHX is scarce, but it might be because of the cationic behavior of CHX, as it binds to both mineralized and unmineralized dentine [105–108]. As CHX consists of chlorine, it prevents its hydrolytic activity by binding to the zinc of the catalytic domain of MMPs [104]. When CHX digluconate (0.5 wt.%, 2.0 wt.% and 4.0 wt.%) was incorporated in experimental etch-and-rinse adhesives, no adverse effects on the degree of conversion were found [110]. However, Cadenaro et al. [111] demonstrated that, although the degree of conversion was not affected by the incorporation of 1% or 5% CHX into the adhesive resins, the elastic modulus was significantly decreased. Furthermore, bond strength was preserved for 12 months without affecting the ultimate tensile strength, solubility, DC, and water sorption after the incorporation of CHX in Etch and Rinse adhesives [112]. Da Silva et al. [113] incorporated Galardin, Batimastat, GM1489, and CHX as enzyme inhibitors in Etch and Rinse adhesives and found similar results in CHX and GM1489.

5.1.2. Quaternary Ammonium Methacrylates Compounds (QAMS)

The mechanism of action of quaternary ammonium compounds is similar to CHX, as both are positively charged. The most prominent and already tested MMP inhibitor in the quaternary ammonium compounds is benzalkonium chloride (BAC). BAC consists of various alkyl chains and is a combination of alkylbenzyl-dimethylammonium chlorides. These compounds showed favorable bond strength results over time in comparison to CHX, as they strongly bind to demineralized dentine [114–116]. To evaluate the mechanical properties of the unfilled resins after incorporating quaternary ammonium methacrylates, Hoshika et al. [117] found that the degree of conversion of these resins was improved, but wet toughness and ultimate tensile strength were decreased by the addition of 10% QAMs, whereas when 15% QAMs were added, it decreased the mechanical properties of the resins significantly [117]. Therefore, it was documented that, although the degree of cure will be increased, the ultimate tensile strength and Young's modulus were decreased in QAM containing adhesive [117]. 12-meth-acryloyl-oxydodecyl-pyridinium bromide (MDPB) is a quaternary ammonium methacrylate QAM and has been incorporated into several contemporary adhesive systems. It is well-known for its antimicrobial properties in the adhesive resins, MDBP polymerize with methacrylate; therefore, leaching of this compound was restricted and can serve as a microbe [118]. It has an excellent capability of preventing MMP activity [118–121]. Further research is needed to demonstrate that adhesives based on QAM compounds increase the strength of the resin-dentine interface by effectively inhibiting MMPs at the resin dentine interface and also to focus on the development of quaternary ammonium functionalities [122].

5.2. Ethylene Diamine Tetra Acetic Acid (EDTA)

For decades, due to the chelating properties, ethylene-diamine tetra acetic acid (EDTA) has been used in endodontics. EDTA binds to Zn^{2+} ions from the catalytic site of the MMPs and removes the Ca^{2+} from the collagen matrices [123,124]. However, a long application time and the reversibility caused by water solubility are the main drawbacks [125].

5.3. Bisphosphonates

Bisphosphonates are proteases inhibitors that act by chelating zinc and calcium ions from several enzymes [126]. In particular, good instant outcomes were observed with polyvinylphosphonic acid (PVPA), but with uncertain durability [127]. Tezvergil-Mulutuay et al. [127] used polyvinylphosphonic acid (PVPA) on recombinant MMP-9 and found that it efficiently inhibited this gene with less hydroxyproline release. The bonding between PVPA and collagen is electrostatics. PVPA can become trapped in collagen through 1-ethyl-3-(3-dimethyl aminopropyl) carbodiimide (EDC) [128]. This phenomenon makes PVPA more advantageous in terms of bond durability when compared with CHX. All these outcomes suggested that, to enhance the durability and longevity of resin dentine bonds, PVPA can be incorporated. However, there is a dearth of literature about bisphosphonates as MMP inhibitors; thus, future work should be performed.

5.4. Tetracycline

Tetracyclines along with their analogues doxycycline and minocycline are broad-spectrum antibiotics used as effective MMP inhibitors, having cationic chelating properties. [129,130]. Doxycycline decreases dentine matrix degradation intensely [131]. Chemically-modified tetracyclines (CMT-3, Metastat) are exceptionally efficient in reducing MMPs in dentine caries and can retain their MMP-inhibiting capacity, although the antimicrobial activity is not ideal [132]. They act on enzymes by inhibiting their activity and secretions. In addition, they are also involved in calcium chelation [133]. However, improvements in the dentine-resin bond by the tetracycline have not yet been evaluated. These compounds need further investigation due to their potent MMP inhibiting potential. Though, during photo-oxidation, these compounds can induce the purple stain of teeth and are therefore not considered suitable for clinical use.

5.5. Collagen Cross Linkers

In dentine, collagen matrix cross-linking is not only a natural mechanism used to increase the mechanical properties of dentine, but it also is used to make dentine less susceptible to enzymatic and hydrolytic degradation [134,135]. Several studies, therefore, have reported the incorporation of numerous chemical constituents that have cross-linking properties in dental adhesives [134,135]. Cross-linking agents stabilize the collagen structure and make it more resilient to enzymatic degradation by attaching to the amino-carbonyl groups of the collagen [134,135]. These cross-linkers prevent the hybrid layer degradation through several mechanisms, as mentioned below:

a. Firstly, cross linkers disable the degeneration process by changing the molecular mobility of the MMPs [134,135].
b. Secondly, they render this collagen less prone to hydrolysis by creating supplementary cross-links between the molecules of collagen [134,135].
c. Lastly, many different mechanisms have been involved in the inhibition of MMPs and CTs by these cross-linkers.

These mechanisms could be evading the cysteine switch oxidation, which in the substrate cleavage sites will be hidden by these cross-linkers, and the dysregulation of endogenous protease expression, within collagen protection of cleavage sites, by cross-linkers and the inactivation/silencing of proteases activity [136].

5.5.1. Aldehydes

Aldehyde (glutaraldehyde) is widely reported in the literature as a strong cross-linking agent for its use in dentistry [135,137–140]. However, it is rarely used in clinical dentistry because of its cytotoxicity [139,141]. It forms a covalent bond between the amino groups of peptidyl lysine and hydroxylysine residues within the collagen [139,141]. To improve and enhance the mechanical durability of dentine, the simplest unsaturated aldehyde known as Acrolein (acrolein (2-propenal)) was used as an additional primer strength preservation of dentine resin interface [142]. As with the glutaraldehyde, Acrolein is also proved to be cytotoxic and inadequate for clinical practice.

5.5.2. Grape Seed Extracts

The agents with low cytotoxicity have been widely investigated, such as crosslinkers derived from natural grape seeds known as proanthocyanidin [143–145] and carbodiimides [103,146]. In contrast to the aldehydes, they can be easily used in daily clinical practice due to the fact that they are biocompatible and are not cytotoxic [134]. In dentistry, 1-ethyl-3-(3-dimethylamino-propyl) carbodiimide (EDC) is the most extensively studied carbodiimide [143–145]. Ionized carboxyl groups in proteins react with the functional group of EDC and form an intermediate compound known as O-acylisourea. This intermediate product forms a steady covalent-amide bond between the two proteins by reacting with lysine and hydroxylysine amino groups to aid amino cross-linkages with collagen. EDC has a dual function; firstly, in collagen, EDC cross-links with both telopeptide and helical domains, whereas, at the same time, EDC prevents telopeptidase activity [147]. A similar bond strength was found in the resin dentine interface preservation for both EDC and GD [148], but in contrast, EDC showed much lower cytotoxicity. The stiffness of both the demineralized dentine matrix as well as the hybrid layer has been well documented with EDC [149–151]. It was demonstrated that, when EDC was applied on acid etch dentine and stored in artificial saliva for 1 year, it deactivated MMPs efficiently. Dentine powder was studied in this research with the help of zymographic assays [148], and three-dimensional in situ zymographic images were obtained by confocal microscopy [103,146]. However, the use of EDC is limited to clinical practice as it needs a relatively long time (1 h) to cross-link to collagen [152]. Similarly, the cytotoxic effects of 5% GA and EDC in varying concentrations on odontoblast-like cells with dentine barriers were studied by Scheffel et al. [153], and they found that lower concentrations of EDC (0.1, 0.3, and 0.5 M) and 5% GA did not produce trans-dentinal cytotoxic effects on odontoblast-like cells.

5.5.3. Plant-Derived Cross-Linking Agents

The most widely reported plant-derived cross-linking agents used in dentistry are genipin, tannins (polyphenolic compounds), oligomeric proanthocyanidins, and curcumin due to their high potency and low cytotoxicity [102,139,145,154]. These cross-linking agents reinforced the hybrid layer and improved the resin-dentine bonds by reacting with degradation products, resulting in the late polymerization of adhesive resin [155]. These antioxidant substances that possibly prevent MMP activity in the dentine also promote non-enzymatic collagen cross-linking. The elastic strength increases with an increase in the degree of cross-linking of the collagen [156]. Dentine biomodification increases the mechanical properties of demineralized dentine through nonenzymatic collagen cross-linking [157,158]. Proanthocyanidins are derived from grape seeds and improve the tenacity and elastic modulus of demineralized dentine [159]. However, they are considered unsuitable clinically due to their lengthier (10–60 min) application duration [138]. Moreover, brown pigmentation in dentine, as well as polymerization inhibition of resin monomers due to a decrease in the degree of conversion, has been observed with these agents [160,161]. The cross-linking effect is also accomplished by photochemical treatment and physical agents apart from chemical compounds [160,161]. These treatments and agents could be drying, heating [162], and ultraviolet A (UVA) [163] and gamma irradiation. Glutaraldehyde [116], EDC [146,147,164], and similar cross-linking agents [165,166] derived from plants have the capabilities to react with collagen-degrading enzymes.

5.6. Residual or Unbound Water Removal within the Hybrid Layer

One of the greatest challenges of highly cross-linked resin polymers is that they undergo phase changes due to their poor solubility in water [64]. Therefore, manufacturers prepare commercial adhesive systems in a variety of solvents such as ethanol to warrant the single phase of resin adhesives during clinical application. When adhesives containing solvent are applied on moist acid-etched dentine, microscopic phase changes are observed in the adhesives. Pashley et al. [167] demonstrated the solution to this problem by replacing the water rinse with ethanol in the wet-bonding technique, which resulted in dentine saturation by ethanol, not with water. Similarly, Tay et al. [168] reported that, when bis-GMA was applied to ethanol-saturated dentine, excellent resin–dentine bonding was achieved. The danger of phase separation can be avoided completely by using ethanol containing adhesives to dentine treated with ethanol by reducing the residual water at the dentine-resin interface [169]. Due to water absence in the hybrid layer, the collagen matrix may not be cleaved by matrix proteases. However, in the water-wet bonding approach, hydrophobic infiltration of resin adhesives is much less than in ethanol wet-bonding [20,170]. It is well documented that demineralized dentine and ethanol can replace and remove unbound water [171]. Jee et al. [98] determined in their study whether bound water can be replaced by collagen matrices and found that the tightly bound first layer and the second layer of water cannot be replaced by ethanol in collagen, although most of the bound water in the outermost layer was substituted by ethanol.

5.7. Calcium-Chelation Dry Bonding

The acid etchant (37% phosphoric acid) used during the adhesive process completely demineralized the collagen fibrils because of its molecular weight, which is around 100 Da, which can easily pervade throughout collagen fibrils [172,173]. Molecules with a molecular weight smaller than 600 Da can enter collagen fibrils easily, and all molecules with a molecular weight larger than 40 KDa cannot permeate into collagen fibrils easily [169,170]. Keeping this point in mind, studies have used calcium chelator (sodium polyacrylate) instead of phosphoric acid [172,173]. They used 15 wt.% calcium chelators with a molecular weight of about 225,000 Da, which is large enough to permeate collagen. After 30 s of chelation, water rinsing and air-drying were carried out, due to which the reaction was stopped. This will lead to the preservation of inter-fibrillar gaps for monomer diffusion inward and those absorbed into the hybrid layer, and only apatite mineral was removed from the extrafibrillar space [174]. In this approach, collagen fibrils remained too stiff to shrink or collapse even after the removal of residual water, as the collagen fibrils remained completely mineralized [174].

5.8. Biomodification of Dentine

To enhance the physical and mechanical properties of dental hard tissue, advancements in biomodification have been made by modifying the biochemistry of these dental hard tissues by incorporating or inducing physical agents [134]. One of the most significant physical agents used as biomodification is photo-oxidative techniques [134]. This technique utilized ultraviolet light, which requires the most reactive and unstable type of oxygen singlet for activation. This type of oxygen singlet can be provided by vitamin B2 (riboflavin). Cross-linking occurs between the proline and hydroxyproline of side chains (carbonyl groups) and the glycine of a collagen chain (amino group). These oxygen singlets form covalent bonds when activated by ultraviolet light [134].

5.9. Ethanol Wet Bonding

The prime factor involved in the adhesive bond strength durability is the hydrolytic degradation of adhesives. Hydrophilic and ionic monomers have been added to these adhesives to ensure the proper hybridization of wet collagen matrix [8,175,176]. Mechanical properties of these hydrophilic adhesives were found to be compromised, as these polymers contain ester linkages and they are susceptible to water sorption and/or hydrolysis [8,175,176]. A 12-month in vivo study conducted by Brackett et al. [177] concluded that, despite preserving the adhesives with CHX, water-related loss of nano-fillers was observed. In a few studies, loss of bond strength was found despite containing CHX and other enzyme inhibitors [8,175,176]. This loss of strength in adhesive resins might be due to polymerized hydrophilic adhesives, water sorption, or adhesive monomer degradation.

These studies concluded that if water is eliminated from the bonded interface, then water hydrolysis of peptide bonds in collagen and ester-bonds in adhesive polymers might possibly be reduced. This has been the main objective of introducing the concept of ethanol-wet bonding [177]. The phenomenon behind this concept is that acid-etched demineralized dentine matrices that were dehydrated by the ethanol aided the penetration of higher hydrophobic monomers into the interfacial dentine and reduced the collagen hydrophilicity [177]. Ethanol-wet bonding wheedles the infiltration of hydrophobic monomers to demineralize collagen with restricted matrix shrinkage [167]. Water sorption/solubility and resin plasticization decrease because of the infiltration of hydrophobic monomers. In addition, it has been suggested that improved durability of the resin bond occurs, resulting in decreased enzyme-catalyzed hydrolytic collagen degradation because of the elimination of residual water [35,178], and the hybrid layer generated with ethanol-wet bonding also resulted in outstanding durability of resin bond strength and an almost complete absence of nano-leakage [179].

5.10. Resin Matrix Reinforcement with Remineralizing Fillers

As in the previous section, it was discussed that not only enzymatic degradation will degrade the hybrid layer, but chemical degradation also played a significant role in the degradation of adhesive durability. Fillers and nanoparticles are included in the top priority list to use as reinforcing adhesives materials [180]. Several studies have demonstrated an increased bond strength and enhanced mechanical and physical properties of the adhesives after the incorporation of copper [180], carboxylic acid-functionalized titanium dioxide [181], silver micro-fillers [182], and zinc oxide [183] nanoparticles. To improve the bond strength of the commercial three-step etch-and-rinse adhesive system (Scotchbond™ 3M ESPE, St Paul, MN, USA), Zirconia nanoparticles were incorporated by Lohbauer et al. [184] into the primer or adhesive. The incorporation of these particles resulted in improved resistance to the hydrolytic process, which might increase the durability of the dentine-resin bond. The rate of the bond degradation will reduce when hydrolysis is diminished due to the reduced water uptake that retarded the proteases activity, leading to the formation of a stronger hybrid layer [161]. Nanotubes are a hexagonal network of carbon atoms that are extremely strong and stiff and have excellent thermal and electrical properties, and they were also incorporated as fillers to resin-based restorative materials to reinforce the resin matrix [185] and thus the resin-dentine bond strength [186,187]. The incorporation up to 20 wt.% of nanotubes in the etch-and-rinse adhesive system and up to 10 wt.% in Self Etch adhesives system have resulted in increased bond strength [183]. The most fascinating quality of nanotubes is the possibility of expanding the cylindrical hollow structure as a medium for the encapsulation of therapeutic molecules as well as protease inhibitors [188]. Feitosa et al. [188] reported that an inhibitor of MMPs (doxycycline) was encapsulated into nanotubes and then incorporated into an adhesive resin, and it was able to inhibit MMP-1 activity without compromising bond strength results. Similarly, nanotubes can be used as a vehicle for the encapsulation of biomimetic agents to prevent bond degradation due to the release of MMP inhibitors, antioxidants, and collagen crosslinkers [188]. Although dental adhesive reinforcement with nanoparticles is an excellent strategy, the beneficial effects due to nanoparticle incorporation are very vulnerable due to agglomeration or the inhomogeneous dispersion of nanofillers in the resin phase, which may reduce the bond strength and physical stability of the adhesive materials [188]. Usually, there is a threshold for filler loading into adhesives, and it depends on the composition of the adhesives and filler type [189].

5.11. Laser Treatment Prior to Bonding

For bond strength enhancement, laser irradiation of enamel/dentine has been utilized [190,191]. Erbium-doped Yttrium Aluminum Garnet (Er,YAG) [190,191] and plasma-based lasers [192,193] are the two key sources of lasers that are employed in dentistry. The key factors that contribute to the success of dental bonding are the smear layer removal, organic content, and water evaporation, and an increased surface area was obtained by Er,YAG laser irradiation [191]. Preceding bonding when dentine is treated with non-thermal atmospheric pressure plasma (NTAPP) laser, it was found that it aided with improved immediate bond strength as well as long-term bond stability after ageing [192]. This might be because carboxyl and carbonyl groups are grafted by the laser application onto the dentinal substrate, which will enhance the chemical and mechanical interaction of the resin monomers. Moreover, positive effects were found when oxidizing agents were used on the dentine before laser application [190], and the authors found a greater resin-dentine bond strength when the Er,YAG laser application on dentine was performed after bleaching. It was demonstrated that increased releases of free radicals were found after the laser during bleaching and made the surface of dentine such that it received adhesives monomers in a better tactic. This will increase the durability of the dentine and resin bond strength. Correspondingly, an efficient method to enhance the resin dentine bond strength is utilizing a non-thermal argon plasma laser for 30 s on sodium hypochlorite-treated dentine [193]. This increase in strength might be due to increased hydrogen bonding interaction between collagen fibrils and adhesive resins after the dentine etching [193].

6. Biomimetic Remineralization

In recent decades, biomimetic has developed as a multi-disciplinary approach in dentistry. There are several biomimetic approaches utilized in the field of restorative dentistry; an example is a tooth that was restored using bioinspired peptides, bioactive biomaterials, and biomimetic tissue regeneration to achieve remineralization [194]. To improve the strength properties of adhesive materials, developments in the contemporary adhesive materials and understanding at the nanoscale of biomaterial–tissue interaction are continuously investigated and evaluated [194]. The most ideal, novel, and exciting approach to prevent the collagen fibrils from degradation is biomimetic remineralization. Biomimetic remineralization involves the leaching of ion-releasing materials that simulate the natural remineralization process [195,196]. This process removes the residual water from water-rich regions and intrafibrillar spaces of the hybrid layer by inactivating proteases and reducing collagen degradation. These phenomena increase and restore the strength of the hybrid layer by replacing fibrils with apatite crystallites as well as preserve the durability of the resin-dentine bond interface by preventing the exposed collagen from external challenges [195,196]. There are two types of biomimetic remineralization that occur in adhesive dentistry:

(a) The first approach cannot occur in demineralized dentine where apatite crystals are absent. In this type of remineralization, the remaining mineral crystals act as templates for the regrowth of apatite crystals [197].
(b) The second type of biomimetic remineralization involves incorporating polyanions (polyacrylic acid/polyaspartic acid) and apatite nucleation, resulting in biomimetic remineralization [197].

The carious lesions may lead to the exposure of the collagen fibrils due to the loss of minerals from the dentine. This will lead to the degradation of the collagen fibrils and the deterioration of the mechanical properties of the dentine [196]. Moreover, during restorative and adhesive procedures, various type of techniques and agents are used, such as acid etching, acidic monomers, and chelating agents, that cause the demineralization of dentine and enamel. In addition, partial infiltration of collagen fibrils with resin monomers will cause the micro-permeability and nano-leakage of hybrid layers, especially in the contemporary adhesive system [197]. In addition these systems are not able to remove collagen fibrils, further compromising the properties of the polymeric adhesive system. Due to the incomplete or partial infiltration of resin monomers, these hybrid layers consist of numerous water-filled regions [197]. Insoluble collagen fibrils are slowly solubilized by these that are water-filled. Remineralizing reagents such as nanometer-sized apatite crystallites can be incorporated into these water-filled voids. Polyanions act as templates for specific calcium-binding to promote the nucleation of appetites [197]. Moreover, intrafibrillar and interfibrillar remineralization of dentin collagen fibrils were very well demonstrated after the application of non-collagenous protein (NCP) analogues [197].

6.1. Amorphous Calcium Phosphate (ACP)

When collagen fibrils were completely demineralized, amorphous calcium phosphate (ACP) was used to achieve the biomimetic remineralization of adhesive resins [198]. To induce growth and nucleation of appetite, this ACP will bind with collagen and serve as a template. In order to produce crystals of apatite with the length that is the best fit in the collagen gap, clusters of polyanions are incorporated around the ACP [198]. This phenomenon is a self-limiting process that takes place inside the water trees (water-filled voids) in the hybrid layer. ACP will slowly and gradually release the apatite in these water trees to mineralize the dentine and enamel. Nevertheless, clinically, this process does not yet seem to be useful, as in vitro lateral diffusion mechanisms were used for the incorporation of the crystals in demineralized dentine [29]. From the practical point of view, this in vitro concept must be translated into clinical trials; although it still has a long way to go, once it is clinically acceptable, these open doors will pave a new revolution in adhesive dentistry. The ossification of MMPs and the remineralization of collagen fibrils occur when hydroxyapatite permeates into these water-filled spaces. Tay and Pashley [199] used ACP in Portland cement for the remineralization of the collagen network by the deposition of apatite crystals and found a meta-stable amorphous calcium-phosphate in these cements. Numerous studies utilized such kinds of biomimetic analogs in the adhesive resins [97,200,201], and they found that the extrafibrillar mineral deposits were formed when mineralization was achieved without the biomimetic analogues. Similarly, biomimetic analogues were directly bonded to the collagen network instead of biomimetic phosphoproteins into the solution and dentinal collagen complete remineralization in a few months [202,203].

6.2. Phosphoproteins Analogues

Phosphoproteins analogues play a very significant role in the maturation and mineralization and process for biomimetic remineralization. Without these analogues, incomplete maturation and mineralization occurs [97,200,201,204]. When adhesives (Etch and Rinse [202]), Self-Etch adhesives [203], and primers were incapacitated with phosphoprotein biomimetic analogue investigators observed after 3 months of storage, Etch and Rinse adhesives were found to be increased in microtensile bond strength, and after 6 months of storage, Self Etch adhesive increased in bond strength. Because of many limitations, this technique is not a routine practice in dentistry. When the biomimetic approach was applied to Etch and Rinse adhesives, both intrafibrillar and extrafibrillar compartments were occupied by the apatite crystals, whereas when this approach was applied in Self Etch adhesives, only intrafibrillar spaces were deposited by the apatite crystals [205,206]. For the remineralization of completely demineralized dentine, usually 3–4 months are required.

6.3. Polyvinylphosphoric Acid Analogue

Better mechanical properties of the dentine can be obtained using polyvinyl phosphonic acid as biomimetic analogue. This strategy inhibits the endogenous MMPs of dentine through biomimetic remineralization and prevents collagen degradation [128]. It was a documented fact that the strength of the resin dentine interface decreases after 12 months of storage prior to biomimetic remineralization, but the strength after the remineralization long term storage will prevent the durability of the adhesive resin bond [207]. Although these treatments need to be optimized and take additional time for application, they also promise a durable bond over time due to the inactivation of the endogenous proteases of dentine [60].

6.4. Bioactive Silicates

One of the strategies to induce biomimetic remineralization is to combine hydrophilic, biodegradable polymers with sodium/calcium phosphosilicate bioglass (bioactive silicates) to decrease collagen degradation [208]. These silicate adhesives showed biological activity when encountered with biological fluid and will produce ionic dissolution products; therefore, they act directly at the level of the hybrid layer. However, this spectacle needs to be further discovered, because in a study, researchers found a decrease in mechanical properties of adhesives and an increase in the bioactivity. Because of the decrease in mechanical properties, bond durability will be compromised [209]. Sauro et al. [202] induced biomimetic remineralization by incorporating phosphoproteins in acid etch dentine and reported that the ion-releasing resin adhesive resulted in biomimetic remineralization due to the release of calcium silicate from adhesive resins [202].

6.5. Fluoride

Tooth remineralization also occurs because of an important element known as fluorine [210]. Thus, an adhesive containing fluoride is superior for the hybrid layer strengthening, prevention of secondary caries, and preservation of degradation of the resin dentine bond. Nonetheless, the use of fluoride as a biomimetic analogue is questionable, as biomimetic remineralization requires the seeding of apatite crystallites to remineralize the collagen matrix [60]. Authors in another study incorporated fluoride in resin adhesives and emphasized the significance of fluoride-containing adhesives. They found that, after water storage, the bond strength to dentine was significantly increased [211]. However, there is a dearth of literature in this context, and further investigation will be needed [60,161,209]. Overall, regarding adhesive dentistry, all biomimetic remineralization/biomimetic "smart" materials are new and require further investigations.

7. Conclusions

Long-term degradation of dentine-bonded interfaces due to the aging of these adhesives is a major drawback. To enhance the adhesion of resin dentine durability, experimental strategies have been developed with varying success rates. This includes improving the dentine resistance of the collagen matrix to the degradation by enzymes and by preventing collagenolytic activity intrinsically. Using nanotechnologies and other innovative techniques, improvements in the conventional materials and a translation into contemporary materials are needed. These improvements could include the inhibition of MMPs and cathepsins, strengthening of the collagen scaffold, antimicrobial properties, collagen strengthening, and regenerative processes of dental hard tissue. In the next coming years, advancements in technology used to resist collagenolytic hydrolysis and form stable resin–dentine bonds will be available.

8. Limitations

After the minimal cavity preparation, when a thicker layer of dentine is affected by the carious lesion, non-homogenous water saturated matrices of resin infiltration occur. The regions where water is in excess and resin is scant will damage over the period of 1–2 years. Moreover, they will undergo fatigue failure due to extreme cyclic strain. To prevent such degradation, demineralized collagen fibers should be remineralized by mineral deposition both intra- and extrafibrillarly. Currently, the dental industry and researchers are more focused on the formulation of the passive adhesive system rather than the development of materials with more clinical longevity that can easily be used in dental practice. These new passive light-curable resin-based materials might cause a toxic reaction by the release of the components.

Author Contributions: Conceptualization, F.A., M.A.F. and M.S.Z.; methodology, F.A., M.A.F., M.S.Z. and Z.K.; software and data extraction validation, F.A., M.A.F. and M.S.Z.; manuscript writing, F.A.,

M.A.F., M.S.Z., Z.K. and N.K.; revision and editing, N.K. and P.J.P. All authors have read and agreed to the published version of the manuscript.

Funding: This research received no external funding.

Institutional Review Board Statement: Not applicable.

Informed Consent Statement: Not applicable.

Data Availability Statement: Not applicable.

Conflicts of Interest: The authors declare no conflict of interest.

References

1. Vaidyanathan, T.; Vaidyanathan, J. Recent advances in the theory and mechanism of adhesive resin bonding to dentine; a critical review. *J. Biomed. Mater. Res. Part B Appl. Biomater.* **2009**, *88*, 558–578. [CrossRef] [PubMed]
2. Suresh, M.; Abraham, T.A.; Venkatachalam, P.; Venkatesh, A.; Malarvizhi, D.; Vivekanandhan, P. Bonding System for Restorative Materials–A Review. *Indian J. Public Health* **2019**, *10*, 1557–1561. [CrossRef]
3. Nazarian, A. The progression of dental adhesives. *ADA CERP* **2011**, *1*, 1–10.
4. Perdigão, J. Current perspectives on dental adhesion:(1) Dentin adhesion–not there yet. *Jpn Dent. Sci. Rev.* **2020**, *56*, 190–207. [CrossRef]
5. Loguercio, A.D.; Bittencourt, D.D.; Baratieri, L.N.; Reis, A. A 36-month evaluation of self-etch and etch-and-rinse adhesives in noncarious cervical lesions. *J. Am. Dent. Assoc.* **2007**, *138*, 507–514. [CrossRef]
6. Breschi, L.; Mazzoni, A.; Ruggeri, A.; Cadenaro, M.; Di Lenarda, R.; Dorigo, E.D.S. Dental adhesion review; aging and stability of the bonded interface. *Dent. Mater.* **2008**, *24*, 90–101. [CrossRef] [PubMed]
7. Perdigao, J. New developments in dental adhesion. *Dent. Clin. N. Am.* **2007**, *51*, 333–357. [CrossRef] [PubMed]
8. Milia, E.; Cumbo, E.; Cardoso, J.A.; Gallina, G. Current dental adhesives systems. A narrative review. *Curr. Pharma.Des.* **2012**, *18*, 5542–5552. [CrossRef] [PubMed]
9. Susin, A.H.; Vasconcellos, W.A.; Saad, J.R.C.; Oliveira Junior, O.B.D. Tensile bond strength of self-etching versus total-etching adhesive systems under different dentinal substrate conditions. *Braz. Oral Res.* **2007**, *21*, 81–86. [CrossRef] [PubMed]
10. Perdigão, J. Dentine bonding—Variables related to the clinical situation and the substrate treatment. *Dent. Mater J.* **2010**, *26*, 24–37. [CrossRef] [PubMed]
11. Belon, C.; Allonas, X.; Croutxé-Barghorn, C.; Lalevée, J. Overcoming the oxygen inhibition in the photopolymerization of acrylates; A study of the beneficial effect of triphenylphosphine. *J. Polym. Sci. A Polym.* **2010**, *48*, 2462–2469. [CrossRef]
12. Atai, M.; Watts, D.C. A new kinetic model for the photopolymerization shrinkage-strain of dental composites and resin-monomers. *Dent. Mater. J.* **2006**, *22*, 785–791. [CrossRef] [PubMed]
13. Kharouf, N.; Rapp, G.; Mancino, D.; Hemmerlé, J.; Haikel, Y.; Reitzer, F. Effect of etching the coronal dentin with the rubbing technique on the microtensile bond strength of a universal adhesive system. *Dent. Med. Probl.* **2019**, *56*, 343–348. [CrossRef] [PubMed]
14. Bolhuis, P.B.; de Gee, A.J.; Kleverlaan, C.J.; El Zohairy, A.A.; Feilzer, A.J. Contraction stress and bond strength to dentinfor compatible and incompatible combinations of bonding systems and chemical and light-cured core build-up resin composites. *Dent. Mater. J.* **2006**, *22*, 223–233. [CrossRef] [PubMed]
15. Finger, W.J.; Osada, T.; Tani, C.; Endo, T. Compatibility between self-etching adhesive and self-curing resin by addition of anion exchange resin. *Dent. Mater. J.* **2005**, *21*, 1044–1050. [CrossRef]
16. Salz, U.; Zimmermann, J.; Salzer, T. Self-curing; self-etching adhesive cement systems. *J. Adhes. Dent.* **2005**, *7*, 7–17. [PubMed]
17. Zhang, J.R.; Xu, L.; Liao, Y.Y.; Deng, J.C.; Tang, R.Y. Advances in Radical Oxidative C—H Alkylation of N-Heteroarenes. *Chin. J.Chem.* **2017**, *35*, 271–279. [CrossRef]
18. Ohyama, K.; Maki, S.; Sato, K.; Kato, Y. Comparative in vitro metabolism of the suspected pro-oestrogenic compound; methoxychlor in precision-cut liver slices from male and female rats. *Xenobiotica* **2005**, *35*, 331–342. [CrossRef] [PubMed]
19. Lapp, C.A.; Schuster, G.S. Effects of DMAEMA and 4-methoxyphenol on gingival fibroblast growth; metabolism; and response to interleukin-1. *J. Biomed. Mater. Res.* **2002**, *60*, 30–35. [CrossRef] [PubMed]
20. Hiraishi, N.; Nishiyama, N.; Ikemura, K.; Yau, J.; King, N.; Tagami, J. Water concentration in self-etching primers affects their aggressiveness and bonding efficacy to dentine. *J. Dent. Res.* **2005**, *84*, 653–658. [CrossRef] [PubMed]
21. Ikeda, T.; De Munck, J.; Shirai, K.; Hikita, K.; Inoue, S.; Sano, H. Effect of evaporation of primer components on ultimate tensile strengths of primer–adhesive mixture. *Dent. Mater. J.* **2005**, *21*, 1051–1058. [CrossRef] [PubMed]
22. Kim, J.S.; Cho, B.H.; Lee, I.B.; Um, C.M.; Lim, B.S.; Oh, M.H. Effect of the hydrophilic nanofiller loading on the mechanical properties and the microtensile bond strength of an ethanol-based one-bottle dentine adhesive. *J. Biomed. Mater. Res. Part B Appl. Biomater.* **2005**, *72*, 284–291. [CrossRef] [PubMed]
23. Conde, M.C.; Zanchi, C.H.; Rodrigues-Junior, S.A.; Carreno, N.L.; Ogliari, F.A.; Piva, E. Nanofiller loading level; Influence on selected properties of an adhesive resin. *J. Dent.* **2009**, *37*, 331–335. [CrossRef] [PubMed]

24. Zhang, Z.; Yu, J.; Yao, C.; Yang, H.; Huang, C. New perspective to improve dentine–adhesive interface stability by using dimethyl sulfoxide wet-bonding and epigallocatechin-3-gallate. *Dent. Mater. J.* **2020**, *36*, 1452–1463. [CrossRef] [PubMed]
25. Kharouf, N.; Eid, A.; Hardan, L.; Bourgi, R.; Arntz, Y.; Jmal, H.; Foschi, F.; Sauro, S.; Ball, V.; Haikel, Y.; et al. Antibacterial and bonding properties of universal adhesive dental polymers doped with pyrogallol. *Polymers* **2021**, *13*, 1538. [CrossRef] [PubMed]
26. Manuja Nair, J.P.; Kumar, S.; Chakravarthy, Y.; Vel Krishna, S. Comparative evaluation of the bonding efficacy of sixth and seventh generation bonding agents; An In-Vitro study. *J. Conserve. Dent.* **2014**, *17*, 27. [CrossRef]
27. Sezinando, A.; Perdigão, J.; Regalheiro, R. Dentin bond strengths of four adhesion strategies after thermal fatigue and 6-month water storage. *J. Esthet. Restor. Dent.* **2012**, *24*, 345–355. [CrossRef]
28. Bedran-Russo, A.; Leme-Kraus, A.A.; Vidal, C.M.; Teixeira, E.C. An overview of dental adhesive systems and the dynamic tooth–adhesive interface. *Dent. Clin.* **2017**, *61*, 713–731. [CrossRef]
29. Van Landuyt, K.L.; Snauwaert, J.; De Munck, J.; Peumans, M.; Yoshida, Y.; Poitevin, A.; Coutinho, E.; Suzuki, K.; Lambrechts, P.; Van Meerbeek, B. Systematic review of the chemical composition of contemporary dental adhesives. *Biomaterials* **2007**, *28*, 3757–3785. [CrossRef]
30. Mazzoni, A.; Carrilho, M.; Papa, V.; Tjäderhane, L.; Gobbi, P.; Nucci, C.; Di Lenarda, R.; Mazzotti, G.; Tay, F.R.; Pashley, D.H.; et al. MMP-2 assay within the hybrid layer created by a two-step etch-and-rinse adhesive; biochemical and immunohistochemical analysis. *J. Dent. Res.* **2011**, *39*, 470–477. [CrossRef] [PubMed]
31. Maravic, T.; Mazzoni, A.; Comba, A.; Scotti, N.; Checchi, V.; Breschi, L. How stable is dentine as a substrate for bonding? *Curr. Oral Health Rep.* **2017**, *4*, 248–257. [CrossRef]
32. Koibuchi, H.; Yasuda, N.; Nakabayashi, N. Bonding to dentin with a self-etching primer; the effect of smear layers. *Dent. Mater. J.* **2001**, *17*, 122–126. [CrossRef]
33. Gateva, N.; Gusiyska, A.; Stanimirov, P.; Raychev, I.; Kabaktchieva, R. Biodegradation and Dentin Bonding Effectiveness of one "Universal" Self-etch Adhesive Used in Multi-mode Manner. *J. IMAB* **2017**, *23*, 1510–1515. [CrossRef]
34. Faye, B.; Sarr, M.; Bane, K.; Aidara, A.W.; Niang, S.O.; Kane, A.W. One-year clinical evaluation of the bonding effectiveness of a one-step; self-etch adhesive in noncarious cervical lesion therapy. *Int. J. Dent.* **2015**, *2015*, 25–32. [CrossRef] [PubMed]
35. Celik, C.; Ozgunaltay, G.; Dayangaç, B. Bond strength of different adhesive systems to dental hard tissues. *Oper. Dent.* **2007**, *32*, 166–172.
36. Pashley, D.H.; Tay, F.R.; Breschi, L.; Tjäderhane, L.; Carvalho, R.M.; Carrilho, M.; Tezvergil-Mutluay, A. State of the art etch-and-rinse adhesives. *Dent. Mater. J.* **2011**, *27*, 1–16. [CrossRef]
37. Gonçalves, M.; Corona, S.A.M.; Palma-Dibb, R.G.; Pécora, J.D. Influence of pulse repetition rate of Er; YAG laser and dentin depth on tensile bond strength of dentin–resin interface. *J. Biomed. Mater. Res. Part A* **2008**, *86*, 477–482. [CrossRef] [PubMed]
38. Manappallil, J.J. *Basic Dental Materials*; JP Medical Ltd.: Clayton, Panama, 2015.
39. Ranka, C.P.; Somasundaram, R. An update review of self etching primers in orthodontics. *JAMDSR* **2020**, *8*, 87–91.
40. Broome, J.; Duke, E.; Norling, B. Shear bond strengths of composite resins with three different adhesives. *J. Dent. Res.* **1985**, *64*, 244.
41. Bowen, R.; Cobb, E.; Rapson, J. Adhesive bonding of various materials to hard tooth tissues; improvement in bond strength to dentine. *J. Dent. Res.* **1982**, *61*, 1070–1076. [CrossRef]
42. Meharry, M.R.; Moazzami, S.M.; Li, Y. Comparison of enamel and dentin shear bond strengths of current dental bonding adhesives from three bond generations. *Oper. Dent.* **2013**, *38*, E237–E245. [CrossRef] [PubMed]
43. Hamama, H.H.; Burrow, M.F.; Yiu, C. Effect of dentine conditioning on adhesion of resin-modified glass ionomer adhesives. *Aust. Dent. J.* **2014**, *59*, 193–200. [CrossRef] [PubMed]
44. Toshniwal, N.; Singh, N.; Dhanjani, V.; Mote, N.; Mani, S. Self etching system v/s conventional bonding; Advantages; disadvantages. *Int. J. Appl. Dent. Sci.* **2019**, *5*, 379–383.
45. De Araujo, E.M., Jr.; Baratieri, L.N.; Monteiro, S., Jr.; Vieira, L.C.; De Andrada, M.A. Direct adhesive restoration of anterior teeth: Part 2. Clinical protocol. *PPAD* **2003**, *15*, 351–357.
46. Carvalho, R.M.; Manso, A.P.; Geraldeli, S.; Tay, F.R.; Pashley, D.H. Durability of bonds and clinical success of adhesive restorations. *Dent. Mater. J* **2012**, *28*, 72–86. [CrossRef] [PubMed]
47. Van Meerbeek, B.; Yoshihara, K.; Yoshida, Y.; Mine, A.J.; De Munck, J.; Van Landuyt, K.L. State of the art of self-etch adhesives. *Dent. Mater.* **2011**, *27*, 17–28. [CrossRef] [PubMed]
48. Alex, G. Adhesive considerations in the placement of direct composite restorations. *Oral Health* **2008**, *98*, 109–119.
49. De Munck, J.D.; Van Landuyt, K.; Peumans, M.; Poitevin, A.; Lambrechts, P.; Braem, M.; van Meerbeek, B. A critical review of the durability of adhesion to tooth tissue; methods and results. *J. Dent. Res.* **2005**, *84*, 118–132. [CrossRef]
50. Reis, A.; Albuquerque, M.; Pegoraro, M.; Mattei, G.; de Oliveira Bauer, J.R.; Grande, R.H.; Klein-Junior, C.A.; Baumhardt-Neto, R.; Loguercio, A.D. Can the durability of one-step self-etch adhesives be improved by double application or by an extra layer of hydrophobic resin? *J. Dent.* **2008**, *36*, 309–315. [CrossRef]
51. Vicente, A.; Bravo, L.A.; Romero, M. Self-etching primer and a non-rinse conditioner versus phosphoric acid: Alternative methods for bonding brackets. *Eur. J. Orthod.* **2006**, *28*, 173–178. [CrossRef]
52. Moszner, N.; Salz, U.; Zimmermann, J. Chemical aspects of self-etching enamel–dentine adhesives; a systematic review. *Dent Mater. J.* **2005**, *21*, 895–910. [CrossRef] [PubMed]

53. Nishiyama, N.; Tay, F.; Fujita, K.; Pashley, D.H.; Ikemura, K.; Hiraishi, N. Hydrolysis of functional monomers in a single-bottle self-etching primer—correlation of 13C NMR and TEM findings. *J. Dent. Res.* **2006**, *85*, 422–426. [CrossRef] [PubMed]
54. Ito, S.; Hashimoto, M.; Wadgaonkar, B.; Svizero, N.; Carvalho, R.M.; Yiu, C.; Rueggeberg, F.A.; Foulger, S.; Saito, T.; Nishitani, Y.; et al. Effects of resin hydrophilicity on water sorption and changes in modulus of elasticity. *Biomaterials* **2005**, *26*, 6449–6459. [CrossRef] [PubMed]
55. Yaseen, S.M.; Reddy, V.S. Comparative evaluation of shear bond strength of two self-etching adhesives (sixth and seventh generation) on dentine of primary and permanent teeth; An in vitro study. *J. Indian Soc. Pedod. Prev. Dent.* **2009**, *27*, 33. [CrossRef]
56. Van Landuyt, K.L.; Kanumilli, P.; De Munck, J.; Peumans, M.; Lambrechts, P.; Van Meerbeek, B. Bond strength of a mild self-etch adhesive with and without prior acid-etching. *J. Dent.* **2006**, *34*, 77–85. [CrossRef]
57. Başaran, G.; Özer, T.; Devecioğlu Kama, J. Comparison of a recently developed nanofiller self-etching primer adhesive with other self-etching primers and conventional acid etching. *Eur. J. Orthod.* **2009**, *31*, 271–275. [CrossRef]
58. Kasraei, S.; Atai, M.; Khamverdi, Z.; Nejad, S.K. The effect of nanofiller addition to an experimental dentine adhesive on microtensile bond strength to human dentine. *FID* **2009**, *6*, 36–41.
59. Karami, N.M.; Javadinezhad, S.; Homayonzadeh, M. Sealant Microleakage in saliva-contaminated enamel: Comparison between three adhesive systems. *J. Dent. Sch. Shahid Beheshti Univ. Med. Sci.* **2010**, *4*, 197–204.
60. Liu, Y.; Tjäderhane, L.; Breschi, L.; Mazzoni, A.; Li, N.; Mao, J. Limitations in bonding to dentine and experimental strategies to prevent bond degradation. *J. Dent. Res.* **2011**, *90*, 953–968. [CrossRef]
61. Carvalho, R.M. Commentary. Assessment of in vitro methods used to promote adhesive interface degradation: A critical review. *J. Esthet. Rest. Dent.* **2007**, *19*, 354–358. [CrossRef]
62. Kermanshahi, S.; Santerre, J.; Cvitkovitch, D.; Finer, Y. Biodegradation of resin-dentine interfaces increases bacterial microleakage. *J. Dent. Res.* **2010**, *89*, 996–1001. [CrossRef]
63. Hashimoto, M. A Review—Micromorphological evidence of degradation in resin-dentine bonds and potential preventional solutions. *J. Biomed. Mater. Res. Part B Appl. Biomater.* **2010**, *92*, 268–280. [CrossRef] [PubMed]
64. Hosaka, K.; Nishitani, Y.; Tagami, J.; Yoshiyama, M.; Brackett, W.W.; Agee, K.A.; Tay, F.R.; Pashley, D.H. Durability of resin-dentin bonds to water-vs. ethanol-saturated dentin. *J. Dent. Res.* **2009**, *88*, 146–151. [CrossRef] [PubMed]
65. Loguercio, A.D.; Moura, S.K.; Pellizzaro, A.; Dal-Bianco, K.; Patzlaff, R.T.; Grande, R.H.M. Durability of enamel bonding using two-step self-etch systems on ground and unground enamel. *Oper. Dent.* **2008**, *33*, 79–88. [CrossRef] [PubMed]
66. Malacarne, J.; Carvalho, R.M.; Mario, F.; Svizero, N.; Pashley, D.H.; Tay, F.R. Water sorption/solubility of dental adhesive resins. *Dent. Mater. J.* **2006**, *22*, 973–980. [CrossRef] [PubMed]
67. Yiu, C.K.; Pashley, E.L.; Hiraishi, N.; King, N.M.; Goracci, C.; Ferrari, M.; Carvalho, R.M.; Pashley, D.H.; Tay, F.R. Solvent and water retention in dental adhesive blends after evaporation. *Biomaterials* **2005**, *26*, 6863–6872. [CrossRef] [PubMed]
68. Ye, Q.; Park, J.; Topp, E.; Wang, Y.; Misra, A.; Spencer, P. In vitro performance of nano-heterogeneous dentine adhesive. *J. Dent. Res.* **2008**, *87*, 829–833. [CrossRef] [PubMed]
69. Wang, Y.; Spencer, P.; Yao, X.; Ye, Q. Effect of coinitiator and water on the photoreactivity and photopolymerization of HEMA/camphoquinone-based reactant mixtures. *J. Biomed. Mater. Res. A* **2006**, *78*, 721–728. [CrossRef]
70. Ilie, N.; Hickel, R. Can CQ be completely replaced by alternative initiators in dental adhesives? *Dent. Mater. J.* **2008**, *27*, 221–228. [CrossRef]
71. Cadenaro, M.; Antoniolli, F.; Codan, B.; Agee, K.; Tay, F.R.; Dorigo, E.D.S. Influence of different initiators on the degree of conversion of experimental adhesive blends in relation to their hydrophilicity and solvent content. *Dent. Mater. J.* **2010**, *26*, 288–294. [CrossRef] [PubMed]
72. Amaral, F.L.; Colucci, V.; Palma-Dibb, R.G.; Corona, S.A. Assessment of in vitro methods used to promote adhesive interface degradation: A critical review. *J. Esthet. Restor. Dent.* **2007**, *19*, 340–353. [CrossRef] [PubMed]
73. Shokati, B.; Tam, L.E.; Santerre, J.P.; Finer, Y. Effect of salivary esterase on the integrity and fracture toughness of the dentine-resin interface. *J. Biomed. Mater. Res, B. Appl. Biomater.* **2010**, *94*, 230–237.
74. Kostoryz, E.L.; Dharmala, K.; Ye, Q.; Wang, Y.; Huber, J.; Park, J.G. Enzymatic biodegradation of HEMA/bisGMA adhesives formulated with different water content. *J. Biomed. Mater. Res. Part B Appl. Biomater.* **2009**, *88*, 394–401. [CrossRef]
75. Boushell, L.W.; Kaku, M.; Mochida, Y.; Bagnell, R.; Yamauchi, M. Immunohistochemical localization of matrixmetalloproteinase-2 in human coronal dentin. *Arch.Oral.Biol.* **2008**, *53*, 109–116. [CrossRef]
76. Hannas, A.R.; Pereira, J.C.; Granjeiro, J.M.; Tjäderhane, L. The role of matrix metalloproteinases in the oral environment. *Acta Odonto Scand* **2007**, *65*, 1–3. [CrossRef] [PubMed]
77. Tjäderhane, L.; Nascimento, F.D.; Breschi, L.; Mazzoni, A.; Tersariol, I.L.; Geraldeli, S.; Tezvergil-Mutluay, A.; Carrilho, M.R.; Carvalho, R.M.; Tay, F.R.; et al. Optimizing dentin bond durability; control of collagen degradation by matrix metalloproteinases and cysteine cathepsins. *Dent. Mater.* **2013**, *29*, 116–135. [CrossRef]
78. Bahudhanapati, H. *In Search of MMP Specific Inhibitors: Protein Engineering of TIMPs*; Florida Atlantic University: Boca Raton, FL, USA, 2009.
79. Shimada, Y.; Ichinose, S.; Sadr, A.; Burrow, M.; Tagami, J. Localization of matrix metalloproteinases (MMPs-2; 8; 9 and 20) in normal and carious dentine. *Aust. Dent. J.* **2009**, *54*, 347–354. [CrossRef]
80. Sulkala, M.; Tervahartiala, T.; Sorsa, T.; Larmas, M.; Salo, T.; Tjäderhane, L. Matrix metalloproteinase-8 (MMP-8) is the major collagenase in human dentin. *Arch. Oral. Biol.* **2007**, *52*, 121–127. [CrossRef]

81. Skurska, A.; Dolinska, E.; Pietruska, M.; Pietruski, J.K.; Dymicka, V.; Kemona, H.; Arweiler, N.B.; Milewski, R.; Sculean, A. Effect of nonsurgical periodontal treatment in conjunction with either systemic administration of amoxicillin and metronidazole or additional photodynamic therapy on the concentration of matrix metalloproteinases 8 and 9 in gingival crevicular fluid in patients with aggressive periodontitis. *BMC Oral Health* **2015**, *15*, 1–5.
82. Mazzoni, A.; Pashley, D.H.; Tay, F.R.; Gobbi, P.; Orsini, G.; Ruggeri, A., Jr.; Carrilho, M.; Tjäderhane, L.; di Lenarda, R.; Breschi, L. Immunohistochemical identification of MMP-2 and MMP-9 in human dentine; correlative FEI-SEM/TEM analysis. *J. Biomed. Mater. Res. Part A* **2009**, *88*, 697–703. [CrossRef]
83. Mazzoni, A.; Pashley, D.H.; Nishitani, Y.; Breschi, L.; Mannello, F.; Tjäderhane, L. Reactivation of inactivated endogenous proteolytic activities in phosphoric acid-etched dentine by etch-and-rinse adhesives. *Biomaterials* **2006**, *27*, 4470–4476. [CrossRef] [PubMed]
84. Nishitani, Y.; Yoshiyama, M.; Wadgaonkar, B.; Breschi, L.; Mannello, F.; Mazzoni, A.; Carvalho, R.M.; Tjäderhane, L.; Tay, F.R.; Pashley, D.H. Activation of gelatinolytic/collagenolytic activity in dentine by self-etching adhesives. *Eur. J. Oral Sci.* **2006**, *114*, 160–166. [CrossRef]
85. Mazzoni, A.; Scaffa, P.; Carrilho, M.; Tjäderhane, L.; Di Lenarda, R.; Polimeni, A. Effects of etch-and-rinse and self-etch adhesives on dentine MMP-2 and MMP-9. *J. Dent. Res.* **2013**, *92*, 82–86. [CrossRef] [PubMed]
86. Iwasa, M.; Tsubota, K.; Shimamura, Y.; Ando, S.; Miyazaki, M.; Platt, J.A. pH changes upon mixing of single-step self-etching adhesives with powdered dentine. *J. Adhes. Dent.* **2011**, *13*, 207. [PubMed]
87. Almahdy, A.; Koller, G.; Sauro, S.; Bartsch, J.W.; Sherriff, M.; Watson, T.F.; Banerjee, A. Effects of MMP inhibitors incorpo-rated within dental adhesives. *J. Dent. Res.* **2012**, *91*, 605–611. [CrossRef] [PubMed]
88. Lehmann, N.; Debret, R.; Roméas, A.; Magloire, H.; Degrange, M.; Bleicher, F. Self-etching increases matrix metalloproteinase expression in the dentine-pulp complex. *J. Dent. Res.* **2009**, *88*, 77–82. [CrossRef]
89. Betancourt, D.; Baldion, P.; Castellanos, J. Resin-dentine bonding interface; Mechanisms of degradation and strategies for stabilization of the hybrid layer. *Int. J. Biomater.* **2019**, *2019*, 5268342. [CrossRef]
90. Tersariol, I.L.; Geraldeli, S.; Minciotti, C.L.; Nascimento, F.D.; Pääkkönen, V.; Martins, M.T.; Carrilho, M.R.; Pashley, D.H.; Tay, F.R.; Salo, T.; et al. Cysteine cathepsins in human dentine-pulp complex. *J. Endod.* **2010**, *36*, 475–481. [CrossRef]
91. Nascimento, F.; Minciotti, C.; Geraldeli, S.; Carrilho, M.; Pashley, D.H.; Tay, F.R.; Nader, H.B.; Salo, T.; Tjäderhane, L.; Tersariol, I.L.S. Cysteine cathepsins in human carious dentine. *J. Dent. Res.* **2011**, *90*, 506–511. [CrossRef]
92. Mahalaxmi, S.; Madhubala, M.; Jayaraman, M.; Sathyakumar, S. Evaluation of matrix metalloproteinase and cysteine cathepsin activity in dentine hybrid layer by gelatin zymography. *Indian J. Dent. Res.* **2016**, *27*, 652.
93. Li, Z.; Kienetz, M.; Cherney, M.M.; James, M.N.; Brömme, D. The crystal and molecular structures of a cathepsin K; chondroitin sulfate complex. *J. Mol. Biol.* **2008**, *383*, 78–91. [CrossRef] [PubMed]
94. Vidal, C.; Tjäderhane, L.; Scaffa, P.; Tersariol, I.; Pashley, D.; Nader, H. Abundance of MMPs and cysteine cathepsins in caries-affected dentine. *J. Dent. Res.* **2014**, *93*, 269–274. [CrossRef] [PubMed]
95. Scaffa, P.M.; Breschi, L.; Mazzoni, A.; Vidal, C.D.; Curci, R.; Apolonio, F.; Gobbi, P.; Pashley, D.; Tjäderhane, L.; dos Santos Tersariol, I.L.; et al. Co-distribution of cysteine cathepsins and matrix metalloproteases in human dentine. *Arch. Oral. Biol.* **2017**, *74*, 101–107. [CrossRef] [PubMed]
96. Baldión, P.A.; Cortés, C.J. Mathematical models of polymer-dentine physicochemical interactions and their biological effects. *Sci. J. Rev.* **2016**, *5*, 319–330.
97. Kim, Y.K.; Gu, L.-S.; Bryan, T.E.; Kim, J.R.; Chen, L.; Liu, Y. Mineralisation of reconstituted collagen using polyvinylphosphonic acid/polyacrylic acid templating matrix protein analogues in the presence of calcium; phosphate and hydroxyl ions. *Biomaterials* **2010**, *31*, 6618–6627. [CrossRef]
98. Jee, S.E.; Zhou, J.; Tan, J.; Breschi, L.; Tay, F.R.; Grégoire, G. Investigation of ethanol infiltration into demineralized dentine collagen fibrils using molecular dynamics simulations. *Acta Biomate* **2016**, *36*, 175–185. [CrossRef]
99. Bertassoni, L.E. Dentine on the nanoscale; Hierarchical organization; mechanical behavior and bioinspired engineering. *Dent. Mater. J.* **2017**, *33*, 637–649. [CrossRef]
100. Fung, D.T.; Wang, V.M.; Laudier, D.M.; Shine, J.H.; Basta-Pljakic, J.; Jepsen, K.J. Subrupture tendon fatigue damage. *J. Orthop. Res.* **2009**, *27*, 264–273. [CrossRef]
101. Breschi, L.; Martin, P.; Mazzoni, A.; Nato, F.; Carrilho, M.; Tjäderhane, L.; Visintini, E.; Cadenaro, M.; Tay, F.R.; Dorigo, E.D.; et al. Use of a specific MMP-inhibitor (galardin) for preservation of hybrid layer. *Dent. Mater. J.* **2010**, *26*, 571–578. [CrossRef] [PubMed]
102. Seseogullari-Dirihan, R.; Mutluay, M.; Vallittu, P.; Pashley, D.H.; Tezvergil-Mutluay, A. Effect of pretreatment with collagen crosslinkers on dentine protease activity. *Dent. Mater. J.* **2015**, *31*, 941–947. [CrossRef]
103. Mazzoni, A.; Angeloni, V.; Sartori, N.; Duarte, S., Jr.; Maravic, T.; Tjäderhane, L. Substantivity of carbodiimide inhibition on dentinal enzyme activity over time. *J. Dent. Res.* **2017**, *96*, 902–908. [CrossRef] [PubMed]
104. Carrilho, M.R.; Carvalho, R.M.; Sousa, E.N.; Nicolau, J.; Breschi, L.; Mazzoni, A. Substantivity of chlorhexidine to human dentine. *Dent. Mater. J.* **2010**, *26*, 779–785. [CrossRef] [PubMed]
105. Carrilho, M.; Carvalho, R.M.; De Goes, M.; Di Hipolito, V.; Geraldeli, S.; Tay, F.R. Chlorhexidine preserves dentine bond in vitro. *J. Dent. Res.* **2007**, *86*, 90–94. [CrossRef] [PubMed]
106. Breschi, L.; Mazzoni, A.; Nato, F.; Carrilho, M.; Visintini, E.; Tjäderhane, L.; Ruggeri, A., Jr.; Tay, F.R.; Dorigo, E.D.; Pashley, D.H. Chlorhexidine stabilizes the adhesive interface; a 2-year in vitro study. *Dent. Mater. J.* **2010**, *26*, 320–325. [CrossRef]

107. Loguercio, A.D.; Hass, V.; Gutierrez, M.F.; Luque-Martinez, I.V.; Szezs, A.; Stanislawczuk, R. Five-year effects of chlorhexidine on the in vitro durability of resin/dentine interfaces. *J. Adhes. Dent.* **2016**, *18*, 35–42.
108. Zheng, P.; Zaruba, M.; Attin, T.; Wiegand, A. Effect of different matrix metalloproteinase inhibitors on microtensile bond strength of an etch-and-rinse and a self-etching adhesive to dentine. *Oper. Dent. J.* **2015**, *40*, 80–86. [CrossRef] [PubMed]
109. Hebling, J.; Pashley, D.H.; Tjäderhane, L.; Tay, F.R. Chlorhexidine arrests subclinical degradation of dentin hybrid layers in vivo. *J. Dent. Res.* **2005**, *84*, 741–746. [CrossRef]
110. Moharam, L.-M.; Salem, H.-N.; Elgamily, H.-M. The effect of incorporating different concentrations of chlorhexidine digluconate on the degree of conversion of an experimental adhesive resin. *J. Clinc. Exp.* **2018**, *10*, e371. [CrossRef]
111. Cadenaro, M.; Pashley, D.H.; Marchesi, G.; Carrilho, M.; Antoniolli, F.; Mazzoni. Influence of chlorhexidine on the degree of conversion and E-modulus of experimental adhesive blends. *Dent. Mater. J.* **2009**, *25*, 1269–1274. [CrossRef]
112. Stanislawczuk, R.; Pereira, F.; Muñoz, M.A.; Luque, I.; Farago, P.V.; Reis, A. Effects of chlorhexidine-containing adhesives on the durability of resin–dentine interfaces. *J. Dent.* **2014**, *42*, 39–47. [CrossRef]
113. Da Silva, E.M.; de Sá Rodrigues, C.U.F.; de Oliveira Matos, M.P.; de Carvalho, T.R.; dos Santos, G.B.; Amaral, C.M. Experimental etch-and-rinse adhesive systems containing MMP-inhibitors; Physicochemical characterization and resin-dentine bonding stability. *J. Dent.* **2015**, *43*, 1491–1497. [CrossRef] [PubMed]
114. Tezvergil-Mutluay, A.; Mutluay, M.M.; Gu, L.S.; Zhang, K.; Agee, K.A.; Carvalho, R.M.; Manso, A.; Carrilho, M.; Tay, F.R.; Breschi, L.; et al. The anti-MMP activity of benzalkonium chloride. *J. Dent.* **2011**, *39*, 57–64. [CrossRef]
115. Sabatini, C.; Patel, S.K. Matrix metalloproteinase inhibitory properties of benzalkonium chloride stabilizes adhesive interfaces. *Eur. J. Oral. Sci.* **2013**, *121*, 610–616. [CrossRef] [PubMed]
116. Sabatini, C.; Ortiz, P.A.; Pashley, D.H. Preservation of resin–dentine interfaces treated with benzalkonium chloride adhesive blends. *Eur. J. Oral Sci.* **2015**, *123*, 108–115. [CrossRef] [PubMed]
117. Hoshika, T.; Nishitani, Y.; Yoshiyama, M.; Key, W.O., III; Brantley, W.; Agee, K.A.; Breschi, L.; Cadenaro, M.; Tay, F.R.; Rueggeberg, F.; et al. Effects of quaternary ammonium-methacrylates on the mechanical properties of unfilled resins. *Dent. Mater. J.* **2014**, *30*, 1213–1223. [CrossRef]
118. Tezvergil-Mutluay, A.; Agee, K.A.; Mazzoni, A.; Carvalho, R.M.; Carrilho, M.; Tersariol, I.L. Can quaternary ammonium methacrylates inhibit matrix MMPs and cathepsins? *Dent. Mater. J.* **2015**, *31*, e25–e32. [CrossRef] [PubMed]
119. Li, F.; Chen, J.; Chai, Z.; Zhang, L.; Xiao, Y.; Fang, M.; Ma, S. Effects of a dental adhesive incorporating antibacterial monomer on the growth; adherence and membrane integrity of Streptococcus mutans. *J. Dent.* **2009**, *37*, 289–296. [CrossRef]
120. Imazato, S.; Tay, F.R.; Kaneshiro, A.V.; Takahashi, Y.; Ebisu, S. An in vivo evaluation of bonding ability of comprehensive antibacterial adhesive system incorporating MDPB. *Dent. Mater. J.* **2007**, *23*, 170–176. [CrossRef]
121. Pashley, D.H.; Tay, F.R.; Imazato, S. How to increase the durability of resin-dentine bonds. *Compend. Contin. Educ. Dent.* **2011**, *32*, 60–64.
122. Antonucci, J.M.; Zeiger, D.N.; Tang, K.; Lin-Gibson, S.; Fowler, B.O.; Lin, N.J. Synthesis and characterization of dimethacrylates containing quaternary ammonium functionalities for dental applications. *Dent. Mater. J.* **2012**, *28*, 219–228. [CrossRef]
123. Thompson, J.M.; Agee, K.; Sidow, S.J.; McNally, K.; Lindsey, K.; Borke, J. Inhibition of endogenous dentine matrix metalloproteinases by ethylenediaminetetraacetic acid. *J. Endod.* **2012**, *38*, 62–65. [CrossRef]
124. Osorio, R.; Erhardt, M.C.; Pimenta, L.A.; Osorio, E.; Toledano, M. EDTA treatment improves resin-dentin bonds' resistance to degradation. *J. Dent. Res.* **2005**, *84*, 736–740. [CrossRef] [PubMed]
125. Carrilho, M.R.; Tay, F.R.; Donnelly, A.M.; Agee, K.A.; Tjäderhane, L.; Mazzoni, A. Host-derived loss of dentine matrix stiffness associated with solubilization of collagen. *J. Biomed. Mater. Res. Part B Appl. Biomater.* **2009**, *90*, 373–380.
126. Heikkilä, P.; Teronen, O.; Moilanen, M.; Konttinen, Y.T.; Hanemaaijer, R.; Laitinen, M.; Maisi, P.; van der Pluijm, G.; Bartlett, J.D.; Salo, T.; et al. Bisphosphonates inhibit stromelysin-1 (MMP-3); matrix metalloelastase (MMP-12); collagenase-3 (MMP-13) and enamelysin (MMP-20); but not urokinase-type plasminogen activator; and diminish invasion and migration of human malignant and endothelial cell lines. *Anti-Cancer Drugs* **2002**, *13*, 245–254. [CrossRef] [PubMed]
127. Tezvergil-Mutluay, A.; Agee, K.A.; Hoshika, T.; Tay, F.R.; Pashley, D.H. The inhibitory effect of polyvinylphosphonic acid on functional matrix metalloproteinase activities in human demineralized dentine. *Acta Biomater.* **2010**, *6*, 4136–4142. [CrossRef]
128. Gu, L.-S.; Kim, Y.K.; Liu, Y.; Takahashi, K.; Arun, S.; Wimmer, C.E. Immobilization of a phosphonated analog of matrix phosphoproteins within cross-linked collagen as a templating mechanism for biomimetic mineralization. *Acta Biomater.* **2011**, *7*, 268–277. [CrossRef]
129. Mazzoni, A.; Mannello, F.; Tay, F.R.; Tonti, G.A.; Papa, S.; Mazzotti, G.; Di Lenarda, R.; Pashley, D.H.; Breschi, L. Zymographic analysis and characterization of MMP-2 and-9 forms in human sound dentin. *J. Dent. Res.* **2007**, *86*, 436–440. [CrossRef]
130. Sorsa, T.; Tjäderhane, L.; Konttinen, Y.T.; Lauhio, A.; Salo, T.; Lee, H.M. Matrix metalloproteinases; contribution to pathogenesis; diagnosis and treatment of periodontal inflammation. *Ann. Med.* **2006**, *38*, 306–321. [CrossRef]
131. Osorio, R.; Yamauti, M.; Osorio, E.; Ruiz-Requena, M.; Pashley, D.H.; Tay, F. Zinc reduces collagen degradation in demineralized human dentine explants. *J. Dent.* **2011**, *39*, 148–153. [CrossRef]
132. Chaussain-Miller, C.; Fioretti, F.; Goldberg, M.; Menashi, S. The role of matrix metalloproteinases (MMPs) in human caries. *J. Dent. Res.* **2006**, *85*, 22–32. [CrossRef] [PubMed]
133. Tallant, C.; Marrero, A.; Gomis-Rüth, F.X. Matrix metalloproteinases; fold and function of their catalytic domains. *Biochim. Biophys. Acta (BBA)-Mol. Cell Res.* **2010**, *1803*, 20–28. [CrossRef]

134. Bedran-Russo, A.K.; Pauli, G.F.; Chen, S.N.; McAlpine, J.; Castellan, C.S.; Phansalkar, R.S.; Aguiar, T.R.; Vidal, C.M.; Napotilano, J.G.; Nam, J.W.; et al. Dentine biomodification; strategies; renewable resources and clinical applications. *Dent. Mater. J.* **2014**, *30*, 62–76. [CrossRef] [PubMed]
135. Xu, C.; Wang, Y. Cross-linked demineralized dentine maintains its mechanical stability when challenged by bacterial collagenase. *J. Biomed. Mater. Res. Part B Appl. Biomater.* **2011**, *96*, 242–248. [CrossRef] [PubMed]
136. Naso, F.; Gandaglia, A.; Bottio, T.; Tarzia, V.; Nottle, M.B.; d'Apice, A.J.; Cowan, P.J.; Cozzi, E.; Galli, C.; Lagutina, I.; et al. First quantification of alpha-G al epitope in current glutaraldehyde-fixed heart valve bioprostheses. *Xenotransplantation* **2013**, *20*, 252–261. [CrossRef]
137. Chen, C.; Mao, C.; Sun, J.; Chen, Y.; Wang, W.; Pan, H. Glutaraldehyde-induced remineralization improves the mechanical properties and biostability of dentine collagen. *Mater. Sci. Eng. C* **2016**, *67*, 657–665. [CrossRef]
138. Hass, V.; Luque-Martinez, I.V.; Gutierrez, M.F.; Moreira, C.G.; Gotti, V.B.; Feitosa, V.P. Collagen cross-linkers on dentine bonding; stability of the adhesive interfaces; degree of conversion of the adhesive; cytotoxicity and in situ MMP inhibition. *Dent. Mater. J.* **2016**, *32*, 732–741. [CrossRef] [PubMed]
139. Seseogullari-Dirihan, R.; Apollonio, F.; Mazzoni, A.; Tjaderhane, L.; Pashley, D.; Breschi, L. Use of crosslinkers to inactivate dentine MMPs. *Dent. Mater. J.* **2016**, *32*, 423–432. [CrossRef]
140. Zhou, J.; Chiba, A.; Scheffel, D.L.; Hebling, J.; Agee, K.; Tagami, J.; Tan, J.; Abuelenain, D.; Nawareg, M.A.; Hassan, A.H.; et al. Cross-linked dry bonding; A new etch-and-rinse technique. *Dent. Mater. J.* **2016**, *32*, 1124–1132. [CrossRef]
141. Maravic, T.; Breschi, L.; Comba, A.; Cunha, S.R.; Angeloni, V.; Nucci, C. Experimental use of an acrolein-based primer as collagen cross-linker for dentine bonding. *J. Dent.* **2018**, *68*, 85–90. [CrossRef]
142. Castellan, C.S.; Bedran-Russo, A.K.; Karol, S.; Pereira, P.N. Long-term stability of dentin matrix following treatment with various natural collagen cross-linkers. *J. Mech. Behav. Biomed. Mater.* **2011**, *4*, 1343–1350. [CrossRef]
143. Bedran-Russo, A.K.B.; Pereira, P.N.; Duarte, W.R.; Drummond, J.L.; Yamauchi, M. Application of crosslinkers to dentine collagen enhances the ultimate tensile strength. *J. Biomed. Mater. Res. Part A* **2007**, *80*, 268–272. [CrossRef] [PubMed]
144. Epasinghe, D.J.; Yiu, C.K.Y.; Burrow, M.F. Effect of proanthocyanidin incorporation into dental adhesive on durability of resin–dentine bond. *Inter. J. Adhes Adhes.* **2015**, *63*, 145–151. [CrossRef]
145. Liu, Y.; Dusevich, V.; Wang, Y. Proanthocyanidins rapidly stabilize the demineralized dentine layer. *J. Dent. Res.* **2013**, *92*, 746–752. [CrossRef] [PubMed]
146. Mazzoni, A.; Apolonio, F.M.; Saboia, V.P.; Santi, S.; Angeloni, V.; Checchi, V.; Curci, R.; Di Lenarda, R.; Tay, F.R.; Pashley, D.; et al. Carbodiimide inactivation of MMPs and effect on dentine bonding. *J. Dent. Res.* **2014**, *93*, 263–268. [CrossRef]
147. Tezvergil-Mutluay, A.; Mutluay, M.; Agee, K.; Seseogullari-Dirihan, R.; Hoshika, T.; Cadenaro, M. Carbodiimide cross-linking inactivates soluble and matrix-bound MMPs in vitro. *J. Dent. Res.* **2012**, *91*, 192–196. [CrossRef]
148. Mazzoni, A.; Angeloni, V.; Apolonio, F.M.; Scotti, N.; Tjäderhane, L.; Tezvergil-Mutluay, A. Effect of carbodiimide (EDC) on the bond stability of etch-and-rinse adhesive systems. *Dent. Mater. J.* **2013**, *29*, 1040–1047. [CrossRef]
149. Bedran-Russo, A.K.B.; Vidal, C.M.; Dos Santos, P.H.; Castellan, C.S. Long-term effect of carbodiimide on dentine matrix and resin-dentine bonds. *Biomed. Mater. Res. Part B* **2010**, *94*, 250–255.
150. Ryou, H.; Turco, G.; Breschi, L.; Tay, F.R.; Pashley, D.H.; Arola, D. On the stiffness of demineralized dentine matrices. *Dent. Mater. J.* **2016**, *32*, 161–170. [CrossRef]
151. Zhang, Z.; Mutluay, M.; Tezvergil-Mutluay, A.; Tay, F.R.; Pashley, D.H.; Arola, D. Effects of EDC crosslinking on the stiffness of dentine hybrid layers evaluated by nanoDMA over time. *Dent. Mater. J.* **2017**, *33*, 904–914. [CrossRef]
152. Tezvergil-Mutluay, A.; Agee, K.; Uchiyama, T.; Imazato, S.; Mutluay, M.; Cadenaro, M. The inhibitory effects of quaternary ammonium methacrylates on soluble and matrix-bound MMPs. *J. Dent. Res.* **2011**, *90*, 535–540. [CrossRef]
153. Scheffel, D.L.S.; Bianchi, L.; Soares, D.G.; Basso, F.G.; Sabatini, C.; de Souza Costa, C. Transdentinal cytotoxicity of carbodiimide (EDC) and glutaraldehyde on odontoblast-like cells. *Oper. Dent.* **2015**, *40*, 44–54. [CrossRef] [PubMed]
154. Kharouf, N.; Haikel, Y.; Ball, V. Polyphenols in dental applications. *J. Bioeng.* **2020**, *7*, 72. [CrossRef] [PubMed]
155. Dikmen, B.; Gurbuz, O.; Ozsoy, A.; Eren, M.M.; Cilingir, A.; Yucel, T. Effect of different antioxidants on the microtensile bond strength of an adhesive system to sodium hypochlorite-treated dentine. *J. Adhes. Dent.* **2015**, *17*, 499–504.
156. Silver, F.H.; Horvath, I.; Foran, D.J. Viscoelasticity of the vessel wall: The role of collagen and elastic fibers. *Crit. Rev. Biomed. Eng.* **2001**, *29*, 279–302. [CrossRef] [PubMed]
157. Liu, Y.; Bai, X.; Li, S.; Liu, Y.; Keightley, A.; Wang, Y. Molecular weight and galloylation affect grape seed extract constituents' ability to cross-link dentine collagen in clinically relevant time. *Dent. Mater. J.* **2015**, *31*, 814–821. [CrossRef]
158. Dos Santos, P.H.; Karol, S.; Bedran-Russo, A.K. Long-term nano-mechanical properties of biomodified dentine–resin interface components. *J. Biomech.* **2011**, *44*, 1691–1694. [CrossRef] [PubMed]
159. Liu, R.-R.; Fang, M.; Zhang, L.; Tang, C.-F.; Dou, Q.; Chen, J.-H. Anti-proteolytic capacity and bonding durability of proanthocyanidin-biomodified demineralized dentine matrix. *Int. J. Oral Sci.* **2014**, *6*, 168–174. [CrossRef] [PubMed]
160. Green, B.; Yao, X.; Ganguly, A.; Xu, C.; Dusevich, V.; Walker, M.P. Grape seed proanthocyanidins increase collagen biodegradation resistance in the dentine/adhesive interface when included in an adhesive. *J. Dent.* **2010**, *38*, 908–915. [CrossRef]
161. Frassetto, A.; Breschi, L.; Turco, G.; Marchesi, G.; Di Lenarda, R.; Tay, F.R. Mechanisms of degradation of the hybrid layer in adhesive dentistry and therapeutic agents to improve bond durability—A literature review. *Dent. Mater. J.* **2016**, *32*, 41–53. [CrossRef]

162. Hayashi, M.; Okamura, K.; Koychev, E.; Furuya, Y.; Sugeta, A.; Ota, T. Effects of rehydration on dentine strengthened by heating or UV irradiation. *J. Dent. Res.* **2010**, *89*, 154–158. [CrossRef]
163. Seseogullari-Dirihan, R.; Tjäderhane, L.; Pashley, D.H.; Tezvergil-Mutluay, A. Effect of ultraviolet A-induced crosslinking on dentine collagen matrix. *Dent. Mater. J.* **2015**, *31*, 1225–1231. [CrossRef] [PubMed]
164. Mazzoni, A.; Nascimento, F.; Carrilho, M.; Tersariol, I.; Papa, V.; Tjäderhane, L. MMP activity in the hybrid layer detected with in situ zymography. *J. Dent. Res.* **2012**, *91*, 467–472. [CrossRef] [PubMed]
165. De Munck, J.; Van den Steen, P.E.; Mine, A.; Van Landuyt, K.L.; Poitevin, A.; Opdenakker, G.; Van Meerbeek, B. COMMENTARY. Inhibition of enzymatic degradation of adhesive-dentin interfaces. *J. Dent. Res.* **2009**, *88*, 1101–1106. [CrossRef] [PubMed]
166. Zhang, Y.; Gu, Y.; Lee, H.-M.; Hambardjieva, E.; Vranková, K.M.; Golub, L. Design; synthesis and biological activity of new polyenolic inhibitors of matrix metalloproteinases; a focus on chemically-modified curcumins. *Curr. Med. Chem.* **2012**, *19*, 4348–4358. [CrossRef] [PubMed]
167. Pashley, D.H.; Tay, F.R.; Carvalho, R.M.; Rueggeberg, F.A.; Agee, K.A.; Carrilho, M. From dry bonding to water-wet bonding to ethanol-wet bonding. A review of the interactions between dentine matrix and solvated resins using a macromodel of the hybrid layer. *Am. J. Dent.* **2007**, *20*, 7. [PubMed]
168. Tay, F.; Pashley, D.H.; Kapur, R.; Carrilho, M.; Hur, Y.; Garrett, L. Bonding BisGMA to dentine—A proof of concept for hydrophobic dentine bonding. *J. Dent. Res.* **2007**, *86*, 1034–1039. [CrossRef]
169. Ayar, M.K. A review of ethanol wet-bonding; Principles and techniques. *Eur. Dent. J.* **2016**, *10*, 155. [CrossRef] [PubMed]
170. Shin, T.P.; Yao, X.; Huenergardt, R.; Walker, M.P.; Wang, Y. Morphological and chemical characterization of bonding hydrophobic adhesive to dentine using ethanol wet bonding technique. *Dent. Mater. J.* **2009**, *25*, 1050–1057. [CrossRef] [PubMed]
171. Agee, K.A.; Prakki, A.; Abu-Haimed, T.; Naguib, G.H.; Nawareg, M.A.; Tezvergil-Mutluay, A. Water distribution in dentine matrices; bound, vs. unbound water. *Dent. Mater. J.* **2015**, *31*, 205–216. [CrossRef]
172. Chiba, A.; Zhou, J.; Nakajima, M.; Tan, J.; Tagami, J.; Scheffel, D. The effects of ethanol on the size-exclusion characteristics of type I dentine collagen to adhesive resin monomers. *Acta Biomater.* **2016**, *33*, 235–241. [CrossRef]
173. Takahashi, M.; Nakajima, M.; Tagami, J.; Scheffel, D.; Carvalho, R.; Mazzoni, A. The importance of size-exclusion characteristics of type I collagen in bonding to dentine matrices. *Acta Biomater.* **2013**, *9*, 9522–9528. [CrossRef] [PubMed]
174. Mai, S.; Wei, C.-C.; Gu, L.-S.; Tian, F.-C.; Arola, D.D.; Chen, J.-H. Extrafibrillar collagen demineralization-based chelate-and-rinse technique bridges the gap between wet and dry dentine bonding. *Acta Biomater.* **2017**, *57*, 435–448. [CrossRef] [PubMed]
175. De Oliveira Carrilho, M.R.; Tay, F.R.; Pashley, D.H.; Tjäderhane, L.; Carvalho, R.M. Mechanical stability of resin–dentine bond components. *Dent. Mater. J.* **2005**, *21*, 232–241. [CrossRef] [PubMed]
176. Reis, A.F.; Carrilho, M.R.; Ghaname, E.; Pereira, P.N.; Giannini, M.; Nikaido, T. Effects of water-storage on the physical and ultramorphological features of adhesives and primer/adhesive mixtures. *Dent. Mater. J.* **2010**, *29*, 1011170090. [CrossRef]
177. Brackett, M.G.; Li, N.; Brackett, W.W.; Sword, R.J.; Qi, Y.; Niu, L. The critical barrier to progress in dentine bonding with the etch-and-rinse technique. *J. Dent.* **2011**, *39*, 238–248. [CrossRef] [PubMed]
178. Sadek, F.; Braga, R.; Muench, A.; Liu, Y.; Pashley, D.H.; Tay, F. Ethanol wet-bonding challenges current anti-degradation strategy. *J. Dent. Res.* **2010**, *89*, 1499–1504. [CrossRef]
179. Sadek, F.T.; Castellan, C.S.; Braga, R.R.; Mai, S.; Tjäderhane, L.; Pashley, D.H. One-year stability of resin–dentine bonds created with a hydrophobic ethanol-wet bonding technique. *Dent. Mater. J.* **2010**, *26*, 380–386. [CrossRef] [PubMed]
180. Gutiérrez, M.; Malaquias, P.; Matos, T.; Szesz, A.; Souza, S.; Bermudez, J. Mechanical and microbiological properties and drug release modeling of an etch-and-rinse adhesive containing copper nanoparticles. *Dent. Mater. J.* **2017**, *33*, 309–320. [CrossRef]
181. Sun, J.; Petersen, E.J.; Watson, S.S.; Sims, C.M.; Kassman, A.; Frukhtbeyn, S.; Skrtic, D.; Ok, M.T.; Jacobs, D.S.; Reipa, V.; et al. Biophysical characterization of functionalized titania nanoparticles and their application in dental adhesives. *Acta Biomater.* **2017**, *53*, 585–597. [CrossRef]
182. Torres-Mendez, F.; Martinez-Castanon, G.-A.; Torres-Gallegos, I.; Zavala-Alonso, N.-V.; Patino-Marin, N.; Nino-Martinez, N. Effects of silver nanoparticles on the bonding of three adhesive systems to fluorotic enamel. *Dent. Mater.* **2017**, *36*, 266–274. [CrossRef]
183. Barcellos, D.C.; Fonseca, B.M.; Pucci, C.R.; das Neves Cavalcanti, B.; Persici, E.D.S.; de Paiva Gonçalves, S.E. Zn-doped etch-and-rinse model dentine adhesives; Dentine bond integrity; biocompatibility; and properties. *Dent. Mater.* **2016**, *32*, 940–950. [CrossRef]
184. Lohbauer, U.; Wagner, A.; Belli, R.; Stoetzel, C.; Hilpert, A.; Kurland, H.-D. Zirconia nanoparticles prepared by laser vaporization as fillers for dental adhesives. *Acta Biomater.* **2010**, *6*, 4539–4546. [CrossRef]
185. Agarwal, U.S.; Nisal, A.; Joseph, R. PET-SWNT nanocomposites through ultrasound assisted dissolution-evaporation. *Eur. Polym. J.* **2007**, *43*, 2279–2285.
186. Alkatheeri, M.S.; Palasuk, J.; Eckert, G.J.; Platt, J.A.; Bottino, M.C. Halloysite nanotube incorporation into adhesive systems—Effect on bond strength to human dentine. *Clinic. Oral. Investig.* **2015**, *19*, 1905–1912. [CrossRef]
187. Feitosa, S.A.; Münchow, E.A.; Al-Zain, A.O.; Kamocki, K.; Platt, J.A.; Bottino, M.C. Synthesis and characterization of novel halloysite-incorporated adhesive resins. *J. Dent.* **2015**, *43*, 1316–1322. [CrossRef] [PubMed]
188. Feitosa, S.; Palasuk, J.; Kamocki, K.; Geraldeli, S.; Gregory, R.; Platt, J. Doxycycline-encapsulated nanotube-modified dentine adhesives. *J. Dent. Res.* **2014**, *93*, 1270–1276. [CrossRef]

189. Leitune, V.C.B.; Collares, F.M.; Trommer, R.M.; Andrioli, D.G.; Bergmann, C.P.; Samuel, S.M.W. The addition of nanostructured hydroxyapatite to an experimental adhesive resin. *J. Dent.* **2013**, *41*, 321–327. [CrossRef]
190. Curylofo, F.A.; Messias, D.C.F.; Silva-Sousa, Y.T.C.; Souza-Gabriel, A.E. Bond strength of restorative material to dentine submitted to bleaching and Er; YAG laser post-treatment. *Photomed Laser Surg.* **2014**, *32*, 495–499. [CrossRef]
191. Kasraei, S.; Yarmohammadi, E.; Ghazizadeh, M.V. Microshear bond strength of OptiBond all-in-one self-adhesive agent to Er; YAG laser treated enamel after thermocycling and water storage. *Lasers. Med. Sci.* **2016**, *7*, 152. [CrossRef]
192. Han, G.J.; Kim, J.H.; Chung, S.N.; Chun, B.H.; Kim, C.K.; Seo, D.G. Effects of non-thermal atmospheric pressure pulsed plasma on the adhesion and durability of resin composite to dentine. *Eur. J. Oral Sci.* **2014**, *122*, 417–423. [CrossRef]
193. de Abreu, J.L.B.; Prado, M.; Simão, R.A.; da Silva, E.M.; Dias, K.R.H.C. Effect of non-thermal argon plasma on bond strength of a self-etch adhesive system to NaOCl-treated dentine. *Braz. Dent.* **2016**, *27*, 446–451. [CrossRef] [PubMed]
194. Zafar, M.S.; Amin, F.; Fareed, M.A.; Ghabbani, H.; Riaz, S.; Khurshid, Z. Biomimetic aspects of restorative dentistry biomaterials. *Biomimetics* **2020**, *5*, 34. [CrossRef] [PubMed]
195. Bertassoni, L.E.; Habelitz, S.; Pugach, M.; Soares, P.C.; Marshall, S.J.; Marshall Jr, G.W. Evaluation of surface structural and mechanical changes following remineralization of dentine. *Scanning Microsc.* **2010**, *32*, 312–319. [CrossRef] [PubMed]
196. He, L.; Hao, Y.; Zhen, L.; Liu, H.; Shao, M.; Xu, X.; Liang, K.; Gao, Y.; Yuan, H.; Li, J.; et al. Biomineralization of dentine. *J. Struct. Biol.* **2019**, *207*, 115–122. [CrossRef]
197. Rokidi, S.; Koutsoukos, P.G. Crystal growth of calcium phosphates from aqueous solutions in the presence of strontium. *Chem. Eng. Sci.* **2012**, *77*, 157–164. [CrossRef]
198. Gower, L.B. Biomimetic model systems for investigating the amorphous precursor pathway and its role in biomineralization. *Chem. Rev.* **2008**, *108*, 4551–4627. [CrossRef]
199. Tay, F.R.; Pashley, D.H. Guided tissue remineralisation of partially demineralised human dentine. *Biomaterials* **2008**, *29*, 1127–1137. [CrossRef]
200. Kim, J.; Arola, D.D.; Gu, L.; Kim, Y.K.; Mai, S.; Liu, Y. Functional biomimetic analogs help remineralize apatite-depleted demineralized resin-infiltrated dentine via a bottom–up approach. *Acta Biomater.* **2010**, *6*, 2740–2750. [CrossRef] [PubMed]
201. Gu, L.; Kim, Y.; Liu, Y.; Ryou, H.; Wimmer, C.; Dai, L. Biomimetic analogs for collagen biomineralization. *J. Dent. Res.* **2011**, *90*, 82–87. [CrossRef]
202. Sauro, S.; Osorio, R.; Watson, T.F.; Toledano, M. Influence of phosphoproteins' biomimetic analogs on remineralization of mineral-depleted resin–dentine interfaces created with ion-releasing resin-based systems. *Dent. Mater.* **2015**, *31*, 759–777. [CrossRef]
203. Abuna, G.; Feitosa, V.P.; Correr, A.B.; Cama, G.; Giannini, M.; Sinhoreti, M.A. Bonding performance of experimental bioactive/biomimetic self-etch adhesives doped with calcium-phosphate fillers and biomimetic analogs of phosphoproteins. *J. Dent.* **2016**, *52*, 79–86. [CrossRef]
204. Breschi, L.; Maravic, T.; Cunha, S.R.; Comba, A.; Cadenaro, M.; Tjäderhane, L. Dentine bonding systems; From dentine collagen structure to bond preservation and clinical applications. *Dent. Mater.* **2018**, *34*, 78–96. [CrossRef] [PubMed]
205. Mai, S.; Kim, Y.K.; Kim, J.; Yiu, C.K.; Ling, J.; Pashley, D.H. In vitro remineralization of severely compromised bonded dentine. *J. Dent. Res.* **2010**, *89*, 405–410. [CrossRef]
206. Kim, J.; Vaughn, R.M.; Gu, L.; Rockman, R.A.; Arola, D.D.; Schafer, T.E. Imperfect hybrid layers created by an aggressive one-step self-etch adhesive in primary dentine are amendable to biomimetic remineralization in vitro. *J. Biomed. Mater. Res. Part A* **2010**, *93*, 1225–1234.
207. Kim, Y.K.; Mai, S.; Mazzoni, A.; Liu, Y.; Tezvergil-Mutluay, A.; Takahashi, K. Biomimetic remineralization as a progressive dehydration mechanism of collagen matrices–implications in the aging of resin–dentine bonds. *Acta Biomater.* **2010**, *6*, 3729–3739. [CrossRef]
208. Osorio, R.; Yamauti, M.; Sauro, S.; Watson, T.F.; Toledano, M. Experimental resin cements containing bioactive fillers reduce matrix metalloproteinase–mediated dentine collagen degradation. *J. Endod.* **2012**, *38*, 1227–1232. [CrossRef]
209. Tjäderhane, L.; Nascimento, F.D.; Breschi, L.; Mazzoni, A.; Tersariol, I.L.; Geraldeli, S. Strategies to prevent hydrolytic degradation of the hybrid layer—A review. *Dent. Mater.* **2013**, *29*, 999–1011. [CrossRef] [PubMed]
210. Kanduti, D.; Sterbenk, P.; Artnik, B. Fluoride: A review of use and effects on health. *Mater. Socio-Med.* **2016**, *28*, 133. [CrossRef] [PubMed]
211. Shinohara, M.S.; Mario, F.; Schneider, L.F.J.; Ferracane, J.L.; Pereira, P.N.; Di Hipólito, V. Fluoride-containing adhesive: Durability on dentine bonding. *Dent. Mater.* **2009**, *25*, 1383–1391. [CrossRef]

coatings

MDPI

Article

Do Different Types of Adhesive Agents Effect Enamel Demineralization for Orthodontic Bonding? An In Vitro Study

Raif Murat Demircioglu [1], Orhan Cicek [1,*], Fusun Comert [2] and Hande Erener [3]

[1] Department of Orthodontics, Faculty of Dentistry, Zonguldak Bulent Ecevit University, 67100 Zonguldak, Turkey
[2] Department of Medical Microbiology, Faculty of Medicine, Zonguldak Bulent Ecevit University, 67630 Zonguldak, Turkey
[3] Department of Orthodontics, Faculty of Dentistry, Tekirdag Namık Kemal University, 59030 Tekirdağ, Turkey
* Correspondence: orhancicek@beun.edu.tr; Tel.: +90-372-2613557; Fax: +90-372-2613603

Abstract: (1) Objective: The aim of this study was to compare the demineralization around brackets bonded with different types of adhesive agents in a cariogenic suspension environment. (2) Methods: In the study, 60 extracted upper first premolar teeth were divided into three groups with 20 teeth in each group. In Group 1, Transbond XT Primer + Transbond XT Light Cure Adhesive (3M Unitek, Monrovia, CA, USA), in Group 2, GC Ortho Connect Light Cure Adhesive (GC Crop, Tokyo, Japan) and in Group 3, Transbond™ Plus Self Etching Primer + Transbond XT Light Cure Adhesive (3M Unitek, Monrovia, CA, USA) adhesive agents were used. In Group 1 and 2, buccal enamel surfaces were etched for 30 s, washed for 15 s and dried for 15 s. All groups were bonded with Gemini metal (3M Unitek, Monrovia, CA, USA) brackets. Gingival, occlusal and proximal enamel surfaces of the brackets were measured with a DIAGNOdent pen (KaVo, Biberach, Germany), and demineralization values were recorded. Measurements were performed after bracketing (T0) and after 28 days in a cariogenic environment (T1), which was renewed every 48 h. The Kolmogorov–Smirnov test was used to determine whether or not the data were homogeneously distributed, the Wilcoxon test was used for comparisons within groups, and the Mann–Whitney U and Kruskal–Wallis tests were used for comparisons between groups. (3) Results: In all groups, demineralization values on all enamel surfaces of the brackets were found to be statistically significantly higher in the T1 period than in the T0 period ($p < 0.05$). In the T1 period, demineralization values of occlusal enamel surfaces in Groups 1 and 2 were found to be significantly higher than in Group 3 ($p < 0.05$). The amount of increase in occlusal enamel surface demineralization value between T0 and T1 periods in Groups 1 and 2 was significantly higher than in Group 3 ($p < 0.05$). There was no statistically significant difference in demineralization values of proximal and gingival enamel surfaces between the groups in the T1 period ($p > 0.05$). (4) Conclusion: Significantly less occlusal enamel surface demineralization was observed in teeth in which the Transbond™ Plus Self Etching Primer adhesive agent was not applied with acid etching.

Keywords: orthodontics; bracket; adhesive; bond; artificial saliva; *Streptococcus mutans*; cariogenic environment; demineralization; DIAGNOdent

1. Introduction

In fixed appliances, the bands and brackets bonded to the teeth create retention areas for dental plaque, bacteria and food on the tooth surfaces [1].

For tooth caries to occur, there must be cariogenic bacteria, a sensitive tooth surface, time for the lesion to develop and nutrients for the bacteria. *Streptococcus mutans* creates strong biofilms on tooth surfaces, quickly metabolizes a wide range of carbohydrates from the host diet and endures numerous (and frequent) environmental challenges encountered in oral biofilms. *Streptococcus mutans* is a cariogenic organism. Cariogenic bacteria are the primary agents of initial caries; they adhere to the enamel, produce and tolerate acid and thrive in a sucrose-rich environment [2].

Excessive and frequent consumption of refined carbohydrates, along with the failure to remove plaque from retentive areas, accelerates the demineralization of tooth enamel. This situation disrupts the balance of remineralization and demineralization. Orthodontic bands and brackets placed on the teeth used in orthodontic treatment cause new retentive areas for plaque on the flat surfaces of the teeth with low caries prevalence. No matter how much patients pay attention to their oral hygiene during orthodontic treatment, auxiliary attachments such as fixed functional appliances, orthodontic arch wires, springs, loops, auxiliary arches and ligatures used during treatment create areas that are difficult to reach and clean [1]. It prevents the oral hygiene of the patients from remaining at the healthy level. Cariogenic activity increases with orthodontic treatment [3].

Increased demineralization causes the development of clinically visible white spot lesions on teeth. The clinical appearance of white spot lesions has a chalky white opacity and is defined as subsurface enamel pores resulting from demineralization [4]. Decalcified and porous enamel's changes in light scattering give it a white appearance [5]. When using fixed appliances for orthodontic treatment, white spot lesions are a frequent and unpleasant side effect. Within four weeks, or the time between two appointments for orthodontic treatment, these initial carious lesions can start developing [6]. The literature has shown that between 2% and 97% of patients undergoing fixed orthodontic treatment have white spot lesions [7–9]. The use of contemporary detection methods indicates a higher prevalence of white spot lesions than with the naked eye (97%) [8].

White spot lesions must be diagnosed as soon as possible in order to apply preventative measures and detect tooth integrity before it is compromised [4]. Without early detection and preventative measures, white spot lesions can advance rapidly and result in irreversible material loss on the teeth. The need for restoration arises in teeth with material loss [4]. In patients undergoing orthodontic treatment, fluoride varnishes are frequently used to increase mineralization and prevent demineralization before and after bonding. However, there are studies showing that the use of fluoride varnish before bonding reduces shear bonding strength (SBS) [10].

During the traditional acid etching method, the enamel preparation steps (acid etching, rinsing, drying and application of bonding agent) should be done properly. Loss of surface enamel and weakening of subsurface enamel can be seen in total-etch systems [11]. It can cause the enamel surface to split or break during debonding due to strong acidic conditioning liquid or prolonged etching [11]. It is quite often repeated that the use of self-etch primers produces a milder etching pattern than 37% phosphoric acid does [12,13]. The application of self-etch primers reduces the amount of adhesive remaining after debonding, thus reducing the invasive procedures required to clean the enamel surface [14].

This study was aimed at evaluating and comparing the enamel demineralization around the brackets bonded to the extracted human maxillary first premolars by using three different types of orthodontic adhesive agents and measuring with a laser fluorescence method, DIAGNOdent pen, in an artificial cariogenic suspension environment.

The null hypothesis was that no difference exists between demineralization around brackets bonded with different adhesive agents in the cariogenic suspension.

2. Materials and Methods

2.1. Preparation of Samples and Bonding

The study was carried out using 60 upper first premolars extracted for orthodontic treatment from patients referred to Zonguldak Bulent Ecevit University, Department of Orthodontics. Ethics committee approval was obtained prior the study, dated 9 February 2022 and numbered 2022/03, from Non-Invasive Clinical Research Ethics Committee of Zonguldak Bulent Ecevit University.

The teeth included in the study had no fluorosis on the enamel, caries, fillings, restorations, cracks or fractures [15–17]. The evaluation of the freshly extracted teeth was done with the naked eye. The patient's age, gender and the quadrant in which the teeth were extracted were neglected.

The teeth were stored in a 0.1% thymol solution until study [18]. The storage period of the teeth did not exceed six months [19]. The sample size of the study, in which the effect size was calculated using the mean and standard deviation of the groups, was performed by the G*Power 3.1.9.7 program. α error probability was set at 0.05. The power of the study ($1 - \alpha$ error prob) was set at 0.95. According to these data, the actual power of the study was calculated as 96%, and total sample size should have been 54. Sixty maxillary first premolars were divided into three groups, each group consisting of 20 teeth. Before bonding, roots of teeth were removed from the crowns with the use of a separator disc under water cooling along the enamel–cementum border of the teeth. The pulp chambers exposed after the incision were cleaned with a probe, and the pulp chambers were filled with a flowable composite [20]. A 3M ESPE Elipar S10 curing light (1200 Mw/cm^2 and a wavelength of 430–480 nm) was used for 20 s for the polymerization of the flowable composite. The flowable composite was polished with polishing discs to prevent a microbial retention area. The buccal surfaces of the teeth were cleaned with a rubber band and pumice before bonding (see Figure 1a). A 4×4 mm windowed acetate sheet was used to seal off the area where the bracket would be bonded to the buccal enamel surfaces. Using an acetate sheet limited the enamel surface that could be etched and adhered. Thus, the potential retentive enamel surface area caused by acid etching was reduced.

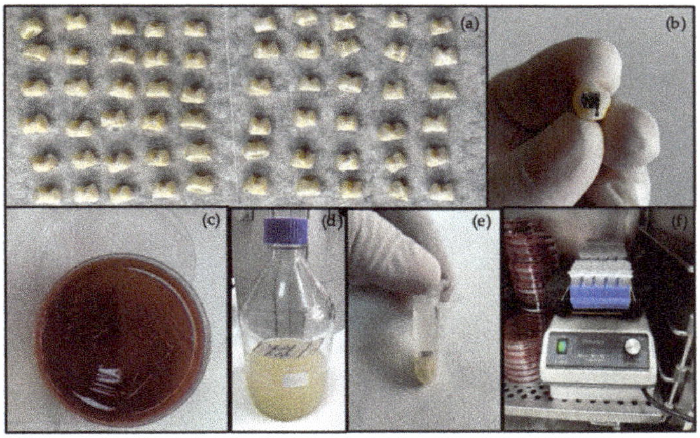

Figure 1. (**a**) Teeth whose roots were removed with a separator disc before bonding; (**b**) bonded tooth sample; (**c**) *Streptococcus mutans* medium; (**d**) artificial saliva solution; (**e**) specimen ready to be placed in the incubator; (**f**) specimens placed in an incubator on a Thermolyne Maxi-Mix III Type 65800 Rotary Shaker (Thermo Scientific, Iowa, IA, USA).

In Group 1, 37% orthophosphoric acid gel for acid etching was applied to the buccal enamel surfaces encircled by the acetate sheet for 30 s. After acid etching, the enamel surface was washed for 15 s and dried for 15 s. Transbond XT Primer (3M Unitek, Monrovia, CA, USA) was applied in a thin layer to the etched enamel surface. Then, brackets loaded with Transbond XT Light Cure Adhesive Paste (3M Unitek, Monrovia, CA, USA) were carefully placed on the teeth in the correct position.

In Group 2, 37% orthophosphoric acid gel for acid etching was applied to the buccal enamel surfaces encircled by the acetate sheet for 30 s. After acid etching, the enamel surface was washed for 15 s and dried for 15 s. Then, brackets loaded with GC Ortho Connect Light Cure Adhesive Paste (GC Crop, Tokyo, Japan) were carefully placed in the correct position.

In Group 3, a thin layer of Transbond Plus Self Etching Primer (3M Unitek, Monrovia, CA, USA) was applied to clean plaque-free enamel surfaces on which acetate-sheet-bounded brackets would be located. Air was applied lightly to the surface with an air syringe. Then, brackets loaded with Transbond XT Light Cure Adhesive Paste (3M Unitek, Monrovia, CA, USA) were carefully placed on the teeth in the correct position.

All samples were bonded with Gemini metal (3M Unitek, Monrovia, CA, USA) brackets (see Figure 1b). Adhesive flashes during bracket placement were removed with a probe and a 3M ESPE Elipar S10 (3M ESPE Dental Products) curing light source with a light intensity of 1200 Mw/cm^2 and wavelength of 430–480 nm used for adhesive paste polymerization. During the polymerization, a total of 20 s of light was applied from the mesial and distal sides of brackets for 10 s. The ingredients of the adhesive agents used in the study are given in Table 1.

Table 1. Composition on ingredients of adhesives.

	Ingredients	wt%	Manufacturer
3M™ Unitek™ Transbond™ XT Primer	Bisphenol A Diglycidyl Ether Dimethacrylate (BISGMA) Triethylene Glycol Dimethacrylate (TEGDMA) 4-(Dimethylamino)-Benzeneethanol	45–55 45–55 <0.5	3M Unitek, Monrovia, CA, USA
3M Unitek Transbond XT Light Cure Adhesive	Silane Treated Quartz Bisphenol A Diglycidyl Ether Dimethacrylate (BISGMA) Bisphenol A *Bis* (2-Hydroxyethyl Ether) Dimethacrylate Silane Treated Silica Diphenyliodonium Hexafluorophosphate	70–80 10–20 5–10 <2 <0.2	3M Unitek, Monrovia, CA, USA
3M™ Unitek™ Transbond™ Plus Self Etch Primer Part A	2-Propenoic acid, 2-methyl-, 2-hydroxyethyl ester, reaction products with phosphorus oxide (P_2O_5) DL-Camphorquinone *N,N*-Dimethylbenzocaine 4-Methoxyphenol Hydroquinone	>95 <2 <2 <0.2 <0.1	3M Unitek, Monrovia, CA, USA
3M™ Unitek™ Transbond™ Plus Self Etch Primer Part B	Water Dipotassium Hexafluorotitanate	>98 <2	3M Unitek, Monrovia, CA, USA
GC Ortho Connect Light Cure Adhesive	Esterification products of 4,4′-isopropylidenediphenol, ethoxylated and 2-methylprop-2-enoic acid Urethane Dimethacrylate (UDMA) methacryloyloxydecyl dihydrogen phosphate 6-*tert*-butyl-2,4-xylenol diphenyl(2,4,6-trimethylbenzoyl)phosphine oxide	25–50 25–50 2.5–5 0.25–0.5 0.2–0.5	GC Crop, Tokyo, Japan

2.2. Measurements of T0 with DIAGNOdent Pen

The DIAGNOdent pen (KaVo, Biberach, Germany) device used in the study was calibrated for each tooth before measurement according to manufacturer's instructions.

Measurements were performed on the occlusal, gingival and proximal enamel surfaces around the brackets by holding the B tip perpendicular to these surfaces. After moving the B tip of the DIAGNOdent pen 3–4 times on the occlusal and gingival surfaces, up and down on the proximal surfaces, the value read on the screen of the device was recorded as T0 [21]. The DIAGNOdent pen device recognizes its readings as fluorescence arbitrary units (a.u.) and generates a score. A single proximal demineralization value was recorded for each tooth by averaging the measurements performed on the mesial and distal proximal surfaces. Measurements were repeated twice, and all measurements were made by the same researcher (R.M.D.). The measurement room was illuminated with a 6400 K white artificial light.

2.3. Preparation of Artificial Saliva and Cariogenic Suspension

Artificial saliva was prepared with the same formula as Toz Ertop et al. did [19]. Artificial saliva was prepared with 0.4 g of sodium chloride (NaCl), 0.4 g of potassium chloride (KCl), 0.8 g of calcium chloride ($CaCl_2 \cdot 2H_2O$), 0.78 g of sodium dihydrogen phosphate ($NaH_2PO_4 \cdot 2H_2O$), 0.005 g of sodium sulfate ($NaS \cdot 9H_2O$) and 1 g urea of in

1000 mL of deionized water [19,22] (see Figure 1c). After the prepared artificial saliva solution was sterilized in an autoclave, 140 mg of mucin (Mucin Type II; SigmaAldrich Chemie GmbH, Deisenhofen, Germany) was added per 100 mL of artificial saliva solution. Adding mucin was aimed at accelerating the development of pellicle formation [19,23].

Bacteria stock culture were taken from Zonguldak Bulent Ecevıt University, Faculty of Medicine, Department of Medical Microbiology. *Streptococcus mutans* culture incubated on blood agar from a stock culture was used to prepare the cariogenic suspension (see Figure 1d). A bacterial suspension equivalent to 0.5 McFarland (10^8 cfu/mL) turbidity was prepared in brain–heart infusion broth with bacteria taken from the medium. The sucrose solution was prepared as 1 g/10 mL distilled water and passed through a sterile 0.22 μm syringe filter. An artificial cariogenic suspension with a turbidity of 10^6 cfu/mL was obtained by adding 0.5 mL of sucrose solution and 0.5 mL of bacterial suspension for each 49 mL of artificial saliva solution [19].

2.4. Cariogenic Suspension Environment

Bonded tooth samples and U-bottom centrifuge tubes to be used were sterilized in an autoclave to prevent contamination. Each tooth sample was placed in a tube. Two milliliters of artificial cariogenic suspension was added to the tubes (see Figure 1e). The tubes were placed in a Thermolyne Maxi-Mix III Type 65800 Rotary Shaker (Thermo Scientific, Iowa, IA, USA) on a tray. The rotation speed was set to 20 rpm. The homogeneous interaction of the artificial cariogenic suspension with all the teeth was done with the use of a rotary shaker. The prepared samples were incubated for 28 days at 37 °C in a 10% CO_2 atmosphere on a rotary shaker placed in incubator (see Figure 1f). During the incubation period, the artificial cariogenic suspension and the used U-bottom centrifuge tubes were renewed every 48 h. After 28 days, the teeth were removed from the cariogenic suspension and washed with distilled water.

2.5. T1 Measurements with DIAGNOdent Pen

After 28 days, enamel demineralization on the occlusal, gingival and proximal enamel surfaces around the brackets was remeasured with the DIAGNOdent pen. Measurement results were recorded as T1. The measurements were made with the same method and environment as for T0 measurements.

2.6. Statistical Analysis

SPSS (Statistical Package for Social Sciences) 28.0 (SPSS Inc., Chicago, IL, USA) was used for statistical analysis. The Kolmogorov–Smirnov test was used to determine whether or not the data were homogeneously distributed, the Wilcoxon test was used for comparisons within groups, the Mann–Whitney U and Kruskal–Wallis tests were used for comparisons between groups. The level of significance chosen for all statistical tests was $p < 0.05$.

3. Results

In the T0 period, all enamel surface demineralization values adjacent to the bracket in all groups did not differ significantly between the groups ($p > 0.05$). In all groups, all enamel surface demineralization values adjacent to the bracket in the T1 period increased significantly compared to those of the T0 period ($p < 0.05$). In all groups, the gingival and proximal enamel surface demineralization values adjacent to the bracket in the T1 period did not show a statistically significant difference between the groups ($p > 0.05$). The change in the demineralization values of the gingival and proximal enamel surface adjacent to the bracket in the T0/T1 period did not show a statistically significantly difference in all groups ($p > 0.05$). The increase in the demineralization values of the gingival enamel surfaces adjacent to the bracket was seen the most in Group 1 and the least in Group 3, but these differences were not statistically significant ($p > 0.05$). The increase in proximal enamel surface demineralization values adjacent to the bracket was seen the most in Group 2 and the least in Group 3, but these changes were not statistically significant ($p > 0.05$).

Occlusal enamel demineralization values adjacent to the bracket in Group 1 and Group 2 in the T1 period showed a statistically significant increase compared to that of Group 3 ($p < 0.05$). No statistically significant difference was found between the demineralization values measured on the occlusal surface adjacent to the bracket in Group 1 and Group 2 in the T1 period ($p > 0.05$). In Group 1 and Group 2, the amount of increase in the demineralization value of the occlusal surface adjacent to the bracket in the T0/T1 period was significantly higher than in Group 3 ($p < 0.05$). In Groups 1 and 2, the amount of increase in the demineralization value of the occlusal surface adjacent to the bracket in the T0/T1 period did not show a statistically significant difference between the groups ($p > 0.05$). Statistical analysis results of demineralization values in T0, T1 and T0/T1 periods are given in Table 2.

Table 2. Demineralization values on the enamel surface adjacent to the brackets at T0, T1 and T0/T1 periods.

		Group 1	Group 2	Group 3	*p*
Occlusal					
T0	Median	3.00	3.00	2.50	NS
T1	Median	6.00 ^{3}	7.00 ^{3}	5.00 ^{1,2}	0.003 ^{K}
T0/T1 difference	Median	4.00 ^{3}	4.00 ^{3}	3.00 ^{1,2}	0.003 ^{K}
Intra-Group difference	*p*	0.000 ^{W}	0.000 ^{W}	0.000 ^{W}	
Proximal					
T0	Median	3.00	2.50	2.75	NS
T1	Median	7.00	7.00	7.00	NS
T0/T1 difference	Median	4.50	4.25	4.00	NS
Intra-Group difference	*p*	0.000 ^{W}	0.000 ^{W}	0.000 ^{W}	
Gingival					
T0	Median	3.00	3.00	3.00	NS
T1	Median	10.00	10.00	10.00	NS
T0/T1 difference	Median	7.50	7.00	7.00	NS
Intra-Group difference	*p*	0.000 ^{W}	0.000 ^{W}	0.000 ^{W}	

^{K}: Kruskal–Wallis (Mann–Whitney U test); ^{W}: Wilcoxon test; T0: before placement in cariogenic environment; T1: 28 days after placement in cariogenic environment; $p < 0.05$: level of significance considered; [1] difference with Transbond XT primer + Transbond XT adhesive group $p < 0.05$; [2] difference with GC Ortho Connect group $p < 0.05$; [3] difference with Transbond Plus primer + Transbond XT adhesive $p < 0.05$; NS: not significant.

4. Discussion

The areas where the appliances used in fixed orthodontic treatment are placed are not generally caries-prone areas [24,25]. Toz Ertop et al. kept the bracketed teeth in a cariogenic suspension that they renewed every 2 days for 28 days and observed demineralization in all teeth [19]. In this study, statistically significant increases in demineralization values were found on all enamel surfaces adjacent to the bracket 28 days after all groups were placed in the cariogenic environment. There is a study showing that the use of biomimetic hydroxyapatite is appropriate to treat demineralization around the bracket and to increase enamel mineralization [26].

Visel et al. compared the effects of a self-etch adhesive system (Transbond Plus) and a conventional total-etch adhesive system (Transbond XT) on demineralization around the bracket in vivo and observed that the enamel samples which were conditioned with the self-etching fluoride-releasing primer (Transbond Plus) displayed the highest degree of remineralization [27]. Montaseer et al. in their study examining the potential protection effect of different treatments against demineralization around orthodontic brackets reported that applying the Transbond Plus Self Etching primer to the enamel surface before demineralization showed significantly less demineralization and more resistance to demineralization than the enamel surfaces that did not undergo any treatment in vitro did [28]. These studies support the fact that less demineralization was observed on the occlusal surfaces adjacent to the enamel in the group in which the self-etch primer was used in our study. We could suggest that time spans for remineralizations were produced as a result of the artificial saliva solution in a cariogenic suspension medium, which was replaced every 48 h.

Kohda et al. stated in their study that the reason why self-etch primer application shows statistically significantly less demineralization than phosphoric acid does could be a result of the higher pH level of self-etch primers and shorter application time in vitro [29]. Narendran and Raghunath reported that the conventional total-etch (Transbond XT) group causes a significantly more irregular structure on the enamel surface and found a deeper penetration of 86.7% compared to that of the self-etch (Transbond Plus) group in their study where they compared the demineralization around orthodontic brackets in vitro. In addition, they stated that the Transbond Plus group underwent less demineralization than the Transbond XT group did [30]. Ghandi et al. performed bracket bonding with a self-etch adhesive system (Transbond Plus) and a conventional total-etch adhesive system (Transbond XT). By decalcifying the bracketed teeth, the resin replicas remaining at the base of the brackets were examined under a scanning electron microscope for micromorphological observation of adhesive penetration in the enamel in vitro. They stated that there was significantly less enamel demineralization and resin infiltration in the self-etch group, and that self-etch adhesive systems were more conservative than conventional total-etch adhesive systems [31]. In our study, lower demineralization values were found in the group in which the self-etch primer was used, and statistically significantly less demineralization was observed on the occlusal surface of the enamel adjacent to the bracket in the same group. This result can be explained by the fact that self-etch primers caused more superficial changes on the enamel surface and had a higher pH level.

Hung et al. found significantly high fluoride release in the teeth they bonded with Transbond Plus SEP and Transbond Plus adhesive, especially in the first 14 days in their in vitro study [32]. Zrinski et al. also supported this finding in vitro and stated that Transbond Plus SEP can release fluorine, but its capacity to store fluorine again is inadequate [33]. Krasniqi et al. compared the antimicrobial effects of different types of adhesive agents on *Streptococcus mutans* and *Lactobacillus acidophilus* bacteria; they found that Transbond Plus SEP was the agent with the largest inhibition area (antibacterial effect) against *Streptococcus mutans* and *Lactobacillus acidophilus* among all groups. In the same in vitro study, they found that the Transbond XT primer and adhesive did not show an antibacterial effect. They thought that this result was related to fluorine release [34].

In this study, statistically significantly less demineralization observed in Group 3 on the occlusal surface adjacent to the bracket and less demineralization in Group 3 on all surfaces adjacent to the bracket is supported by studies indicating that the self-etch adhesive system causes less demineralization [14,27,28]. In this situation, it was thought that the use of a self-etch adhesive system caused a more superficial change in the enamel. Additionally, it may result in less irregular surfaces to which bacteria can adhere, and these outcomes were influenced by fluorine's antibacterial properties and its capacity to release fluorine in Transbond SEP [14,32–34].

Turğut clinically evaluated the white spot lesion formation and bond failure of the self-priming total-etch adhesive system for bonding orthodontic metal brackets in vivo [35]. In that study, 51 patients were bonded with a split-mouth study protocol using a self-priming total-etch adhesive system (GC Ortho Connect) and a conventional total-ecth adhesive system (Transbond XT). As a result of the study, it was stated that there was no significantly difference between the GC Ortho Connect and Transbond XT groups in terms of demineralization formation and bond failure [35]. In our study, enamel demineralization on the gingival, occlusal and proximal enamel surfaces adjacent to the bracket measured at T0, T1 and T0/T1 periods did not show a statistically significant difference between Group

1 and Group 2. However, a statistically significant difference was observed in the occlusal surface of Group 3; based on these findings, the null hypothesis was rejected.

Regarding the limitations of the study, in addition to the composite materials to which bacteria can easily adhere, the patient's diet may also be effective in the demineralization of the hard tissues of the teeth. However, using an artificial cariogenic suspension environment simulated an intact surface layer and a subsurface lesion pattern [36,37]. Considering the inadequacies of the artificial cariogenic suspension environment created in the in vitro environment to fully simulate the oral flora, it is thought that further studies planned under in vivo conditions are needed.

5. Conclusions

1. Significant increases in demineralization occurred on all enamel surfaces adjacent to the bracket 28 days after placement in an artificial cariogenic suspension in all groups.
2. There was no statistically significant difference between Group 1 and Group 2 in the demineralization values of enamel surfaces adjacent to the bracket after 28 days.
3. The null hypothesis was rejected. Demineralization values on the occlusal surfaces of the brackets bonded using a Transbond™ Plus Self Etching Primer adhesive agent were found to be significantly lower than those with other adhesive agents. Since the use of a self-etch primer does not require etching on the enamel surface, it can be assumed that the result was less enamel surface changes. It is possible that self-etching teeth had enamel surfaces that were more resistant to plaque formation. The use of a self-etch primer may have made remineralization more effective. The use of a self-etch primer in bracketing may cause less demineralization on the occlusal surfaces of the teeth in cariogenic attacks.

Author Contributions: Conceptualization, O.C., R.M.D., F.C. and H.E.; methodology, O.C., R.M.D., F.C. and H.E.; software, R.M.D., O.C. and F.C.; validation, O.C., R.M.D. and H.E.; formal analysis, O.C., R.M.D., H.E. and F.C.; investigation, R.M.D., O.C. and F.C.; resources, R.M.D., O.C., H.E. and F.C.; data curation, R.M.D., O.C., H.E. and F.C.; writing—original draft preparation, R.M.D., O.C. and F.C.; writing—review and editing, R.M.D., O.C., H.E. and F.C.; visualization, O.C., R.M.D. and F.C.; supervision, O.C.; project administration, O.C., R.M.D., H.E. and F.C. All authors have read and agreed to the published version of the manuscript.

Funding: This research received no external funding.

Institutional Review Board Statement: The study was conducted in accordance with the Declaration of Helsinki. We would like to thank the Non-Invasive Clinical Research Ethics Committee of Zonguldak Bulent Ecevit University for the ethical approval of this study (dated 9 February 2022 and numbered 2022/03).

Informed Consent Statement: Not applicable.

Data Availability Statement: All data supporting the results of this study are included within the article.

Acknowledgments: This study constitutes a specialty dissertation thesis by Raif Murat DEMIRCIOGLU, Zonguldak Bulent Ecevit University, Department of Orthodontics, Turkey. We would like to thank the Non-Invasive Clinical Research Ethics Committee of Zonguldak Bulent Ecevit University for the ethical approval of this study.

Conflicts of Interest: The authors declare no potential conflicts of interest with respect to the authorship and/or publication of this article. The authors report no commercial, proprietary, or financial interest in the products or companies described in this article.

References

1. Proffit, W.R.; Fields, H.W.; Larson, B.; Sarver, D.M. *Contemporary Orthodontics*, 6th ed.; Elsevier Health Sciences: Philadelphia, PA, USA, 2018; pp. 321–332.
2. Lemos, J.A.; Palmer, S.R.; Zeng, L.; Wen, Z.T.; Kajfasz, J.K.; Freires, I.A.; Abranches, J.; Brady, L.J. The Biology of Streptococcus mutans. *Microbiol. Spectr.* **2019**, *7*, 1–18.
3. Reichardt, E.; Geraci, J.; Sachse, S.; Rödel, J.; Pfister, W.; Löffler, B.; Wagner, Y.; Eigenthaler, M.; Wolf, M. Qualitative and quantitative changes in the oral bacterial flora occur shortly after implementation of fixed orthodontic appliances. *Am. J. Orthod. Dentofacial. Orthop.* **2019**, *156*, 735–744. [CrossRef] [PubMed]
4. Bishara, S.E.; Ostby, A.W. White spot lesions: Formation, prevention, and treatment. *Semin. Orthod.* **2008**, *14*, 174–182. [CrossRef]
5. Bahramian, H.; Argani, P.; Baghalian, A. Comparison of different diagnostic techniques in detecting smooth surface caries in primary molars using the histological gold standard: An in vitro study. *Photodiagnosis Photodyn. Ther.* **2020**, *31*, 101867. [CrossRef]
6. Zentner, A. Structural changes of acid etched enamel examined under confocal laser scanning microscope. *J. Orofac. Orthop.* **1996**, *57*, 202–209. [CrossRef]
7. Øgaard, B.; Larsson, E.; Henriksson, T.; Birkhed, D.; Bishara, S.E. Effects of combined application of antimicrobial and fluoride varnishes in orthodontic patients. *Am. J. Orthod. Dentofac. Orthop.* **2001**, *120*, 28–35. [CrossRef]
8. Boersma, J.G.; van der Veen, M.H.; Lagerweij, M.D.; Bokhout, B.; Prahl-Andersen, B. Caries prevalence measured with QLF after treatment with fixed orthodontic appliances: Influencing factors. *Caries Res.* **2005**, *39*, 41–47. [CrossRef]
9. Hadler-Olsen, S.; Sandvik, K.; El-Agroudi, M.A.; Øgaard, B. The incidence of caries and white spot lesions in orthodontically treated adolescents with a comprehensive caries prophylactic regimen—A prospective study. *Eur. J. Orthod.* **2012**, *34*, 633–639. [CrossRef]
10. Cossellu, G.; Lanteri, V.; Butera, A.; Sarcina, M.; Farronato, G. Effects of six different preventive treatments on the shear bond strength of orthodontic brackets: In vitro study. *Acta Biomater. Odontol. Scand.* **2015**, *1*, 13–17. [CrossRef]
11. Kim, M.-J.; Lim, B.-S.; Chang, W.-G.; Lee, Y.-K.; Rhee, S.-H.; Yang, H.-C. Phosphoric Acid Incorporated with Acidulated Phosphate Fluoride Gel Etchant Effects on Bracket Bonding. *Angle Orthod.* **2005**, *75*, 678–684.
12. Vilchis, R.J.S.; Yamamoto, S.; Kitai, N.; Hotta, M.; Yamamoto, K. Shear bond strength of a new fluoride-releasing orthodontic adhesive. *Dent. Mater. J.* **2007**, *26*, 45–51. [CrossRef]
13. Hosein, I.; Sherriff, M.; Ireland, A.J. Enamel loss during bonding, debonding, and cleanup with use of a self-etching primer. *Am. J. Orthod. Dentofacial. Orthop.* **2004**, *126*, 717–724. [CrossRef]
14. Zope, A.; Zope-Khalekar, Y.; Chitko, S.S.; Kerudi, V.V.; Patil, H.A.; Bonde, P.V.; Jaltare, P.; Dolas, S.G. Comparison of Self-Etch Primers with Conventional Acid Etching System on Orthodontic Brackets. *J. Clin. Diagn. Res.* **2016**, *10*, 19–22. [CrossRef]
15. Alabdullah, M.M.; Nabawia, A.; Ajaj, M.A.; Saltaji, H. Effect of fluoride-releasing resin composite in white spot lesions prevention: A single-centre, split-mouth, randomized controlled trial. *Eur. J. Orthod.* **2017**, *39*, 634–640. [CrossRef]
16. Tan, A.; Çokakoğlu, S. Effects of adhesive flash-free brackets on enamel demineralization and periodontal status. *Angle Orthod.* **2020**, *90*, 339–346. [CrossRef]
17. Bazargani, F.; Magnuson, A.; Löthgren, H.; Kowalczyk, A. Orthodontic bonding with and without primer: A randomized controlled trial. *Eur. J. Orthod.* **2016**, *38*, 503–507. [CrossRef]
18. Ak, İ.C. Evaluation of in vitro the shear bond strength of brackets with different base designs. Ph.D. Thesis, Cukurova University, Department of Orthodontics, Adana, Türkiye, 2018.
19. Ertop, M.T.; Cicek, O.; Erener, H.; Ozkalayci, N.; Cicek, B.D.; Comert, F. Evaluation of the Demineralization Development around Different Types of Orthodontic Brackets. *Materials* **2023**, *16*, 984. [CrossRef]
20. Baroudi, K.; Rodrigues, J.C. Flowable resin composites: A systematic review and clinical considerations. *J. Clin. Diagn. Res.* **2015**, *9*, 18–22. [CrossRef]
21. Diniz, M.; Campos, P.; Sanabe, M.; Duarte, D.; Santos, M.; Guaré, R.; Duque, C.; Lussi, A.; Rodrigues, J. Effectiveness of fluorescence-based methods in monitoring progression of noncavitated caries-like lesions on smooth surfaces. *Oper. Dent.* **2015**, *40*, 230–241. [CrossRef]
22. Fatima, S.; Panda, N.; Reddy, A.V.; Fatima, S. Buccal Mucoadhesive Tablets of Sumatriptan Succinate for Treatment of Sustainable Migraine: Design, Formulation and In Vitro Evaluation. *Int. J. Pharm Res.* **2015**, *4*, 114–126.
23. Aykent, F.; Yondem, I.; Ozyesil, A.G.; Gunal, S.K.; Avunduk, M.C.; Ozkan, S. Effect of different finishing techniques for restorative materials on surface roughness and bacterial adhesion. *J. Prosthet. Dent.* **2010**, *103*, 221–227. [CrossRef]
24. Khalaf, K. Factors affecting the formation, severity and location of white spot lesions during orthodontic treatment with fixed appliances. *J. Oral. Maxillofac. Res.* **2014**, *5*, e4. [CrossRef] [PubMed]
25. Øgaard, B. White spot lesions during orthodontic treatment: Mechanisms and fluoride preventive aspects. *Semin. Orthod.* **2008**, *14*, 183–193. [CrossRef]
26. Scribante, A.; Dermenaki Farahani, M.R.; Marino, G.; Matera, C.; Rodriguez y Baena, R.; Lanteri, V.; Butera, A. Biomimetic Effect of Nano-Hydroxyapatite in Demineralized Enamel before Orthodontic Bonding of Brackets and Attachments: Visual, Adhesion Strength, and Hardness in In Vitro Tests. *BioMed. Res. Int.* **2020**, *2020*, 6747498. [CrossRef] [PubMed]
27. Visel, D.; Jäcker, T.; Jost-Brinkmann, P.G.; Präger, T.M. Demineralization adjacent to orthodontic brackets after application of conventional and self-etching primer systems. *J. Orofac. Orthop.* **2014**, *75*, 358–373. [CrossRef] [PubMed]

28. Montasser, M.A.; El-Wassefy, N.A.; Taha, M. In vitro study of the potential protection of sound enamel against demineralization. *Prog. Orthod.* **2015**, *16*, 12. [CrossRef]
29. Kohda, N.; Iijima, M.; Brantley, W.; Muguruma, T.; Yuasa, T.; Nakagaki, S.; Mizoguchi, I. Effects of bonding materials on the mechanical properties of enamel around orthodontic brackets. *Angle Orthod.* **2011**, *82*, 187–195. [CrossRef]
30. Narendran, S.; Raghunath, N. Comparison of enamel demineralization around orthodontic brackets bonded with conventional etching, self-etch primer and antimicrobial monomer containing self-etch primer. *Int. J. Adv. Res. Innov. Ideas Educ.* **2019**, *2*, 18–22.
31. Gandhi, G.; Kalra, J.P.S.; Goyal, A.; Sharma, A. Microphotographic Assessment of Enamel Surface using Self-Etching Primer and Conventional Phosphoric Acid: An In vitro Study. *Contemp. Clin. Dent.* **2018**, *9*, 15–19. [CrossRef]
32. Hung, C.-Y.; Yu, J.-H.; Su, L.-W.; Uan, J.-Y.; Chen, Y.-C.; Lin, D.-J. Shear Bonding Strength and Thermal Cycling Effect of Fluoride Releasable/Rechargeable Orthodontic Adhesive Resins Containing LiAl-F Layered Double Hydroxide (LDH) Filler. *Materials* **2019**, *12*, 3204. [CrossRef]
33. Zrinski, M.T.; Miljanic, S.; Peros, K.; Turco, G.; Contardo, L.; Spalj, S. Fluoride release and recharge potential of remineralizing orthodontic adhesive systems. *Flouride* **2019**, *52*, 397–403.
34. Krasniqi, S.; Sejdini, M.; Stubljar, D.; Jukic, T.; Ihan, A.; Aliu, K.; Aliu, X. Antimicrobial Effect of Orthodontic Materials on Cariogenic Bacteria Streptococcus mutans and Lactobacillus acidophilus. *Med. Sci. Monit. Basic Res.* **2020**, *26*, e920510. [CrossRef]
35. Turğut, A. Clinical Evaluation of One Step Orthodontic Adhesive Without Primer for Bonding of Metal Brackets in Terms of White Spot Lesion Formation and Bond Strength. Ph.D. Thesis, Pamukkale University, Department of Orthodontics, Denizli, Türkiye, 2020.
36. Femiano, F.; Femiano, R.; Femiano, L.; Nucci, L.; Santaniello, M.; Grassia, V.; Scotti, N.; Aversa, R.; Perrotta, V.; Apicella, A.; et al. Enamel Erosion Reduction through Coupled Sodium Fluoride and Laser Treatments before Exposition in an Acid Environment: An In Vitro Randomized Control SEM Morphometric Analysis. *Appl. Sci.* **2022**, *12*, 1495. [CrossRef]
37. Szalewski, L.; Wójcik, D.; Bogucki, M.; Szkutnik, J.; Różyło-Kalinowska, I. The influence of popular beverages on mechanical properties of composite resins. *Materials* **2021**, *14*, 3097. [CrossRef]

coatings

Article

Evaluation of the Cyclic Fatigue Resistance of Tia Tornado Blue and Tia Tornado Gold in Curved Canals: In Vitro Study

Ziyad Allahem *, Mohammad Bendahmash, Reem Almeaither, Hussam Alfawaz and Abdullah Alqedairi

Department of Restorative Dental Sciences, College of Dentistry, King Saud University, Riyadh 11451, Saudi Arabia; mohammad.wd9@gmail.com (M.B.); reemalmeaither@gmail.com (R.A.); halfawaz1@ksu.edu.sa (H.A.); aalqedairi@ksu.edu.sa (A.A.)

* Correspondence: zallahem@ksu.edu.sa

Abstract: (1) Background: The aim of this study was to investigate the cyclic fatigue behavior of the newly introduced endodontic instrument systems Tia Tornado Blue and Tia Tornado Gold. (2) Methods: The tested rotary instruments were divided according to their type into four groups as follows: Tia Tornado Blue (TTB), Tia Tornado Gold (TTG), Protaper Gold (PTG), and Vortex Blue (VB). The cyclic fatigue resistance of fifteen instruments of each group, totaling 60, was tested. Each instrument was rotated under continuous motion inside an artificial canal at simulated body temperature at the speed recommended by the manufacturer until fracture. The time to fracture was recorded, and the number of cycles to fracture (NCF) was calculated for each instrument. Additionally, the fractured segment length was calculated, and scanning electron microscopic (SEM) images were captured of the fractured surfaces. (3) Results: Statistical analysis revealed that the VB significantly had the highest NCF followed by the PTG, TTG, and TTB ($p < 0.05$). The lengths of the fractured segments were found to be similar among the tested instrument, which ranged from 4.42 to 4.86 mm ($p > 0.05$). SEM images exhibited the typical features of cyclic fatigue. (4) Conclusions: The newly introduced instruments, TTG and TTB, exhibited a significantly lower resistance to cyclic fatigue compared to the PTG and VB rotary instruments.

Keywords: cyclic fatigue; Tia Tornado Blue; Tia Tornado Gold; Protaper Gold; endodontic instrument

1. Introduction

Nickel titanium (NiTi) rotary instruments have been routinely used in endodontics practice since their innovation. Their priority in practice has been mostly due to the high flexibility, lower canal transportation, and reduced apical extrusion compared to stainless steel instruments [1,2]. However, NiTi fractures can happen during use due to many factors, although their prevalence is low and ranges from 0.7% to 3% [3,4].

Factors that have a crucial effect on the fatigue of NiTi instruments are the radius of curvature, the angle of curvature, operator skill, canal geometry, NiTi alloy, the heat treatment of the alloy, instrument size, the point of maximal instrument flexure, sterilization cycles, and manufacturing process [5–11]. Two mechanisms of fracture have been identified include cyclic fatigue and torsional fatigue. Cyclic fatigue is defined by the tension/compression cycles at the point of maximum flexure, while torsional fatigue is determined when an instrument tip or another part of the instrument is locked in a canal while the shank continues to rotate, and the torque goes beyond the elastic limit of the metal of the endodontic instrument [12,13].

According to reported findings, cyclic fatigue has been found to be the most frequent cause of failure [14–16]. Some new modifications have been made regarding endodontic instruments with advancements in the manufacturing processes of NiTi in attempts to reduce procedural errors and change the instrument performance in terms of the cyclic fatigue resistance. These modifications involved changes in heat treatments, alloy composition, different cross-sectional design, thermomechanical processes, and surface treatment of the endodontic instruments [17–20].

In general, conventional NiTi instruments have an austenite structure at 37 °C degree body temperature. When the NiTi instrument is the control memory wire or M-wire, it is in the martensite phase [21]. The M-wire has been introduced to improve the flexibility to overcome complex curved canal anatomy. The M-Wire alloy is made of 508 nitinol wires that have been processed through a proprietary method of treatment consisting of drawing the raw wire under specific tension and heat treatments at various temperatures, thereby resulting in a material that includes some portions in both the martensitic and the premartensitic R phases while maintaining a pseudoelastic state [19,22].

The changes in heat treatment are made by changing the transition temperatures of the NiTi instruments, hence improving their cyclic fatigue resistance and flexibility, in comparison to the conventional superelastic NiTi instruments [13]. At human body temperature, traditional NiTi instruments have a lower finish temperature for austenite transformation (16–31 °C). However, thermally modified instruments have demonstrated increased austenite transformation finish temperatures (50–55 °C), which, in consequence, entails more martensitic components involved at body temperature [23].

Toward increasing the cyclic fatigue resistance, some manufacturers applied to NiTi heat-treated instruments a continuous heating and cooling process, thereby resulting in a surface oxidation layer that gives them a gold or blue appearance [24–26]. Gold and blue heated treatments enhance the fatigue resistance when compared with M-wire and conventional superelastic wires, and they are categorized as endodontic instruments with a superior martensite phase [27].

Vortex Blue (Dentsply Tulsa Dental, Tulsa, OK, USA) undergoes pre- and postmanufacture heat treatment, while Protaper Gold (Dentsply Maillefer, Baillagues, Switzerland) undergoes postmanufacture heat treatment [17].

The Tiadent manufacturer (Tiadent, Houston, TX, USA) provides several versions of rotary instruments with claimed superior strength and resistance to cyclic fatigue such as Tia Tornado Blue (TTB) (TTB; Tiadent, Houston, TX, USA) and Tia Tornado Gold (TTG) (TTB; Tiadent, Houston, TX, USA). The manufacturer also claims that the Tia Tornado Gold (TTG) wires are made from M-wire and are compatible with the ProTaper Gold (PTG) wires. It is imperative to understand the mechanical properties of the latest introduced instruments, as these can affect the clinician's choice in clinical management and predict their performance when preparing challenging root canal systems.

The newly introduced instruments provide a cost-efficient alternative to those commonly used in the dental office by dentist. To date, limited information is available on the mechanical properties of the TTB and TTG instruments. Therefore, the aim of this in vitro study was to investigate the cyclic fatigue behavior of the newly introduced rotary instruments TTB and TTG.

The null hypothesis was that there is no significant difference among the tested instruments in term of cyclic fatigue resistance.

2. Materials and Methods

2.1. Preparation of Artificial Canals

The artificial canals were made using the same technique used in the previous study by Alqedairi et al. [28]. The laser micromachining technique using the LASERTEC 40 (Deckel Maho Gildemeister, Hamburg, Germany), which consists of a Q-switched neodymium-doped yttrium aluminum garnet (Nd: Y3Al5O12 (Nd: YAG)) laser operating at a wavelength of 1064 nm with a maximum average power of 30 W, was used to make artificial canals in stainless steel plates with dimensions of 100 mm × 50 mm × 10 mm.

The artificial canal was modeled using CATIA V5® software (Dassault Systèmes, Version 5, Vélizy, France), and laser path programming was performed with a standard triangle language file of the proprietary machine software. After the process parameters were established, the laser was focused on the block with the aid of a galvanometer scanner, and the canal was then machined layer by layer [29].

The artificial canals were prepared in stainless steel blocks with dimensions corresponding to the dimensions of the instrument tested: +0.1 mm in width and +0.2 mm in depth, with an angle of curvature of 60°, a radius of curvature of 5 mm, and a center of curvature 5 mm from the tip of the instrument.

2.2. Cyclic Fatigue Resistance Testing

The experiment was conducted according to a previous paper by Jamleh et al. [30]. Four different rotary instruments, PTG F2, VB 25.06, TTB 25.06, and TTG TF2, were tested in terms of cyclic fatigue resistance. Sample size calculation was performed based on previous studies [28,31] using G-power 3.1.9.4. Fifteen new instruments of each system were used in this study (Figure 1). All the selected instruments had a tip size of 0.25 mm and length of 25 mm. The experiment model was composed of an artificial canal that was milled in stainless steel block. The artificial canal had a maximum curvature point located 5 mm (D5) from the canal end. The metallic canal was covered with transparent glass to prevent the instrument from slipping out and to aid in visualizing the instrument when fracture occurred. The model was submersed in distilled water bath at temperature of 37 °C. X-Smart Plus (Dentsply Maillefer, Ballaigues, Switzerland) endodontic motor was used, and the instrument (19 mm in length) was inserted in the artificial canal. After the instrument was in place, the motor was operated at different speeds according to the manufacturers' instructions: TTB at 350 rpm, TTG at 300 rpm, PTG at 300 rpm, and VB at 500 rpm. The experiment was video recorded and time to fracture was calculated. The number of cycles to fracture (NCF) was calculated by multiplying the time to fracture in minutes by the tested system's rotational speed [30].

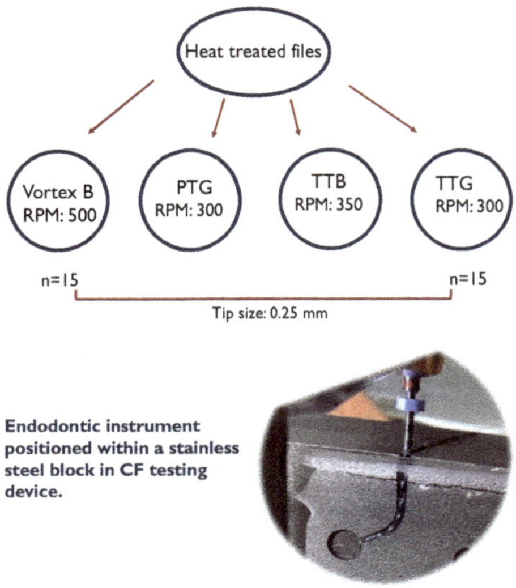

Figure 1. Flow chart of the experiment method for cyclic fatigue of an instrument in the artificial metal canal.

2.3. Scanning Electron Microscopy (SEM)

For the topographic features, one fractured instrument from each instrument system was selected, cleaned using alcohol, and dried at room temperature. Then, the instruments were mounted vertically on 15 mm metal stubs using double-sided carbon tape and subjected to microscopic analysis using SEM (SEM; 6360LV Scanning Electron Microscope;

JEOL, Tokyo, Japan) with 10 kV. The SEM photomicrographs were captured at different magnifications ranging from 150× to 1000×.

2.4. Statistical Analysis

Descriptive statistics were used to summarize the number of cycles to failure for each of the four groups. Normality of data was assessed using histograms, box–whisker plots, and a normality test (Shapiro–Wilk test). Distribution of data appeared to be normal, with normality test p values > 0.05.

Mean and standard deviation values are reported as descriptive statistics, and a one-way ANOVA test with Tukey's HSD was conducted to compare the groups.

3. Results

The mean NCF values, the outer diameter at the D5 and surface area of the tested instruments, are presented in Table 1. Using one-way ANOVA, a statistically significant difference between the groups was found ($p < 0.05$). A large effect size was found ($\eta^2 = 0.93$), thereby suggesting that the group accounted for 93% of the variability in the number of cycles to failure. Tukey's HSD showed that all the groups were significantly different from each other. The VB significantly had the highest NCF followed by the PTG, TTG, and TTB ($p < 0.05$) (Table 1). Moreover, the VB had the highest reliability (M = 1125.00), while the TTB had the lowest (M = 117.93).

Table 1. NCF: the diameter and the surface area of the instruments at the level of fracture of the experimental instrument systems. (n = 15).

Instrument System	Mean (±SD) (NCF)	Outer Diameter at D5 (mm)	Surface Area at D5 (mm^2)
VB	1125.00 ± 174.83 a *	1.9	0.22
TTB	117.93 ± 38.91 b	2	0.19
PTG	831.60 ± 79.88 c	2	0.23
TTG	542.80 ± 84.55 d	1.4	0.21

*: Different letters indicate a statistically significant difference ($p \leq 0.05$). NCF: number of cycles fracture.

There was no significant difference in the length of the fractured pieces among the groups; it ranged from 4.42 to 4.86 mm ($p > 0.05$). The outer diameter and the surface area of the instrument at the level of fracture were measured using a digital microscope (Hirox, Tokyo, Japan) (Table 1).

The SEM images exhibited the typical features of cyclic fatigue behavior (Figure 2). An area of fatigue striations (a) and another with a dimpled surface (b) were noticed. The crack usually initiated at the edge (A1, white arrow) and propagated to the fatigue striations (a). The instrument was weakened by the coalescence of the microvoids (B3, black arrows) produced, after which ductile fracture occurred, which was evident from the dimpled surface (b), until failure occurred. The round dimples indicated normal rupture patterns caused by tensile stresses.

The taper of each instrument was measured using a digital microscope (Digital Microscope, Hirox, Tokyo, Japan) (Figure 3). For the PTG, a steady increase in the degree of progressive taper was noticed from D0 to D5, whereas in the VB, a constant increase of 0.04 mm along the tested segment was recorded. On the other hand, the Tia Tornado Blue showed an irregular taper pattern, as there was no taper from D0 to D2. Then, an abrupt increase in the taper between D2 and D3, as well as D4 and D5, occurred. Lastly, the Tia Tornado Gold showed a similar result as the PTG. However, the increase in the taper was shown to be less steady as measurements were taken further from the tip. Looking at D4 and D5 specifically, the tapers between them in the VB and PTG were the same, which were 0.04. In the TTB and TTG, they were 0.07 and 0.09, respectively.

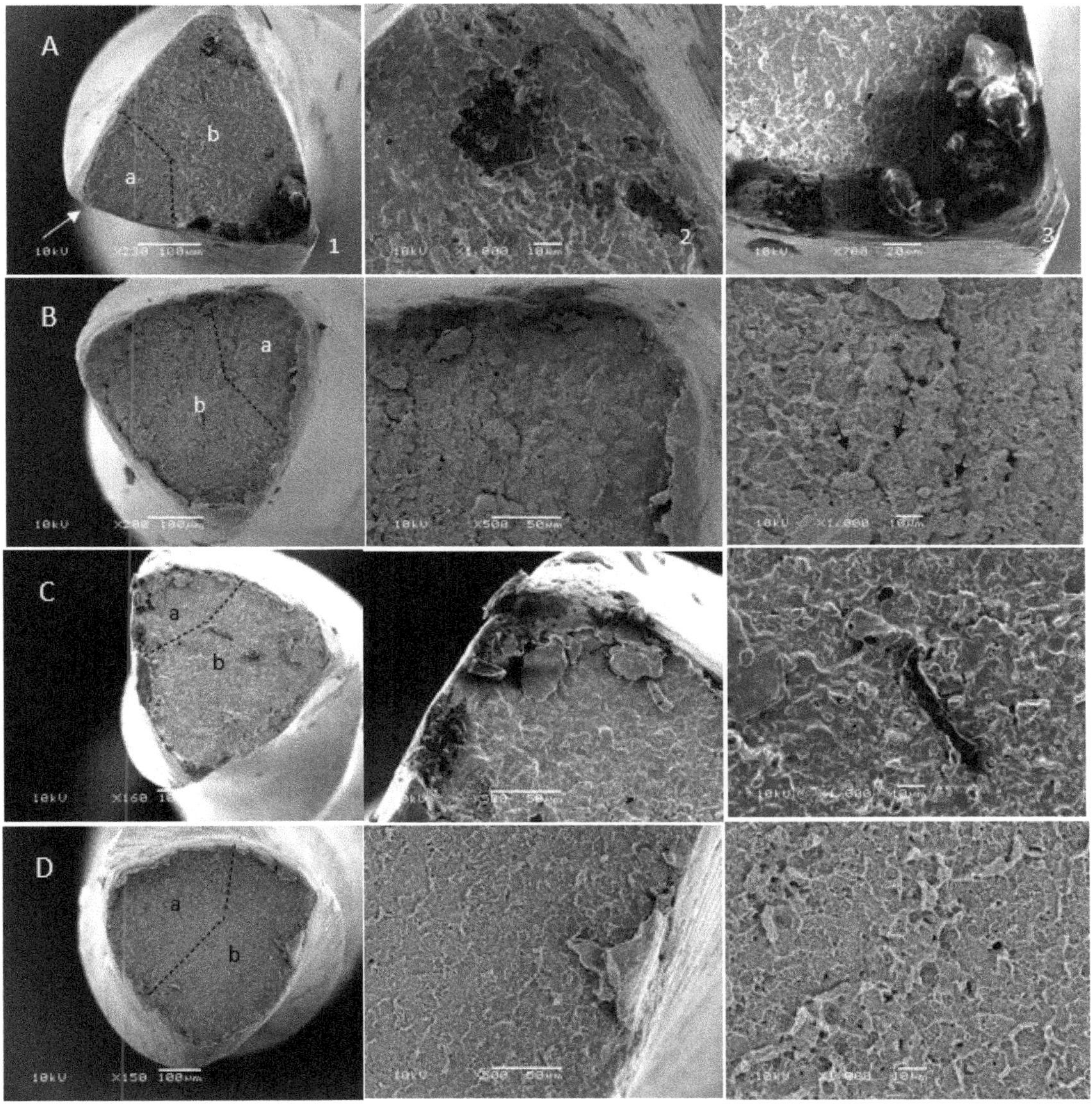

Figure 2. SEM of the cross-sections of the fractured surfaces. (**A**) TTB. (**B**) VB. (**C**) TTG. (**D**) PTG. SEM images exhibited typical features of cyclic fatigue, including (a) striations, (b) dimples, microvoids (black arrows), and initiation of cracks (white arrows).

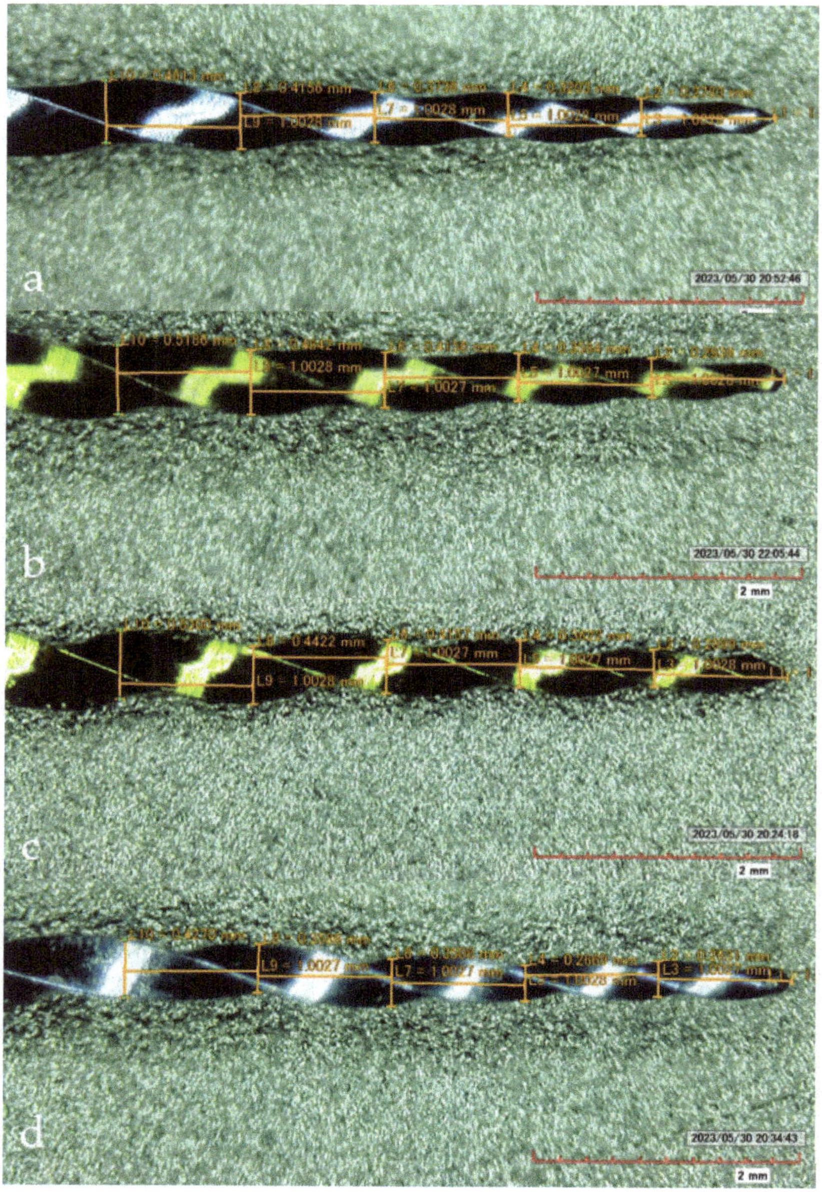

Figure 3. Measurements of each instrument taper from D0–D5 using a digital microscope. (**a**) VB. (**b**) PTG. (**c**) TTG. (**d**) TTB.

4. Discussion

Two new rotary systems, TTB and TTG, with a claim of possession of high fatigue resistance were tested in this study. The cyclic fatigue resistances of these instruments were compared to the VB and PTG, whose values are known in the market and feature enhanced resistance to cyclic fatigue [24–26]. The VB demonstrated a profound superiority of fatigue resistance compared to the other tested instruments. Furthermore, the PTG exhibited better fatigue resistance compared to the TTG and TTB.

A study by Uygun et al. 2020 found that there was no statistically significant difference between the VB and PTG, and this could be due to varied factors during the test like the difference in radius, which was 3 mm, while we used a 5 mm radius in this study [32]. Also, presenting a different result from our study, Sanchez et al. 2020, used a 5 mm radius and reported that the PTG had greater resistance than the CF compared to the VB. However, they used a synthetic oil designed for the lubrication of the artificial root canal systems instead of body temperature distilled water [33]. However, in agreement with our results, a study that used a synthetic oil and the same radius (5 mm) and angle of curvature (60 degrees) resulted in a superior NCF of the VB compared to the PTG [34].

SEM was used to visualize the broken fragments to verify that the topographic features were consistent with the cyclic fatigue failure. Typical cyclic fatigue features were seen with SEM, thereby identifying one or more crack initiation areas, fatigue striation, and a fast fracture zone with dimples [28].

To minimize the variables in this study, instruments with continuous rotation and a similar tip size were chosen. By considering the simulation of the clinical conditions and the complexity of the root canal system, a medium temperature of 37 °C and a 60-degree angle of curvature were used, as they may have crucial effects on cyclic fatigue resistance [5,31]. It was reported that increasing the temperature would have a significant influence on reducing the cyclic fatigue resistance [31]. Moreover, different medium solutions may impact cyclic fatigue resistance. Sodium hypochlorite (NaOCl), which is commonly used in root canal treatment, will lead to lower cyclic resistance, while an endodontic instrument performs with a similar cyclic resistance when placed in an ethylenediaminetetraacetic acid (EDTA) medium compared to distilled water [35]. In this study, body temperature distilled water was used as the medium.

The rotation speed of the instruments in this study was performed according to the manufacturers' instructions, as Pruett et al. reported that the rpm does not affect the cyclic fatigue resistance [5]. Static motion was used to compare the results with other studies, as the majority of studies used this method [3,4]. In static motion, flexural stress is generated mostly within a limited area, which is on the center of the curve and has less variables to change. On the other hand, in the dynamic motion, having the instrument move up and down while the instrument is rotating inside the artificial canal produces more cyclic resistance, as the stress occurs on a wider area of the instrument as the metal transforms while moving [3,4].

Cyclic fatigue is primarily caused by the propagation of faults from the metal surface until it leads to failure [7]. The low quality of the manufacturing can form microcracks and defects on the surface of the instrument [36]. When exposed to compression and tensile forces, the propagation of such defects can lead to metal fatigue, as these microcracks can act as stress concentration points [36]. These failures can be predicted by the stresses that are created by geometrical discontinuities, porosities, inclusions, and overheating during manufacturing [36]. According to Pruett et al., instruments separate at the point of maximum flexure, which is in accordance with our finding where most instruments fractured at 4–5 mm from the tip [5].

Different performance outcomes between the instruments with inferiority, as in the Tia Tornado, could be attributed to the chemical composition of the alloy, design features, and the quality of the manufacturing processes that the manufacturer did not disclose. The instrument cross-section, diameter, and surface area at the level of separation could play a significant role in fatigue behavior [36]. As shown in the results, the increases in taper for the Tia Tornado instruments at the point of failure were larger than those shown in the VB and PTG. It was reported that the taper/mm size affects the fatigue by concentrating the stresses over a short distance with a large taper, thereby making the instrument more prone to fracture [36]. The wider distance between the stress of tension and the stress of compression has been found to inversely affect fatigue failure, since it has a large total stress area [36]. Other factors that affect cyclic fatigue are flute design, circumference shape, flute depth, and the number of spirals [36].

Thermomechanical treatments enhance the instrument's performance in curved canals when exposed to cyclic compressions and tensions [19]. This improvement allows the instrument to show different crystal configurations based on the applied temperature and strain (martensite and austenite) [37]. It provides a martensitic phase at a lower temperature, which gives the instrument the flexibility and the cyclic fatigue resistance required [13,21].

From the results of the current study, the newly introduced instruments, TTG and TTB, need more improvements in terms of the cyclic fatigue before being used as alternatives for the VB and PTG in clinical settings. One of the limitations of this study was not including cutting efficiency and microhardness. However, future suggestions are to limit the variables more, which entails conducting the experiment at a similar speed between instrument groups and under a more simulated clinical condition like using a NaOCl medium instead of distilled water, in addition to clinical studies to determine the clinical outcome of endodontic treatment with the newly introduced instrument to be included in future studies.

5. Conclusions

The newly introduced rotary systems (TTB and TTG) failed to show any improvement in the cyclic fatigue resistance when compared to the VB and PTG. Therefore, the research hypothesis was rejected.

Author Contributions: Conceptualization, Z.A. and A.A.; methodology, A.A. and H.A.; software, M.B. and R.A.; validation, Z.A. and A.A.; formal analysis, R.A. and M.B.; investigation, M.B. and R.A.; resources, Z.A. and R.A.; data curation, Z.A. and M.B.; writing—original draft preparation, M.B. and R.A.; writing—review and editing, Z.A. and A.A.; visualization, H.A.; supervision, Z.A.; project administration, Z.A.; funding acquisition, Z.A. All authors have read and agreed to the published version of the manuscript.

Funding: This research received no external funding.

Institutional Review Board Statement: Not applicable.

Informed Consent Statement: Not applicable.

Data Availability Statement: All data supporting the results of this study are included within the article.

Acknowledgments: The authors thank the King Saud University College of Dentistry Research Center (CDRC), particularly the physical laboratory staff, for their assistance in conducting this study.

Conflicts of Interest: The authors declare no potential conflict of interest with respect to the authorship and/or publication of this article. The authors report no commercial, proprietary, or financial interest in the products or companies described in this article.

References

1. Puleio, F.; Bellezza, U.; Torre, A.; Giordano, F.; Lo Giudice, G. Apical Transportation of Apical Foramen by Different NiTi Alloy Systems: A Systematic Review. *Appl. Sci.* **2023**, *13*, 10555. [CrossRef]
2. Del Fabbro, M.; Afrashtehfar, K.I.; Corbella, S.; El-Kabbaney, A.; Perondi, I.; Taschieri, S. In Vivo and In Vitro Effectiveness of Rotary Nickel-Titanium vs. Manual Stainless Steel Instruments for Root Canal Therapy: Systematic Review and Meta-analysis. *J. Evid. Based Dent. Pract.* **2018**, *18*, 59–69. [CrossRef]
3. Alfouzan, K.; Jamleh, A. Fracture of nickel titanium rotary instrument during root canal treatment and re-treatment: A 5-year retrospective study. *Int. Endod. J.* **2018**, *51*, 157–163. [CrossRef]
4. Wang, N.N.; Ge, J.Y.; Xie, S.J.; Chen, G.; Zhu, M. Analysis of Mtwo rotary instrument separation during endodontic therapy: A retrospective clinical study. *Cell Biochem. Biophys.* **2014**, *70*, 1091–1095. [CrossRef] [PubMed]
5. Pruett, J.P.; Clement, D.J.; Carnes, D.L., Jr. Cyclic fatigue testing of nickel-titanium endodontic instruments. *J. Endod.* **1997**, *23*, 77–85. [CrossRef] [PubMed]
6. Parashos, P.; Linsuwanont, P.; Messer, H.H. A cleaning protocol for rotary nickel-titanium endodontic instruments. *Aust. Dent. J.* **2004**, *49*, 20–27. [CrossRef] [PubMed]
7. Kuhn, G.; Tavernier, B.; Jordan, L. Influence of structure on nickel-titanium endodontic instruments failure. *J. Endod.* **2001**, *27*, 516–520. [CrossRef] [PubMed]

8. Pereira, E.S.J.; Peixoto, I.F.C.; Viana, A.C.D.; Oliveira, I.I.; Gonzalez, B.M.; Buono, V.T.L.; Bahia, M.G.A. Physical and mechanical properties of a thermomechanically treated NiTi wire used in the manufacture of rotary endodontic instruments. *Int. Endod. J.* **2012**, *45*, 469–474. [CrossRef] [PubMed]
9. Alapati, S.B.; Brantley, W.A.; Svec, T.A.; Powers, J.M.; Nusstein, J.M.; Daehn, G.S. SEM observations of nickel-titanium rotary endodontic instruments that fractured during clinical Use. *J. Endod.* **2005**, *31*, 40–43. [CrossRef] [PubMed]
10. Topcuoglu, H.S.; Demirbuga, S.; Duzgun, S.; Topcuoglu, G. Cyclic fatigue resistance of new reciprocating files (Reciproc Blue, WaveOne Gold, and SmartTrack) in two different curved canals. *J. Investig. Clin. Dent.* **2018**, *9*, e12344. [CrossRef] [PubMed]
11. Zubizarreta-Macho, Á.; Alonso-Ezpeleta, Ó.; Albaladejo Martínez, A.; Faus Matoses, V.; Caviedes Brucheli, J.; Agustín-Panadero, R.; Mena Álvarez, J.; Vizmanos Martínez-Berganza, F. Novel Electronic Device to Quantify the Cyclic Fatigue Resistance of Endodontic Reciprocating Files after Using and Sterilization. *Appl. Sci.* **2020**, *10*, 4962. [CrossRef]
12. Hulsmann, M.; Donnermeyer, D.; Schafer, E. A critical appraisal of studies on cyclic fatigue resistance of engine-driven endodontic instruments. *Int. Endod. J.* **2019**, *52*, 1427–1445. [CrossRef] [PubMed]
13. Baird, E.; Huang, X.; Liu, H.; Hieawy, A.; Ruse, N.D.; Wang, Z.; Haapasalo, M.; Shen, Y. A novel model to evaluate the fatigue resistance of NiTi instruments: Rotational and axial movement at body temperature. *Aust. Endod. J.* **2023**, *49*, 301–307. [CrossRef] [PubMed]
14. Plotino, G.; Grande, N.M.; Cordaro, M.; Testarelli, L.; Gambarini, G. A review of cyclic fatigue testing of nickel-titanium rotary instruments. *J. Endod.* **2009**, *35*, 1469–1476. [CrossRef] [PubMed]
15. Cheung, G.S.P.; Peng, B.; Bian, Z.; Shen, Y.; Darvell, B.W. Defects in ProTaper S1 instruments after clinical use: Fractographic examination. *Int. Endod. J.* **2005**, *38*, 802–809. [CrossRef] [PubMed]
16. Shen, Y.; Zhou, H.; Campbell, L.; Wang, Z.; Wang, R.; Du, T.; Haapasalo, M. Fatigue and nanomechanical properties of K3XF nickel-titanium instruments. *Int. Endod. J.* **2014**, *47*, 1160–1167. [CrossRef] [PubMed]
17. Gavini, G.; Santos, M.D.; Caldeira, C.L.; Machado, M.E.D.L.; Freire, L.G.; Iglecias, E.F.; Peters, O.A.; Candeiro, G.T.D.M. Nickel-titanium instruments in endodontics: A concise review of the state of the art. *Braz. Oral Res.* **2018**, *32* (Suppl. S1), e67. [CrossRef] [PubMed]
18. Gambarini, G.; Pongione, G.; Rizzo, F.; Testarelli, L.; Cavalleri, G.; Gerosa, R. Bending properties of nickel-titanium instruments: A comparative study. *Minerva Stomatol.* **2008**, *57*, 393–398.
19. Larsen, C.M.; Watanabe, I.; Glickman, G.N.; He, J. Cyclic fatigue analysis of a new generation of nickel titanium rotary instruments. *J. Endod.* **2009**, *35*, 401–403. [CrossRef]
20. Ataya, M.; Ha, J.H.; Kwak, S.W.; Abu-Tahun, I.H.; El Abed, R.; Kim, H.C. Mechanical Properties of Orifice Preflaring Nickel-titanium Rotary Instrument Heat Treated Using T-Wire Technology. *J. Endod.* **2018**, *44*, 1867–1871. [CrossRef]
21. Shen, Y.; Zhou, H.M.; Zheng, Y.F.; Peng, B.; Haapasalo, M. Current challenges and concepts of the thermomechanical treatment of nickel-titanium instruments. *J. Endod.* **2013**, *39*, 163–172. [CrossRef] [PubMed]
22. Ye, J.; Gao, Y. Metallurgical characterization of M-Wire nickel-titanium shape memory alloy used for endodontic rotary instruments during low-cycle fatigue. *J. Endod.* **2012**, *38*, 105–107. [CrossRef] [PubMed]
23. Yahata, Y.; Yoneyama, T.; Hayashi, Y.; Ebihara, A.; Doi, H.; Hanawa, T.; Suda, H. Effect of heat treatment on transformation temperatures and bending properties of nickel-titanium endodontic instruments. *Int. Endod. J.* **2009**, *42*, 621–626. [CrossRef] [PubMed]
24. Elnaghy, A.M.; Elsaka, S.E. Mechanical properties of ProTaper Gold nickel-titanium rotary instruments. *Int. Endod. J.* **2016**, *49*, 1073–1078. [CrossRef] [PubMed]
25. Hieawy, A.; Haapasalo, M.; Zhou, H.; Wang, Z.J.; Shen, Y. Phase Transformation Behavior and Resistance to Bending and Cyclic Fatigue of ProTaper Gold and ProTaper Universal Instruments. *J. Endod.* **2015**, *41*, 1134–1138. [CrossRef] [PubMed]
26. Plotino, G.; Grande, N.M.; Cotti, E.; Testarelli, L.; Gambarini, G. Blue treatment enhances cyclic fatigue resistance of vortex nickel-titanium rotary files. *J. Endod.* **2014**, *40*, 1451–1453. [CrossRef] [PubMed]
27. Hou, X.M.; Yang, Y.J.; Qian, J. Phase transformation behaviors and mechanical properties of NiTi endodontic files after gold heat treatment and blue heat treatment. *J. Oral. Sci.* **2020**, *63*, 8–13. [CrossRef] [PubMed]
28. Alqedairi, A.; Alfawaz, H.; Bin Rabba, A.; Almutairi, A.; Alnafaiy, S.; Khan Mohammed, M. Failure Analysis and Reliability of Ni–Ti-Based Dental Rotary Files Subjected to Cyclic Fatigue. *Metals* **2018**, *8*, 36. [CrossRef]
29. Mohammed, M.K.; Al-Ahmari, A.; Umer, U. Multiobjective optimization of Nd:YAG direct laser writing of microchannels for microfluidic applications. *Int. J. Adv. Manuf. Technol.* **2015**, *81*, 1363–1377. [CrossRef]
30. Jamleh, A.; Alghaihab, A.; Alfadley, A.; Alfawaz, H.; Alqedairi, A.; Alfouzan, K. Cyclic Fatigue and Torsional Failure of EdgeTaper Platinum Endodontic Files at Simulated Body Temperature. *J. Endod.* **2019**, *45*, 611–614. [CrossRef]
31. Alfawaz, H.; Alqedairi, A.; Alsharekh, H.; Almuzaini, E.; Alzahrani, S.; Jamleh, A. Effects of Sodium Hypochlorite Concentration and Temperature on the Cyclic Fatigue Resistance of Heat-treated Nickel-titanium Rotary Instruments. *J. Endod.* **2018**, *44*, 1563–1566. [CrossRef]
32. Uygun, A.D.; Unal, M.; Falakaloglu, S.; Guven, Y. Comparison of the cyclic fatigue resistance of hyflex EDM, vortex blue, protaper gold, and onecurve nickel-Titanium instruments. *Niger. J. Clin. Pract.* **2020**, *23*, 41–45. [CrossRef]
33. Ruiz-Sánchez, C.; Faus-Llácer, V.; Faus-Matoses, I.; Zubizarreta-Macho, Á.; Sauro, S.; Faus-Matoses, V. The Influence of NiTi Alloy on the Cyclic Fatigue Resistance of Endodontic Files. *J. Clin. Med.* **2020**, *9*, 3755. [CrossRef]

34. Furlan, R.D.; Alcalde, M.P.; Duarte, M.A.; Bramante, C.M.; Piasecki, L.; Vivan, R.R. Cyclic and Torsional Fatigue Resistance of Seven Rotary Systems. *Iran Endod. J.* **2021**, *16*, 78–84.
35. Alfawaz, H.; Alqedairi, A.; Alhamdan, M.; Alkhzim, N.; Alfarraj, S.; Jamleh, A. Effect of NaOCl and EDTA irrigating solutions on the cyclic fatigue resistance of EdgeTaper Platinum instruments. *BMC Oral Health* **2022**, *22*, 195. [CrossRef]
36. McSpadden, J.T. *Mastering Endodontic Instrumentation*; Cloudland Institute: Chattanooga, TN, USA, 2007.
37. Pereira, É.S.J.; Viana, A.C.D.; Buono, V.T.L.; Peters, O.A.; de Azevedo Bahia, M.G. Behavior of nickel-titanium instruments manufactured with different thermal treatments. *J. Endod.* **2015**, *41*, 67–71. [CrossRef] [PubMed]

coatings

Article

Amantadine-Heparin-Polypyrrole as a Promising Drug Delivery Reservoir with a Biological Approach

Sara Kulik [1,*], Sylwia Golba [2,*], Izabela Matuła [2], Ewa Stodolak-Zych [3] and Roksana Kurpanik [3]

[1] Doctoral School, Department of Science and Technology, Institute of Materials Engineering, University of Silesia, Bankowa 14, 40-007 Katowice, Poland
[2] Department of Science and Technology, Institute of Materials Engineering, University of Silesia, Bankowa 14, 40-007 Katowice, Poland; izabela.matula@us.edu.pl
[3] Department of Biomaterials and Composites, Faculty of Materials Science and Ceramics, AGH University of Science and Technology, 30-059 Cracow, Poland; stodolak@agh.edu.pl (E.S.-Z.); kurpanik@agh.edu.pl (R.K.)
* Correspondence: sara.krawczyk@us.edu.pl (S.K.); sylwia.golba@us.edu.pl (S.G.); Tel.: +48-666-847-133 (S.K.)

Abstract: There is an urgent need to alleviate the symptoms of neurodegenerative diseases. The presented work includes the use of electrochemical polymerization (CV) to obtain active polypyrrole layers with incorporated molecules of a neurological drug substance—amantadine hydrochloride. The obtained films were characterized chemically, structurally, and functionally in terms of their use as a drug delivery systems which are neurologically active. FTIR spectra were recorded to identify the incorporation of drug substances into the matrix. The obtained results showed that amantadine and heparin were embedded to the polypyrrole matrix. Scanning electron microscopy (SEM) was used to examine the morphology of the films. The films deposited on the steel substrate showed a compact, smooth structure, where there was no visible organized structure. After release, the film became corrugated. Adhesive tests were conducted with the cross-cut Test Method B to determine the mechanical properties, and the results showed that amantadine improves adhesion for steel substrates. The films were potentially stimulated by chronoamperometry, and UV-Vis spectra were registered to calculate the concentration of AMA in the solution after release. The release curves indicate a 95% efficiency of AMA release over the studied time period and protocol. Later, antibacterial properties were tested. The proposed system was able to provide a daily dose of drugs that ensures a therapeutic effect. This is a significant step towards developing systems capable of delivering a wider range of doses, potentially in line with the full spectrum recommended for therapeutic efficacy. The antibacterial properties of the material allows it to be considered as a material with antibacterial potential in the presence of the *Staphylococcus aureus* (*S. aureus*) strain. The percentage reduction ratio indicates a 90–100% reduction of bacteria in the suspension.

Keywords: conducting polymers; amantadine hydrochloride; polypyrrole; antibacterial test; neurologically active system

1. Introduction

Current demographic trends indicate that society is aging. Statistics from the World Health Organization (WHO) and the United Nations (UN) show that the population of people of elderly age affected by neurodegeneration diseases will increase from less than 1 million in 2020 to 2 million in 2050 [1,2]. Forecasts also indicate that the number of elderly people will be 1.5 billion in 2050, which is twice as many as in 2020 [3]. These changes also result in an increase in the incidence of dementia [4]. Dementia is a neurodegenerative disease that usually begins with a deterioration of cognitive functions—i.e., a deterioration of memory and thinking skills [5].

Neurodegenerative disorders are characterized by the progressive loss of selectively sensitive populations of neurons. The most common neurodegenerative disorders are amyloidoses (insoluble fibrous proteins that have specific structural features that allow them to bind to specific dyes such as Congo red [6,7]) and tauopathies (tau protein associated with microtubules). The protein abnormalities in these disorders have abnormal conformational structures. Although neurodegenerative diseases are usually defined by specific protein accumulations and anatomical sensitivity, there are many underlying processes involved in progressive neuronal dysfunction and death [8], such as oxidative stress, programmed cell death, and neuroinflammation.

Alzheimer's disease is treated with two classes of drugs: cholinesterase inhibitors and glutamate regulators. Cholinesterase inhibitors work by inhibiting the breakdown of the neurotransmitter acetylcholine. Their side effects include vomiting, loss of appetite, nausea, and increased bowel movements. The only approved drug that regulates glutamate levels is the active substance memantine that is used to treat Alzheimer's disease (AD), which is associated with memory and learning. The drug's effectiveness is comparable to that of cholinesterase inhibitors, in that it works in about half of the people who take it and only for a short time [9]. Amantadine (AMA), an aminoadamantane long known for its moderate anti-Parkinsonian effects, has recently been shown to antagonize central N-methyl-D-aspartate (NMDA) receptors at therapeutically relevant concentrations.

A drug delivery system (DDS) is defined as a preparation or device that allows the introduction of a drug substance into the body and improves its efficacy and safety by controlling the rate, time, and location of drug release in the body. This process involves the administration of a drug product, the release of active ingredients by the product, and the subsequent transport of the active ingredients across biological membranes to the site of action. The term therapeutic substance also refers to an agent such as gene therapy that induces the production of an active therapeutic agent in vivo. A drug delivery system is the interface between a patient and a drug. It may be a form of drug to be administered for therapeutic purposes or a device used to administer the substance [10]. One of the most known materials used in DDS are conducting polymers.

Conducting polymers (CP) undergo a reversible redox reaction that causes ions to be transported into and out of the polymer. Typically, depending on the environmental conditions, a potential difference of less than 1 V must be applied to the electrode to release or capture ions. Conducting polymers can operate over a wide temperature range in a liquid electrolyte or in air using a polymer electrolyte [11]. It is assumed in the literature that polypyrrole (PPy), polythiophene (PTh), polyaniline (PANI), and their derivatives exhibit some biocompatibility with living tissues and body fluids. Long-term (90 days) in vitro and in vivo tests have shown little evidence of toxicity or immunological problems [12,13]. In the dedoped state, counterions are removed from the film and cause it to shrink. These properties can be exploited in drug delivery systems where drugs are incorporated into the polymer coating during oxidation and released during film reduction [14]. It is also possible to incorporate a cationic drug after the polymerization process, but the polymer must be doped with anions such as polystyrene sulfonate (PSS). During reduction of the polymer, the cationic drug is introduced to compensate for the negatively charged anions that cause the polymer to swell. During oxidation of the film, the cationic drug contained in it is released, and the film shrinks [15]. Cycles of film activation between redox states can cause cracks and discontinuities in the film, which can lead to an increased release rate of molecules [16]. Anionic drugs can also be introduced after the polymerization process by doping the polymer with a small anion, which can be the active form of the drug. Under these conditions, the coating oxidizes by incorporating anions and is reduced by releasing anions. In the case of introducing a relatively less mobile ion as a doping agent, both cationic and anionic drugs are simultaneously incorporated and removed [17].

PPy films loaded with anionic drugs (dexamethasone phosphate (DMP); meropenem (MER)) were deposited on the surface of indium tin oxide (ITO)-coated glass electrodes by electropolymerization by Shah et al. [18]. As a result of electrochemical oxidation (at a potential of 1 V) and polymerization of pyrrole on the anode, a layer was obtained. In the polypyrrole layer, the cationic charges were compensated with anionic dopants from the electrolyte (in this case, one of the anionic drugs—DMP or MER). The successful deposition of polypyrrole films on the surface of ITO electrodes was noticeable (i.e., the presence of a black film on the surface of the transparent and colorless ITO electrode). It was observed that the release of relevant drugs (DMP or MER) loaded into PPy films was enhanced upon the application of electrochemical stimulus, providing proof of concept that such materials can form a useful protective coating on the surface of implantable medical devices, potentially reducing adverse reactions to their implantation in vivo [19]. Among the conductive polymers, the self-healing hydrogels were revealed [20]. Electrical conductivity was achieved by the polypyrrole, free ions, and the synergistic effect between the PPy particles and the free ions. The resultant hydrogels have potential applications in electronic skin and biomedical implants. These hydrogels improve mechanical properties such as toughness and stretchability [21].

The presented work includes the use of electrochemical polymerization (using two synthetic protocols) to obtain active polypyrrole layers with incorporated molecules of a neurological drug substance—amantadine hydrochloride. To our knowledge and review of source materials in the literature, there is no yet known amantadine-loaded polypyrrole coating. The obtained layers will be characterized in terms of their use as drug delivery systems neurologically active which is used in medicine. The developed procedure for deposition of the amantadine-enriched coating allows for the control of the mass and morphology of the deposited polymer, which directly affects the amount of encapsulated drug. The proposed system can be prepared on various conductive substrates (e.g., medical steel, platinum), which expands the solution's applicability. Optimal electrochemical synthesis conditions were developed to obtain coatings with high drug accumulation capacity and very good adhesion to the substrate (class 4B according to ASTM D3359-23, Standard Test Methods for Rating Adhesion by Tape Test, World Trade Organization Technical Barriers to Trade (TBT) Committee, New York, 2023). The process of amantadine release is confirmed and potential controlled, allowing for the control of the release rate.

2. Materials and Methods

2.1. Synthesis

The synthesis was performed in an aqueous medium containing the following ingredients: pyrrole (98%, Sigma Aldrich, Schnelldorf, Germany), amantadine hydrochloride AMA (>98% TLC, Sigma Aldrich, Schnelldorf, Germany), and necessary commercially available heparin sodium salt from porcine intestinal mucosa (HEP) (>180 USP units/mg, Sigma Aldrich, Schnelldorf, Germany) and sodium dodecyl sulphate (>99% Sigma Aldrich, Schnelldorf, Germany). Prior to synthesis, pyrrole was pre-distilled for purification, resulting in a transparent pale yellow liquid, which was then stored in a freezer. The electrolyte solution was prepared by dissolving SDS in distilled water to achieve a concentration of 0.1 M, surpassing the critical micelle concentration of 0.008 M [22]. The monomer solution was prepared at a concentration of 0.2 M with the molar ratio of the components AMA:HEP:Py set at 1:0.005:2. To avoid inhomogeneities, the synthetic solution was freshly prepared just before the experiment. The stainless steel substrate (316L (Stalglass)) was used due to its relevance for biomedical applications (composition: Cr 16%, Ni 10%, Mo 2%, C 0.08%, P 0.045%, S 0.03%, Si 0.75%, N 0.1%). Stainless steel (square, 8 mm × 8 mm) was used as a working electrode (WE) prepared by polishing with sandpaper (ranging from grit 300 to 600 Klingspor, Bielsko Biała, Poland) and washed with distilled water to remove any powder impurities. Subsequently, the substrate was immediately immersed in a 60% ethanol solution for 24 h to prevent the formation of an oxide layer. All measurements were conducted at room temperature. Indium tin oxide-coated PET (60 Ω/sq, 5 mil) (Sigma Aldrich, Schnelldorf, Germany) was immersed in distilled water and acetone before usage.

2.2. Polymerization of AMA/HEP/Py via CV and CR Methods: Procedures Used to Stimulate the Release of Entrapped AMA and HEP

The measurements were carried out in an electrochemical cell containing a reference Ag/AgCl electrode, a counter electrode (platinum spiral), and a 316 L steel plate or ITO as a working electrode. Cyclic voltammetry (CV) was employed for the synthesis and characterization of the deposited polymeric films. The CV analysis was performed using an Autolab PGSTAT12 potentiostat, and the results were analyzed with Autolab software (version 4.9). During the synthesis, two maximum potentials were applied, E = 0.70 V and E = 0.85 V, with different numbers of cycles (20 cycles and 30 cycles) to detect their influence on the deposition process as well as on drug immobilization efficiency. To stimulate the release of the drug substance, chronoamperometry was used with a potential set at either 0.5 or 0.7 V, held for 150 s, and in long term release for 48 h. The scheme of the releasing procedure is shown in Figure 1.

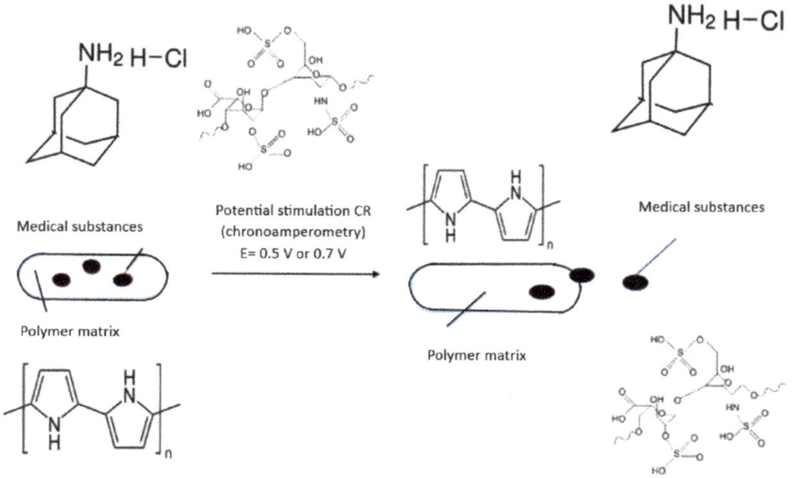

Figure 1. Scheme of potential release of AMA and HEP from PPy matrix.

2.3. *Characterization of AMA/HEP/PPy Films*

Scanning electron microscopy (SEM) was carried out using a Hitachi S4100 microscope equipped with an energy-dispersive X-ray spectrometer (EDS) with an acceleration voltage of 15 keV. For these analyses, the samples were firmly fixed onto a steel sample holder using adhesive carbon tape (SEM Conductive Double sided Carbon Tape, Extra Pure, width: 5 mm, micro-shop). Additionally, the surface of the PPy film was coated with an additional layer of carbon powder using an EMITECH K950X carbon coater to guarantee a conducting surface for the analysis.

The UV-Vis measurements were carried out using a quartz cuvette (Bionovo, Chorzów, Poland) in a Biowave II UV-vis spectrometer (WPA, Biochron, Surrey, UK). For the release stage, the coated samples were fully immersed in a 0.9% NaCl solution and placed along the side wall of the cuvette to allow the diffusion of drug molecules into the bulk of the solution. Aliquots of the solution were taken with a pipette (2.5 mL) and then poured into a measuring cuvette and placed in the UV-Visspectrometer. The solution was then led back to the measuring cell, allowing simultaneous mixing of the solution. The absorption band of AMA was located at 237 nm [23], while the HEP absorption band was observed at 213 nm [24]. The spectra were recorded in real time with readings taken every 15 s, while a constant potential was applied using the AUTOLAB PGSTAT12 potentiostat. The chosen potentials values were based on the respective potentials observed in the CV doping/dedoping cycles. The system worked in air-conditioned room temperature.

2.4. *Antimicrobial Activity of AMA/HEP/PPy Films*

The biological properties of AMA/HEP/PPy film were tested on the basis of the PN-EN ISO 22196:2007 standard (Measurement of antibacterial activity on plastics and other non-porous surfaces, Technical Committee ISO/TC 61, Plastics, Subcommittee SC 6, Ageing, chemical and environmental resistance, UK, 2011).

Preparation of strains Escherichia coli ATCC 8739 and Staphylococcus aureus ATCC 6538P: bacterial cultures were refreshed twice before analysis by sieving onto slants with nutrient agar and incubated for 24 h at 35 °C ± 1 °C.

Preparing materials for tests: The test and control materials were sterilized with 70% ethyl alcohol in an ultrasonic bath.

Inoculation: Microbial cells were suspended in a sterile nutrient broth with an amount of ~10^5 cells/mL. The prepared test and control materials were placed in sterile Petri dishes. A dose of inoculum was applied to the surface of each plant and placed in a culture well.

Incubation: Sterile bags containing inoculated test materials and control materials were incubated for 24 h at 35 °C ± 0.1 °C.

According to the standard PN-EN ISO 22196:2007, the crucial R parameter can be calculated according to the Equation (1). The results give information on whether a material shows microbiological activity.

$$R = (U_t - U_0) - (A_t - U_0) = U_t - A_t \qquad (1)$$

where:

R—value of antimicrobial activity,
U_0—average of the decimal logarithm of the number of live bacterial cells on control samples immediately after inoculation,
U_t—average of the decimal logarithm of the number of live bacterial cells on control samples after 24 h of incubation,
A_t—the average of the decimal logarithm of the number of live bacterial cells on samples tested after 24 h of incubation.

In turn, the percentage value of antimicrobial activity is expressed as Equation (2):

$$R[\%] = \frac{A - B}{A} \cdot 100 \qquad (2)$$

where:

R—percentage of antimicrobial activity,
A—initial number of bacteria in the suspension contacted with the tested materials, CFU,
B—number of bacteria after 24 h of contact between bacteria and the tested material, CFU.

3. Results

3.1. Synthesis and Electrochemical Characterization of AMA/HEP/PPy

PPy was synthesized in the presence of amantadine (AMA) and heparin (HEP) using cyclic voltammetry in a sodium dodecyl sulfate solution. The aim of this study is to prepare a system with a portion of AMA enclosed in the polymer structure. The incorporation of saccharide-type HEP is dictated by its anionic character, which allowed it to serve as a polymeric counterion, balancing the positive charge generated during the growth of PPy chains. At the same time, large HEP molecules can be incorporated into the PPy matrix, interacting with AMA molecules (AMA is a cationic drug). Electropolymerization at a molar ratio of AMA/HEP/PPy 0.1:0.002:0.1 M was induced by cyclic voltammetry in a specify potential range. Two values of maximum potentials were defined: E_1 = 0.70 V and E_2 = 0.85 V. The process parameters are shown in the table below (Table 1). The samples were numbered, taking into account the process parameters and the used substrate.

Table 1. Electrochemical synthesis parameters of AMA/HEP/Py.

Potential	N° Scan	Film	Substrate
E_{1max} = 0.70 V	20	f_1	ITO
	30	f_2	
E_{2max} = 0.85 V	20	f_3	
	30	f_4	
E_{1max} = 0.70 V	20	f_5	Steel
	30	f_6	
E_{2max} = 0.85 V	20	f_7	
	30	f_8	

An example of the cyclic voltammogram of AMA/HEP/PPy is shown in the figure (Figure 2). In order to characterize the deposition process of the AMA/HEP/PPy system, the CV cyclic voltammetry technique was used in the potential range from −0.45 to 0.85 V and from −0.35 to 0.70 V with respect to the Ag/AgCl electrode.

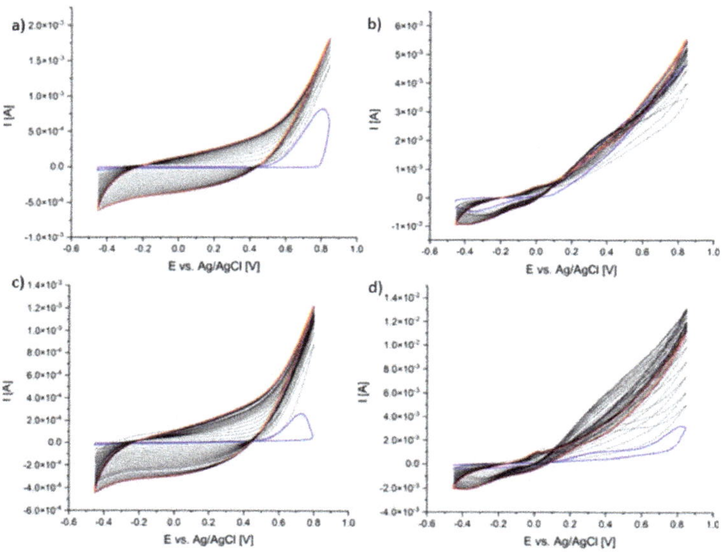

Figure 2. Cyclic voltammograms during synthesis of AMA/HEP/PPy films: (**a**) f_3; (**b**) f_7; (**c**) f_4; (**d**) f_8 (abbreviation explained in Table 1).

The CV curves during the synthesis (Figure 1) of the AMA/HEP/PPy system recorded at a potential of 0.85 V were significantly different in shape from the curves obtained at 0.7 V (Figure S1). The curves of the films on the ITO substrate showed a higher current amplitude. The CV curve corresponding to the polymerization process on the steel substrate showed higher current values by two orders of magnitude for f_7 and by one order of magnitude for f_8 compared to f_5 and f_6. Oxidation peaks were visible at a potential of 0.05 V and reduction at −0.35 V for f_7 and f_8. The electropolymerization process was more balanced for the films on the ITO substrate. The process proceeded according to the scheme of pyrrole polymerization, where the monomer was oxidized, cation-radical coupling occurred, and less soluble chains were formed and grew, which adsorbed on the electrode surface. The electrochemical synthesis process on a steel substrate was less effective, but it produced a more electrochemically active film (doping and dedoping peaks were visible). The higher applied potentials led to intensive oxidation of the monomer, which did not undergo further effective polymerization. Hysteresis appeared on the CV with each subsequent cycle (the cathodic current intensity was higher than the anodic current), which may indicate changes in the nature of the substrate during the process.

The ITO and steel electrodes were fully coated with AMA/HEP/PPy layers prepared according to the given procedures. To characterize the polypyrrole films, cyclic voltammograms of the obtained films were recorded in the SDS environment.

The CV curves of f_5 and f_6 films (Figure S2) do not overlap, and after each cycle, the curves change, which indicates low electrochemical stability. On the surface of the working electrode, intensive ion exchange occurs in the electrolyte solution [25]. The electrolyte solution became colored and turned turbid, which was caused by dissolution of oligomers and migration to the film, which can be observed as a decrease in current intensity. The CV curves (Figure 3) showed higher values of current intensity for all films, compared to the curves in the Figure S2. The widest amplitudes of the curves were visible for the films on the steel substrate, which indicates their charge storage ability. The curves for f_7 and f_8 overlap, indicating better electrochemical stability compared to f_5 and f_6. The table (Table 2) summarizes the values of electrochemical stability and film thickness calculated according to the Equations (3) and (4).

$$S = \frac{Q_{10}^{+}}{Q_{2}^{+}} \cdot 100\% \quad (3)$$

where:

Q_{10}^{+}—the charge of the last doping half cycle [C],
Q_{2}^{+}—the charge of the second doping half cycle [C].

$$g = \frac{q_{pol} \cdot M_{mon}}{n \cdot F \cdot A_w \cdot \rho} \quad (4)$$

where:

g—the thickness of the dry polymer film formed in the electropolymerization process, [cm],
q_{pol}—the total charge of the polymerization process, expressed as the charge of the last polymerization cycle,
M_{mon}—molar mass of monomer [g·mol^{-1}],
n—the number of electrons involved in the oxidation of one monomer unit, (2.3 for pyrrole [23],
F—Faraday's constant, (96.48 C·mol^{-1});
A_w—the surface of the working electrode covered with a polymer film, (cm^2),
ρ—approx. film density (1.48 g·cm^{-3}) [26].

The increase in the potential window during synthesis and the number of cycles influenced the increase in the film thickness, especially for f_4 by about 50%. Increasing the potential window during synthesis influenced as well the electrochemical stability of the obtained films on both substrates. Increasing the number of cycles for coatings on ITO substrate at the applied lower potential of 0.7 V resulted in a decrease in the value of film stability. In the case of coatings obtained at 0.85 V, the stability of the coatings increased. Compared to coatings obtained on ITO, on steel, an increase in the number of potential cycles resulted in a decrease in the stability of coatings for E = 0.7 V and an increase in the stability of coatings at E = 0.85 V. In the polymer formation process, both anions and electrons moved through the film. In the subsequent reduction, the electroneutrality was restored by pushing out anions or by incorporating cations from the electrolyte solution. After applying a positive potential, the neutral film was oxidized and anions were absorbed or cations were rejected. The redox activity of the polymer was regulated by the electron transfer reaction and mass transport process. The decrease in stability values upon potential reduction was related to the difficulty of mass transport or electron transfer.

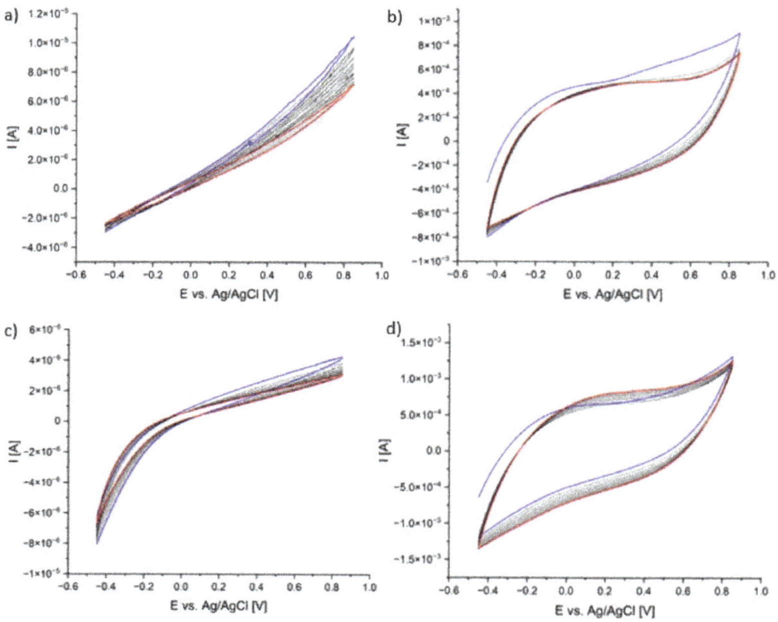

Figure 3. Cyclic voltammograms of AMA/HEP/PPy films: (**a**) f_3; (**b**) f_7; (**c**) f_4; (**d**) f_8 (abbreviation explained in Table 1).

Table 2. Electrochemical stability and thickness of AMA/HEP/PPy film deposited on ITO and steel substrates.

E = 0.70 V, c = 20			E = 0.70 V, c = 30		
Substrate	**S [%]**	**g [µm]**	**Substrate**	**S [%]**	**g [µm]**
ITO f_1	94.94	0.017	ITO f_2	75.84	0.019
Steel f_5	73.29	0.011	Steel f_6	61.41	0.12
E = 0.85 V, c = 20			**E = 0.85 V, c = 30**		
Substrate	**S [%]**	**g [µm]**	**Substrate**	**S [%]**	**g [µm]**
ITO f_3	71.70	0.036	ITO f_4	83.63	0.078
Steel f_7	93.80	0.13	Steel f_8	112.56	0.14

3.2. Morphology of AMA/HEP/PPy Films by SEM

Scanning electron microscopy was used to analyze the morphology of the deposited AMA/HEP/PPy films in pristine post-synthetic form and after the release event (results reported in the following section).

The films deposited on the steel substrate (Figure 4) showed a compact, smooth structure, where there was no visible organized structure. After release, the film became corrugated. Due to the potential release, the HEP molecules with high mass and AMA surface-bonded to PPy came out of the coating, hence the release voids were visible in the SEM image (marked with circles in Figure 3). Due to electrical stimulation, the drug substances bound by electrostatic interactions to oxidized PPy were released. In the case of the f_7 film, after release, clusters were visible revealing a cauliflower-like structure (Figure S3), typical for PPy. The microstructures of the films on the steel substrate were different. During the synthesis, the micelle adsorption was less efficient, hence only trace amounts of the scaffold are visible, changing the structure to a more typical one (cauliflower-like).

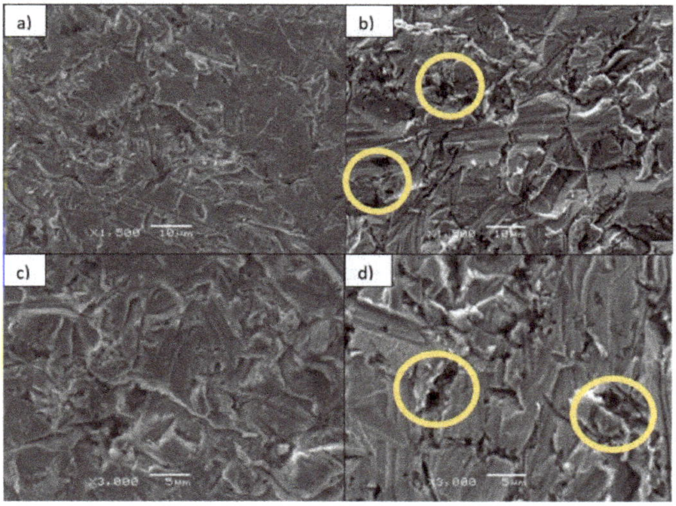

Figure 4. SEM images of the film f_5 AMA/HEP/PPy on steel substrate: (**a**,**c**) before AMA and HEP release; (**b**,**d**) after 0.7 V potential release of AMA and HEP (voids after release of AMA and HEPmarked with yellow circles).

3.3. Adhesion Tests of AMA/HEP/PPy Films

Adhesion tests were performed according to the ASTM D3359-23 standard. The investigation included Test Method B—cross-cut tape test. The test results of the obtained films were classified according to the requirements of the standard. The results are shown in the table (Table 3) and in the figure (Figure S4).

Table 3. Classification of AMA/HEP/PPy film adhesion test results according to ASTM D3359-23 standard (Standard Test Methods for Rating Adhesion by Tape Test, World Trade Organization Technical Barriers to Trade (TBT) Committee, New York, 2023).

Process Parameters	N° Scan	Film	Substate	Classification
E_{1max} = 0.70V	20	f_1	ITO	0B
E_{1max} = 0.70V	30	f_2	ITO	2B
E_{2max} = 0.85V	20	f_3	ITO	1B
E_{2max} = 0.85V	30	f_4	ITO	0B
E_{1max} = 0.70V	20	f_5	Steel	3B
E_{1max} = 0.70V	30	f_6	Steel	2B
E_{2max} = 0.85V	20	f_7	Steel	4B
E_{2max} = 0.85V	30	f_8	Steel	1B

Based on the results, it is evidenced that AMA-added coatings improved the adhesion of the coating to the ITO substrate. The f_3 film flaked along the edges of the cuts, the damaged area was within 60%, and the f_2 coating was damaged within 35% of the grid. The results showed significantly better adhesion for the films on the steel substrate. The f_7 film showed the best adhesion, with only small flakes peeling off at the line intersections (less than 5% of the area). The f_5 coating peeled off within 15% of the cut grid area, and the f_6 coating peeled off within 3%.

3.4. Short-Term Release of AMA and HEP from AMA/HEP/PPy Films

UV-Vis spectra (Figure 5) of solutions were recorded during the release of medicinal substances from AMA/HEP/PPy films. The initial spectrum was marked in blue, and then the next ones were recorded every 15 s. The spectrum after 150 s was marked in red.

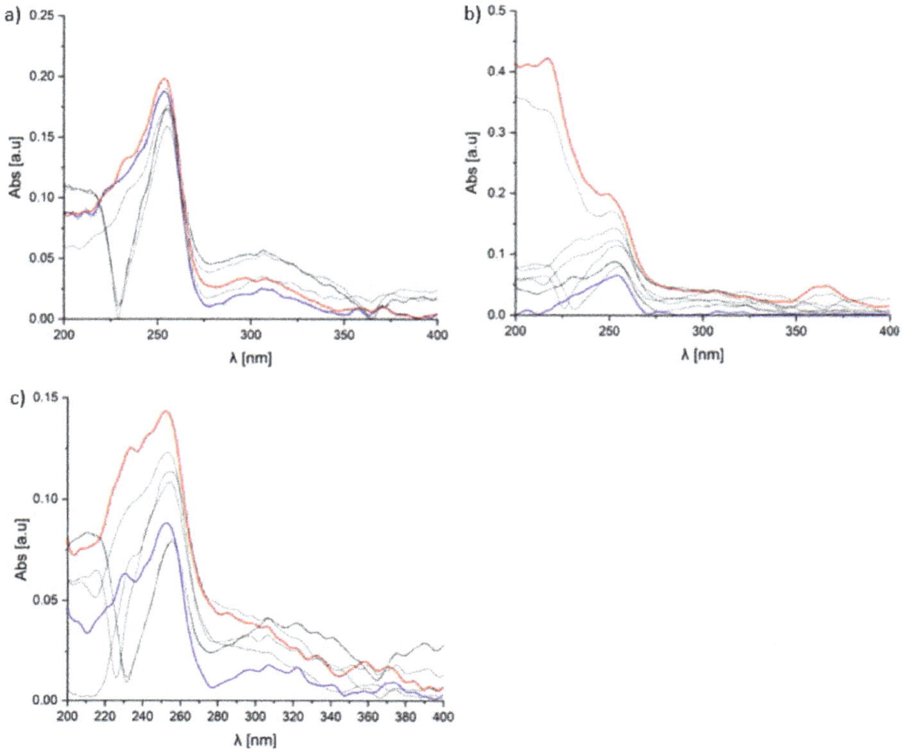

Figure 5. UV-Vis spectra during the release of AMA and HEP from the f_7 AMA/HEP/PPY film: (**a**) at a constant potential of 0.5 V; (**b**) at a constant potential of 0.7 V; (**c**) diffuse.

Each spectrum showed characteristic bands corresponding to AMA λ = 269 nm [26,27] and HEP λ = 213 nm [28], and the absorbance values were converted to concentrations using absorption coefficients $\varepsilon = 3.9 \cdot 10^4$ $cm^{-1} \cdot M^{-1}$ [29] for AMA and $\varepsilon = 5500$ $cm^{-1} \cdot M^{-1}$ for HEP and are listed in the table below (Table 4). The absorbance value in the range of 200–220 nm is related to π-π* electron transitions in C=C bonds or n-π* from nitrogen atoms in amine groups. The highest concentrations of AMA were visible for solutions taken during desorption from the cell, in which the film was placed on a steel substrate. The HEP concentration values were six orders of magnitude lower compared to the AMA concentrations due to its lower molar fraction in the monomer solution and its high molecular weight.

The highest value of AMA concentration in solution was visible after the procedure of 0.7 V release from f_7 film and the lowest for the procedure of diffusion release. It was observed that drug molecules were detached from the matrix more efficiently under the influence of higher potential. Drug concentration in solution after diffusion release showed one order of magnitude lower values compared to potential release. At the structural level, it could be concluded that during the oxidation reaction, the volume of the PPy matrix increased, which led to the release of the cationic drug (AMA) in a process called "cation-driven activation". The expansion and increase in the volume of the polymer was controlled by changing the applied voltage. A higher value of the applied voltage caused intensification of the oxidation process, allowing for control over the release of drug molecules. This proved that the prepared AMA/HEP/PPy coatings were useful for the construction of drug delivery systems.

Table 4. List of concentrations of AMA and HEP after release from PPy matrix obtained from UV-Vis spectra.

Film	C for 0.5 V release [M]		C for 0.7 V release [M]		C for diffusion release [M]	
	HEP	AMA	HEP	AMA	HEP	AMA
f_1	9.09×10^{-6}	8.46×10^{-1}	1.82×10^{-6}	1.77×10^{-1}	3.64×10^{-6}	6.15×10^{-1}
f_2	3.27×10^{-5}	1.62×10^{1}	5.82×10^{-5}	2.23×10^{1}	7.27×10^{-6}	3.85×10^{-1}
f_3	9.09×10^{-5}	1.92×10^{1}	7.09×10^{-5}	3.15×10^{1}	1.09×10^{-5}	6.92×10^{-1}
f_4	9.09×10^{-6}	8.46×10^{-1}	1.82×10^{-5}	1.77×10^{1}	3.64×10^{-6}	6.15×10^{-1}
f_5	7.27×10^{-5}	2.92×10^{1}	6.55×10^{-6}	7.69×10^{-1}	1.09×10^{-5}	1.54×10^{-1}
f_6	2.73×10^{-5}	1.00×10^{1}	8.36×10^{-5}	1.85×10^{1}	1.82×10^{-5}	4.62×10^{-1}
f_7	2.36×10^{-5}	2.00×10^{1}	4.91×10^{-5}	3.85×10^{1}	1.64×10^{-5}	8.46×10^{-1}
f_8	5.45×10^{-5}	1.38×10^{1}	5.82×10^{-5}	2.77×10^{1}	9.09×10^{-6}	6.92×10^{-1}

Based on the obtained data from UV-Vis spectra, the release efficiency of the active substance (RE) was calculated. RE was calculated as the ratio of c_3 to c_1, where c_3 was the concentration of the substance in the solution after release, and c_1 is the concentration of the active substance introduced into the matrix during synthesis. The value of c_1 was calculated from the difference between the initial concentration of the active substance in the solution (c_0) and its concentration after synthesis (c_2). The values were converted to percentages. The table below (Table 5) also included the AMA encapsulation efficiency in polypyrrole films with the uncertainty calculated using the total differential, calculated on the basis of UV-Vis spectra data and Equation (5).

$$EE = \left[\frac{drug\ concentration\ in\ the\ polymer\ film}{drug\ concentration\ in\ the\ solution\ during\ synthesis}\right] \cdot 100\% \quad (5)$$

Table 5. Release efficiency (RE) of AMA and HEP and encapsulation efficiency (EE) of AMA in polypyrrole films.

Film	AMA RE [%]	HEP RE [%]	AMA EE [%]	m AMA Released [mg·cm^{-3}]
f_1	19.10 ± 0.25	8.90 ± 0.25	1.77 ± 0.11	1.79 ± 0.12
f_2	28.44 ± 0.35	24.04 ± 0.32	22.30 ± 0.23	2.26 ± 0.16
f_3	34.61 ± 0.41	33.96 ± 0.51	31.50 ± 0.28	3.19 ± 0.20
f_4	30.12 ± 0.32	26.87 ± 0.28	17.70 ± 0.12	1.79 ± 0.17
f_5	17.02 ± 0.32	12.08 ± 0.23	7.69 ± 0.09	1.84 ± 0.19
f_6	25.11 ± 0.51	21.16 ± 0.42	18.50 ± 0.21	3.64 ± 0.26
f_7	55.26 ± 1.10	47.24 ± 0.93	39.57 ± 0.31	5.13 ± 0.34
f_8	39.36 ± 0.85	19.34 ± 0.14	27.70 ± 0.32	2.39 ± 0.21

The entrapment percentage for AMA was the highest for f_7 (39.5%) and for f_3 (31.5%) (Table 5), which was related to the higher roughness of the layers. The EE of AMA ranged from 17 to 39.5%. This system helped to accumulate high AMA values. Direct comparison with the literature data is impossible due to the novel nature of the proposed system. However, in the literature data, the EE in PLA nanospheres ranged from 19 to 30%, depending on the particle size (81 to 113 nm) [30]. The active substance release efficiency for both AMA and HEP showed high values reaching up to 47%. These values were much higher than those of the literature data, which allowed us to conclude that the proposed AMA/HEP/PPy system can be used for further studies in order to use it as a drug delivery system.

3.5. FTIR Analysis of AMA/HEP/PPy Films Before and After Release of AMA and HEP from the Polypyrrole Matrix

FTIR studies were performed to confirm the incorporation of AMA and HEP into the PPy matrix and after the desorption process, to determine the loss after the release process (Figure 6).

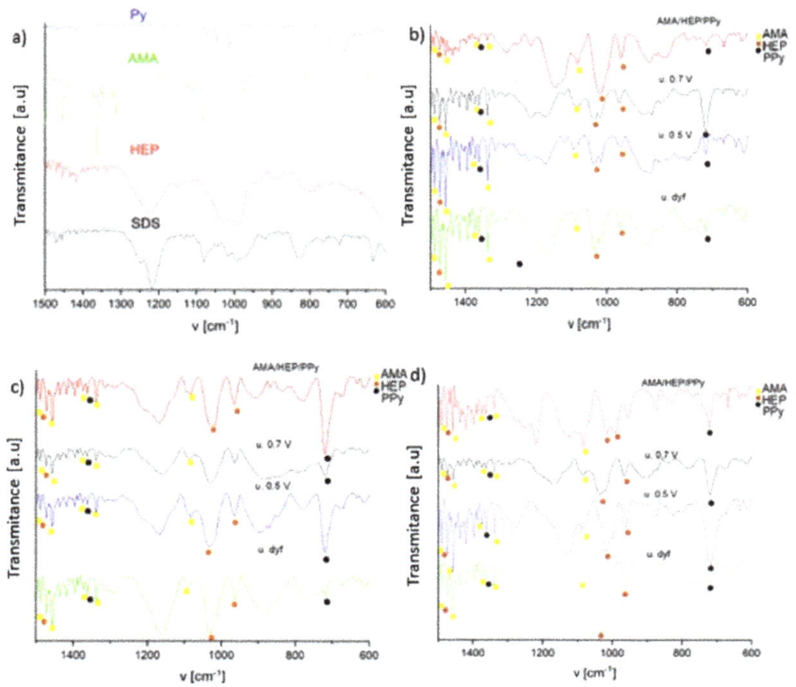

Figure 6. FTIR spectra of AMA/HEP/PPy films synthesized on ITO substrate: (**a**) Reference samples—blue line for pyrrole, green for amantadine, red for heparin, and black for sodium dodecyl sulfate; (**b**) film f_3; (**c**) f_4; (**d**) f_7, where (red line—synthesized film, black—film after 0.7 V release, blue—film after 0.5 V release, green—film after diffusion release) yellow dot—AMA, red—HEP, and black—PPy.

The results are shown in Figure 5. PPy-related bands were identified at 1313 cm^{-1}, corresponding to the C-H in-plane deformation of PPy and N-H bending. Bands at 1074 cm^{-1}, 1047 cm^{-1}, 1012 cm^{-1}, and 811 cm^{-1}, attributed to the C-H wagging mode, along with a signal at 727 cm^{-1} were also detected. Bands at 1546 and 1452 cm^{-1} are associated with the fundamental vibration of the pyrrole ring, while the in-plane vibration of =C–H was observed at 1298 and 1039 cm^{-1}. The vibration of the C–N bond corresponded to the band occurring at around 1196 cm^{-1}. This analysis suggests a successful polymerization process resulting in PPy formation. Bands associated with HEP were detected at the following positions: 1487 cm^{-1}, assigned to C=C stretching; 1430 cm^{-1}, to the symmetric stretching of the carboxy group; and 932 cm^{-1}, assigned to C-O-S stretching and C-O-C glycosidic bonds. The bands belonging to AMA were as follows: 1455 cm^{-1}—band related to bending vibrations of N-H groups; 1354 cm^{-1}—stretching vibrations of C-N bonds; 1110 cm^{-1}; 974 cm^{-1} stretching vibrations of C-C bonds in the cycloalkyl ring [31]. The transmittance intensities of the 1455 cm^{-1} band corresponding to AMA were compared for AMA/HEP/PPy coatings before and after the desorption process. These values are estimated and exemplary, and they refer to the spectrum recorded in the middle part of the sample. The band intensity is the highest for the synthesized films, and the lowest after release at 0.7 V. The changes in the band intensity are visible after the release of AMA and HEP from the polypyrrole matrix. The largest difference in transmittance before desorption to the band after release was detected for the band 1455 cm^{-1} for f_4 and f_7 after stimulation with a potential of 0.7 V.

3.6. Long-Term Release of AMA and HEP from AMA/HEP/PPy Films

For the most promising systems, long-term release studies of drug substances were conducted (time scale of experiment: 48 h). The selection criterion was based on the possibility and effectiveness of providing a daily therapeutic dose, standardly used in the treatment of Alzheimer's disease. The studies were carried out according to the procedure described in Section 2.2. UV-Vis spectra of solutions during the release of drug substances from the films were recorded at a frequency of every 15 min for the first hour, every 20 min for the second hour, every 30 min during the third hour, every 60 min during 12 h, and every 240 min for the rest of the time. The spectra are presented in the figures below (Figure 7). Then, the cumulative release curves (CR—cumulative release) were prepared by plotting the relationship between the concentration of the drug released at time t and the total concentration of the solute (this value corresponds to the maximum concentration achieved during the release procedure from the matrix): $\left(\frac{M_t}{M_\infty}\right)$ = f(t). Selected mathematical models were fitted to the curves based on experimental data, which allowed for the description of the kinetics of the release of substances from the polymer network, including thin polymer films [32]. The mathematical models are summarized and briefly characterized in the table below (Table 6).

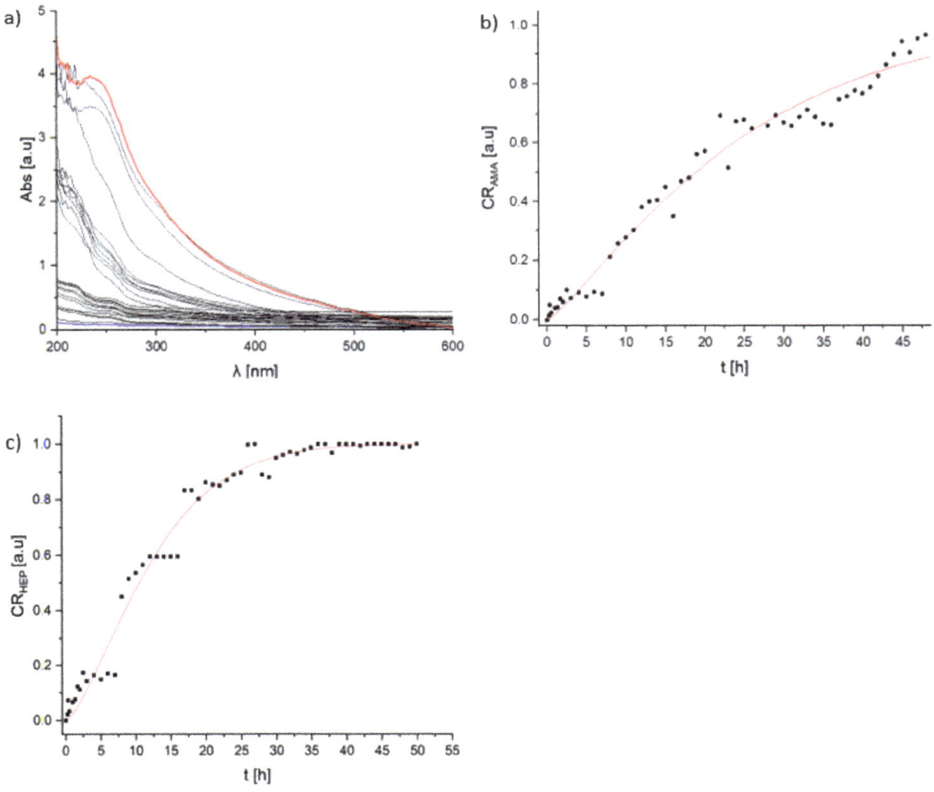

Figure 7. (**a**) UV-Vis spectrum of the solution during the release of AMA and HEP from the f_7 film and the kinetic curves of the release of (**b**) AMA and (**c**) HEP with the simulation curve fitted by the Avrami model.

Table 6. Mathematical models describing controlled release kinetics.

Model	Analytical Relations	Description
Korsmeyer–Peppas (Power)	$\frac{M_t}{M_\infty} = k_p \cdot t^{n_p}$	$\frac{M_t}{M_\infty}$ is the fraction of drug released over time t, k_p is the Power release rate constant, and n_p is the Power release exponent The value of n characterizes the drug release mechanism. To find the exponent n, only part of the release curve was used, where $\frac{M_t}{M_\infty}$ < 0.6. If n = 0.5, the release process is described as diffusion controlled, and if n = 1, the process is non-Fick diffusion, and other controlling factors are present (e.g., swelling)
Avrami	$\frac{M_t}{M_\infty} = 1 - \exp(-k_A \cdot t^{n_A})$	$\frac{M_t}{M_\infty}$ is the fraction of drug released over time t, n_A is an Avrami parameter, and k_A is the Avrami release rate constant. The value n = 1 indicates a first-order kinetic process (perturbed diffusion), and n = 0.54 corresponds to diffusive release kinetics

The first example of a mathematical model aimed to describe drug release from a matrix system was proposed by Higuchi [33]. Initially conceived for planar systems, it was then extended to different geometrics and porous systems [34]. The Avrami model was originally constructed to describe crystallization kinetics [35], but over time, it was adapted to simulate drug release from matrices. The value of n exponent estimates the type of release mechanism: a diffusion-controlled process for n between 0.5 and 1 with higher values indicating mechanisms involving swelling or degradation. Hence, it can provide insight into a mechanism of release process. The Avrami model is applicable for drug release simulation from a thin polymer matrix [36], especially for longer deposition times [37].

The UV-Vis spectra obtained for the f_7 film were compared, and the concentration dependence curves for the active substance were plotted, on the basis of which analogous mathematical modeling was performed, which was the basis for the kinetic description of AMA and HEP from the AMA/HEP/PPy system (Figure 5).

AMA and HEP bands were identified, and absorbance values were recorded for the given wavelengths (Figure 7a). The release curves of AMA (Figure 7b). and HEP (Figure 7c). were plotted, for which 90% CR was achieved for AMA after 48 h, while for HEP, it was after 40 h. The best fit to the release curves was provided by the Avrami model. The calculated parameters of the tested models are listed in the table below (Table 7).

Table 7. Fitting parameters of AMA release kinetics curves obtained from mathematical simulations using the Avrami and Korsmeyer–Peppas models.

System	Film	Drug	Model		Avrami Fitted Parameters	
			R^2 (Avrami)	R^2 (K-P)	k	n
AMA/HEP/PPy (steel)	f_3	AMA	0.970	0.922	0.019 ± 0.036	0.973 ± 0.057
		HEP	0.981	0.969	0.025 ± 0.048	0.969 ± 0.071
AMA/HEP/PPy (ITO)	f_7	AMA	0.961	0.911	0.017 ± 0.029	0.961 ± 0.054
		HEP	0.975	0.911	0.018 ± 0.039	0.953 ± 0.069

The parameters' n values indicate that the release processes for both molecules are described by first-order kinetics. The minor differences between the k and n parameters were obtained for applied fitting models depending on substrate, and still higher values were noted for the f_3 film on a steel substrate than that for the ITO one. It indicates a slightly more restricted release of molecules from the matrix of the samples deposited on the steel substrate. According to the literature, to obtain a therapeutic effect during the use of the drug, patients usually have to take AMA in daily doses of about 274 mg with the possibility of increasing to 400 mg [38,39]. The proposed systems could provide a dosage of 281 mg per day (calculated based on 1 cm^3 of the deposited sample), covering part of the range of concentrations necessary to obtain therapeutic effects.

3.7. Microbiological Tests

In order to verify the possibility of using drug dosing systems in medical applications, the antibacterial activity of the steel plate surface was measured based on PN-EN ISO 22196:2007. The designations of the samples subjected to microbiological analysis are presented in the table below (Table 8).

Table 8. Designations of coatings used in microbiological tests.

Name	System	Parameter of Synthesis
BBT3 (f7)	AMA/HEP/PPy	E = 0.85 V, 20 c
BBT4 (f7)	AMA/HEP/PPy	
A control	steel substrate 316l	-
B control	steel substrate 316l	-

A slight reduction of microorganisms around the materials contacted with the Gram-negative strain occurs in the area of the BBT4 material (marked as 4). However, due to the lack of a homogeneous area, no growth inhibition zone was determined for it. The material (BBT3) contacted with the Gram-positive strain *Staphylococcus aureus* (*S. aureus*) shows a growth inhibition zone after 24 h of incubation. The size of the inhibition zones is presented in the figure (Figure 8), and it was equal to 7 mm according to *S. aureus*.

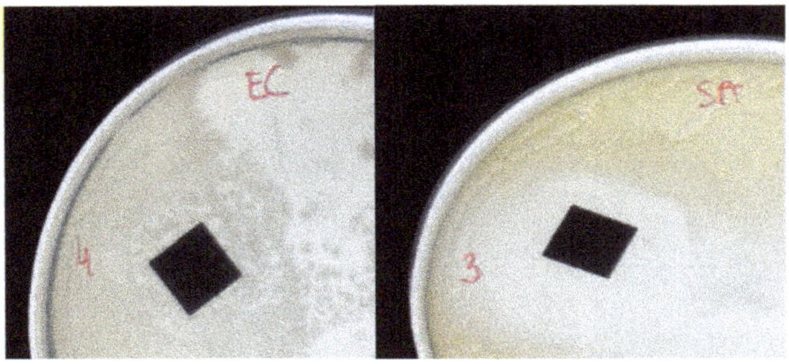

Figure 8. Growth inhibition zone visible around the tested materials on agar media (MHA) after 24 h of bacterial culturing with *Escherichia coli* (*E. coli* EC) and *S. aureus* (SA).

Antimicrobial activity for *S. aureus*: The degree of reduction was determined for the BBT3 material. The obtained results indicate that the materials have a biocidal effect. The reference material (B) did not show any biocidal properties. The data obtained during the experiment are summarized in Figure 9.

The degrees of reduction of *S. aureus* and *E. coli* bacteria after 24 h of contact of the tested materials with microorganisms were equal to 71.42% (reduction log 1.48) for *E. coli* and 93.94% (reduction log 4.13) for *S. aureus*.

The antibacterial properties of the BBT3 material allows it to be considered as a material with antibacterial potential in the presence of the *S. aureus* strain, with reduction ratio indices above 2 (BBT3). The percentage reduction ratio indicates a 90–100% reduction of bacteria in the suspension.

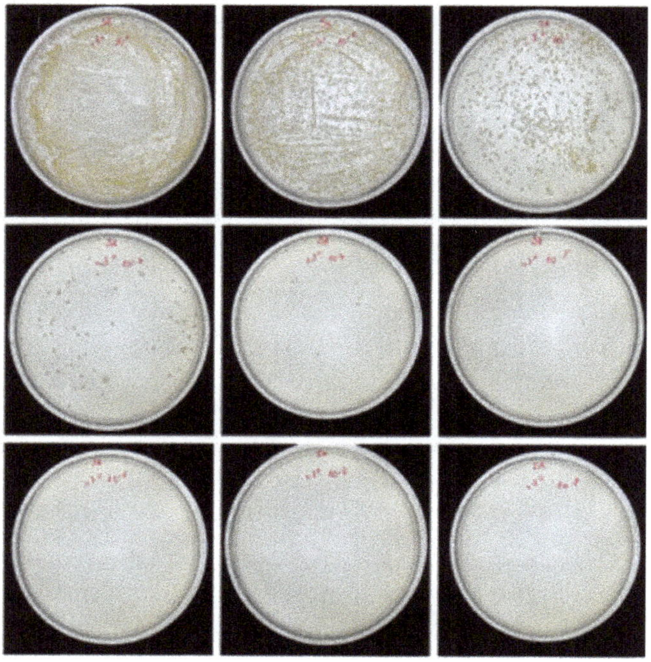

Figure 9. Inoculation of *S.aureus* on PCA medium; reading 24 h after inoculation on the tested material BBT3.

4. Discussion

The increase of the potential window during synthesis and the number of cycles influenced the increase of the layer thickness. The most effective polymerization process and electrochemical stability of the coatings were visible for f_7. Increasing the potential window during synthesis influenced the increase of the process and electrochemical stability of the obtained coatings on both substrates. After the release of AMA and HEP, single globular structures with a diameter of 0.05 μm were visible. AMA increased the adhesion of the coating to the ITO substrate. The results indicate a significantly better adhesion for the films on the steel substrate. The f_7 film showed the best adhesion, as only small flakes of the coating were detached at the intersections of the lines (less than 5% of the area). The f_5 coating detached in 15% of the area of the incision grid and f_6 from the area of 30%. The release efficiency of the active substance for both AMA and HEP showed high values reaching even 47%. These values were much higher than those of the literature data, which allowed us to conclude that the proposed AMA/HEP/PPy system can be used for further studies on its use as a drug delivery system.

5. Conclusions

AMA and HEP were successfully incorporated into the PPy matrix. The films deposited on the steel substrate showed a compact, smooth structure, where there were no visible organized structures. After release, the film became corrugated. AMA-added films improved the adhesion of the coating to the ITO substrate. The f_7 film showed the best adhesion, with only small flakes peeling off at the line intersections (less than 5% of the area). The short-term release of AMA and HEP showed promising results for film f_7. Based on short-term results, long-term tests were conducted. For the AMA/HEP/PPy system, the most effective film in terms of drug release from the matrix (based on short-term release studies) was selected for the steel and ITO substrates, and release curves were plotted. The best fit of the experimental curves was obtained using the Avrami model. The curves showed an n-factor above 0.54, indicating first-order release kinetics with a significant matrix effect on this process. DDS released the drug at a constant rate, thus maintaining the drug concentration in the therapeutic window for a longer period. The release curves indicate a 95% efficiency of AMA release over the studied time period and protocol. The proposed system was able to provide a daily dose that ensures a therapeutic effect. This was a significant step towards developing systems capable of delivering a wider range of doses, potentially in line with the full spectrum recommended for therapeutic efficacy. The antibacterial properties of the material can be considered as a composite with antibacterial potential in the presence of the *S. aureus* strain. The percentage reduction ratio indicates a 90–100% reduction of bacteria in the suspension.

6. Patents

Research based on the databases developed from the patent application: Kulik S., Golba S., Stodolak Zych E., Kurpanik R., Polymer composite based on polypyrrole, implant with a coating of a polymer composite based on polypyrrole and a method of manufacturing an implant with a coating of a polymer composite based on polypyrrole. Application data: 17 September 2024. Confirmed number: P.449842.

Supplementary Materials: The following supporting information can be downloaded at: https://www.mdpi.com/article/10.3390/coatings14111389/s1, Figure S1: Synthesis of AMA/HEP/PPy in 0.1 M SDS: (a) f1; (b) f5; (c) f2; (d) f6. Figure S2: Cyclic voltammogram of AMA/HEP/PPy films in doping/dedoping state of: (a) f1; (b) f5; (c) f2; (d) f6. Figure S3: SEM images of the film f7 AMA/HEP/PPy on steel substrate: (a,c) before AMA and HEP release; (b,d) after 0.7 V potential release of AMA and HEP. Figure S4: Adhesion test of AMA/HEP/PPy films before and after the adhesion test: on ITO substrate (a) f1; (b) f2; (c) f3; (d) f4; (e) f5; (f) f6; (g) f7; (h) f8.

Author Contributions: Conceptualization, S.K., S.G. and R.K.; data curation, S.K. and E.S.-Z.; formal analysis, S.K., S.G., I.M., E.S.-Z. and R.K.; funding acquisition, S.K. and S.G.; investigation, S.K., S.G., I.M., E.S.-Z. and R.K.; methodology, S.K. and E.S.-Z.; project administration, S.K. and S.G.; resources, S.K. and S.G.; software, S.K.; supervision, S.G., E.S.-Z. and R.K.; validation, S.K., S.G. and E.S.-Z.; visualization, S.K.; writing—original draft, S.K.; writing—review & editing, S.G. All authors have read and agreed to the published version of the manuscript.

Funding: The research was funded by the Doctoral School of the University of Silesia in Katowice Research Excellence Initiative (IDB) of the UŚ and the "Freedom of Research" competition—3rd edition of IDB, UŚ.

Institutional Review Board Statement: Not applicable.

Informed Consent Statement: Not applicable.

Data Availability Statement: Dataset available on request from the authors.

Conflicts of Interest: The authors declare no conflicts of interest.

References

1. World Health Organization. Available online: https://www.who.int/health-topics/ageing#tab=tab_1 (accessed on 16 June 2024).
2. United Nations. Available online: https://www.un.org/development/desa/pd/content/world-population-ageing-2017-highlightsasp (accessed on 16 June 2024).
3. Klimova, B.; Kuca, K. Speech and language impairments in dementia. *J. Appl. Biomed.* **2016**, *14*, 97–103. [CrossRef]
4. Klimova, B.; Maresova, P.; Valis, M.; Hort, J.; Kuca, K. Alzheimer's disease and language impairments: Social intervention and medical treatment. *Clin. Interv. Aging* **2015**, *10*, 1401–1407. [CrossRef]
5. Maresova, P.; Hruska, J.; Klimova, B.; Barakovic, S.; Krejcar, O. Activities of Daily Living and Associated Costs in the Most Widespread Neurodegenerative Diseases: A Systematic Review. *Clin. Interv. Aging* **2020**, *15*, 1841–1862. [CrossRef]
6. Fomo, G.; Waryo, T.; Feleni, U.; Baker, P.; Iwuoha, E. Electrochemical Polymerization. In *Functional Polymers—Polymers and Polymeric Composites: A Reference Series*; Jafar Mazumder, M., Sheardown, H., Al-Ahmed, A., Eds.; Springer: Berlin/Heidelberg, Germany, 2019; pp. 105–131. [CrossRef]
7. Skovronsky, D.M.; Lee, V.M.-Y.; Trojanowski, J.Q. Neurodegenerative Diseases: New Concepts of Pathogenesis and Their Therapeutic Implications. *Annu. Rev. Pathol. Mech. Dis.* **2006**, *1*, 151–170. [CrossRef]

8. Schmidt, M.L.; Zhukareva, V.; Newell, K.L.; Lee, V.M.; Trojanowski, J.Q. Tau isoform profile and phosphorylation state in dementia pugilistica recapitulate Alzheimer's disease. *Acta Neuropathol.* **2000**, *101*, 518–524. [CrossRef]
9. Akhtar, A.; Andleeb, A.; Waris, T.S.; Bazzar, M.; Moradi, A.-R.; Awan, N.R.; Yar, M. Neurodegenerative diseases and effective drug delivery: A review of challenges and novel therapeutics. *J. Control. Release* **2020**, *10*, 1152–1167. [CrossRef]
10. Metman, L.V.; Del Dotto, P.; van den Munckhof, P.; Fang, J.; Mouradian, M.M.; Chase, T.N. Amantadine as treatment for dyskinesias and motor fluctuations in Parkinson's disease. *Neurology* **1998**, *50*, 1323–1326. [CrossRef]
11. Jain, K.K. Drug Delivery Systems—An Overview. *Methods Mol. Biol.* **2008**, *437*, 1–50. [CrossRef]
12. Smela, E. Conjugated polymer actuators for biomedical applications. *Adv. Mater.* **2003**, *15*, 481–494. [CrossRef]
13. Ferraz, N.; Strømme, M.; Fellström, B.; Pradhan, S.; Nyholm, L.; Mihranyan, A. In vitro and in vivo toxicity of rinsed and aged nanocellulose–polypyrrole composites. *J. Biomed. Mater. Res.* **2012**, *100*, 2128–2138. [CrossRef]
14. Kamalesh, S.; Tan, P.; Wang, J.; Lee, T.; Kang, E.-T.; Wang, C.-H. Biocompatibility of electroactive polymers in tissues. *J. Biomed. Mater. Res.* **2000**, *52*, 467–478. [CrossRef]
15. Kontturi, K.; Pentti, P.; Sundholm, G. Polypyrrole as a model membrane for drug delivery. *J. Electroanal. Chem.* **1998**, *453*, 231–238. [CrossRef]
16. Gandhi, M.R.; Murray, P.; Spinks, G.M.; Wallace, G.G. Mechanism of electromechanical actuation in polypyrrole. *Synth. Met.* **1995**, *73*, 247–256. [CrossRef]
17. Svirskis, D.; Travas-Sejdic, J.; Rodgers, A.; Garg, S. Electrochemically controlled drug delivery based on intrinsically conducting polymers. *J. Control. Release* **2010**, *146*, 6–15. [CrossRef]
18. Shah, S.A.A.; Firlak, M.; Berrow, S.R.; Halcovitch, N.R.; Baldock, S.J.; Yousafzai, B.M.; Hathout, R.M.; Hardy, J.G. Electrochemically Enhanced Drug Delivery Using Polypyrrole Films. *Materials* **2018**, *11*, 1123. [CrossRef]
19. Alshammary, B.; Walsh, F.C.; Herrasti, P. Electrodeposited conductive polymers for controlled drug release: Polypyrrole. *J. Solid State Electrochem.* **2016**, *20*, 839–859. [CrossRef]
20. Hasan, N.; Bhuyan, M.M.; Jeong, J.-H. Single/Multi-Network Conductive Hydrogels—A Review. *Polymers* **2024**, *16*, 2030. [CrossRef]
21. Wang, X.; Li, X.; Zhao, L.; Li, M.; Li, Y.; Yang, W. Polypyrrole-doped conductive self-healing multifunctional composite hydrogels with a dual crosslinked network, The Royal Society of Chemistry. *Soft. Matter.* **2021**, *17*, 8363–8372. [CrossRef]
22. Hazarika, J.; Kumar, A. Controllable synthesis and characterization of polypyrrole nanoparticles in sodium dodecylsulphate (SDS) micellar solution. *Synth. Met.* **2013**, *175*, 155–162. [CrossRef]
23. Kirschbaum, J. Amantadine. *Anal. Profiles Drug Subst.* **1983**, *12*, 1–36. [CrossRef]
24. Rodríguez-Torres, M.; Díaz-Torres, L.; Romero-Servin, S. Heparin Assisted Photochemical Synthesis of Gold Nanoparticles and Their Performance as SERS Substrates. *Int. J. Mol. Sci* **2014**, *15*, 19239–19252. [CrossRef]
25. Shama, S.A.; Amin, A.S.; Mabrouk, E.S.M.; Omara, H.A. Utility of oxidation–reduction reaction for the spectrophotometric determination of amlodipine besylate. *Arab. J. Chem.* **2009**, *2*, 59–63. [CrossRef]
26. Ryan, E.M.; Breslin, C.B. Formation of polypyrrole with dexamethasone as a dopant: Its cation and anion exchange properties. *J. Electroanal. Chem.* **2018**, *824*, 188–194. [CrossRef]
27. Arroyoa, J.; Akieh-Pirkanniemia, M.; Lisakb, G.; Latonena, R.M.; Bobacka, J. Electrochemically controlled transport of anions across polypyrrole-based membranes. *J. Membr. Sci.* **2019**, *581*, 50–57. [CrossRef]
28. Song, Y.; Wei, J.; Ma, Y.; Zeng, P.; Kong, M. Removal and recovery of amantadine from water by liquid–liquid extraction. *Environ. Earth Sci.* **2015**, *73*, 4931–4938. [CrossRef]
29. Mazibuko, Z.; Indermun, S.; Govender, M.; Kumar, P.; Du Toit, L.C.; Choonara, Y.E.; Pillay, V. Targeted Delivery of Amantadine-loaded Methacrylate Nanosphere-ligands for the Potential Treatment of Amyotrophic Lateral Sclerosis. *J. Pharm. Pharm. Sci.* **2018**, *21*, 94. [CrossRef]
30. Rivas, L.; Sánchez-Cortés, S.; Stanicova, J.; García-Ramos, J.; Miskovsky, P. FT-Raman, FTIR and surface-enhanced Raman spectroscopy of the antiviral and antiparkinsonian drug amantadine. *Vib. Spectrosc.* **1999**, *20*, 179–188. [CrossRef]
31. Wang, L.-Y.; Zhao, M.-Y.; Bu, F.-Z.; Niu, Y.-Y.; Yu, Y.-M.; Li, Y.-T.; Wu, Z.-Y. Cocrystallization of Amantadine Hydrochloride with Resveratrol: The First Drug–Nutraceutical Cocrystal Displaying Synergistic Antiviral Activity. *Cryst. Growth Des.* **2021**, *21*, 2763–2776. [CrossRef]
32. Suman Kumar Valeveti, Shailaja Pashikanti Design, development, and evaluation of transdermal patches containing memantine hydrochloride. *J. Appl. Pharm.* **2023**, *15*, 5. [CrossRef]
33. Higuchi, T.J. Mechanism of sustained medication. Theoretical analysis of rate of release of solid drugs dispersed in solid matrices. *Pharm. Sci.* **1963**, *84*, 1464–1477. [CrossRef]
34. Grassi, M.; Grassi, G. Mathematical modeling and controlled drug delivery: Matrix systems. *Curr. Drug Deliv.* **2005**, *2*, 97–116. [CrossRef] [PubMed]
35. Avrami, M. Kinetics of Phase Change. I General Theory. *J. Chem. Phys.* **1939**, *7*, 1103–1112. [CrossRef]
36. Brady, Á.; Sakathinatan, I.; Barreca, D.; Maccato, C.; McCormac, T. Electrochemical Doping and Surface Studies of Poly(3,4-Ethylenedioxythiophene) Films for the Release of the Anticancer Drug Mitoxantrone. Available online: https://www.researchgate.net/publication/373428270_Electrochemical_Doping_and_Surface_Studies_of_Poly34-Ethylenedioxythiophene_Films_for_the_Release_of_the_Anticancer_Drug_Mitoxantrone (accessed on 1 September 2024). [CrossRef]

37. Alizadeh, N.; Shamaeli, E. Electrochemically controlled release of anticancer drug methotrexateusing nanostructured polypyrrole modified with cetylpyridinium: Release kinetics investigation. *Electrochim. Acta* **2014**, *130*, 488–496. [CrossRef]
38. Drugs.com. Amantadine Dosage. 2023. Available online: https://www.drugs.com/dosage/amantadine.html#Usual_Adult_Dose_for_Parkinson_s_Disease (accessed on 1 September 2024).
39. Mayo Clinic. Amantadine (Oral Route). 2024. Available online: https://www.mayoclinic.org/drugs-supplements/amantadine-oral-route/proper-use/drg-20061695 (accessed on 1 September 2024).

coatings

MDPI

Article

The Combination of Decellularized Cartilage and Amniotic Membrane Matrix Enhances the Production of Extracellular Matrix Elements in Human Chondrocytes

Antonio Rojas-Murillo [1], Jorge Lara-Arias [2], Héctor Leija-Gutiérrez [3], Rodolfo Franco-Márquez [4], Nidia Karina Moncada-Saucedo [5], Abel Guzmán-López [6], Félix Vilchez-Cavazos [2], Elsa Nancy Garza-Treviño [1,*] and Mario Simental-Mendía [2,*]

1 Department of Biochemistry and Molecular Medicine, Universidad Autónoma de Nuevo Leon, Monterrey 66460, Nuevo Leon, Mexico; juan.rojasmrll@uanl.edu.mx
2 Orthopedic Trauma Service, University Hospital "Dr. José Eleuterio González", Universidad Autónoma de Nuevo Leon, Monterrey 66460, Nuevo Leon, Mexico; jorge.larars@uanl.edu.mx (J.L.-A.); jose.vilchezcvz@uanl.edu.mx (F.V.-C.)
3 Center for Research in Physical and Mathematical Sciences, Universidad Autónoma de Nuevo Leon, San Nicolás de los Garza 66455, Nuevo Leon, Mexico; hector.leijagt@uanl.edu.mx
4 Department of Anatomic Pathology and Cytopathology, University Hospital "Dr. José Eleuterio González", Universidad Autónoma de Nuevo Leon, Monterrey 66460, Nuevo Leon, Mexico; rodolfo.francomrqz@uanl.edu.mx
5 Department of Hematology, University Hospital "Dr. José Eleuterio González", Universidad Autónoma de Nuevo Leon, Monterrey 66460, Nuevo Leon, Mexico; nidia.moncadas@uanl.edu.mx
6 Department of Gynecology and Obstetrics, University Hospital "Dr. José Eleuterio González", Universidad Autónoma de Nuevo Leon, Monterrey 66460, Nuevo Leon, Mexico; abel.guzmanlp@uanl.edu.mx
* Correspondence: elsa.garzatr@uanl.edu.mx (E.N.G.-T.); mario.simentalme@uanl.edu.mx (M.S.-M.); Tel.: +52-81-83294173 (E.N.G.-T.)

Abstract: Articular cartilage lesions are challenging to regenerate, prompting the investigation of novel biomaterial-based therapeutic approaches. Extracellular matrix (ECM)-derived biomaterials are a promising option for this purpose; however, to date, the combination of amniotic membrane (AMM) and articular cartilage (ACM) has not been tested. This study evaluated different concentrations of soluble extracts from the decellularized ECM of amniotic membrane (dAMM) and articular cartilage (dACM), both individually and in combination, to determine their ability to maintain the chondrogenic phenotype in human chondrocytes. After the decellularization process 90–99% of the cellular components were removed, it retains nearly 100% of type 2 collagen and 70% of aggrecan (ACAN) for dACM, and approximately 90% of type IV collagen and 75% of ACAN for dAMM. The biological activity of soluble extracts from dACM and dAMM were evaluated on human chondrocytes. After 72 h, 1.5 mg/mL of dACM and 6 mg/mL of dAMM significantly increased ($p < 0.05$) the proliferation and expression of SOX9 and ACAN. Also, the combination of both (1.5 mg/mL dACM and 6 mg/mL dAMM) showed synergistic effects, enhancing chondrocyte proliferation, maintaining chondrogenic lineage, and increasing the production of cartilage ECM components, such as COLII (1.5-fold), SOX9 (2-fold), and ACAN (2-fold). These results suggest that the combined use of dACM and dAMM has potential for cartilage regeneration.

Keywords: articular cartilage; amniotic membrane; fibrin scaffold; tissue engineering; chondrocyte; ECM-derived biomaterials

1. Introduction

Articular cartilage injury poses a significant challenge in orthopedics due to its limited self-repair capacity, often leading to conditions like osteoarthritis if left untreated [1]. In response, innovative therapeutic strategies such as tissue engineering, regenerative medicine, and cell therapy, often in conjunction with biomaterials, have emerged as promising approaches for addressing such injuries [2]. Among these, biomaterials derived from the extracellular matrix (ECM) have garnered attention for their biomimetic properties, closely resembling the native tissue ECM composition [3].

The ECM is a complex structure containing bioactive molecules critical for maintaining tissue integrity and function [4]. ECM-based biomaterials provide an ideal microenvironment for cell adhesion, proliferation, and differentiation, and they contain growth factors and signaling molecules that facilitate tissue repair and regeneration [5,6]. These biomaterials also exhibit excellent biocompatibility and cytocompatibility, essential for various biomedical applications [6]. However, the presence of cellular components in ECM-based materials can trigger immune responses and implant rejection, necessitating decellularization [7].

Decellularized ECMs (dECM) derived from different sources, such as porcine liver [8], porcine kidney [9], rat pancreas [10], spinal cord meninges [11], tendon [12], periodontal ligament [13], and others, including amniotic membrane [14,15] and articular cartilage, have shown promise in tissue regeneration [16]. The decellularization process aims to remove cells while preserving the native ECM architecture and bioactive components [17]. Despite their potential in cartilage repair, challenges remain, such as managing inflammatory responses associated with these matrices and limited availability of healthy tissues for decellularization [18,19].

Decellularized articular cartilage ECM (dACM) has gained attention for its ability to retain tissue-specific proteins and bioactive molecules crucial for cartilage regeneration, even after cellular removal [20]. Biofunctionalized dACM scaffolds have demonstrated success in recruiting host cells and promoting chondrogenic differentiation, showing promise for in situ cartilage regeneration [21]. Similarly, decellularized amniotic membrane ECM (dAMM) is notable for its regenerative capabilities in promoting chondrocyte proliferation and cartilage repair [22]. The unique composition of dAMM, including hyaluronic acid, proteoglycans, and growth factors like EGF and bFGF, supports chondrocyte function and enhances cartilage repair strategies [23,24].

Despite their individual successes, the combined effects of dACM and dAMM have not been thoroughly studied, either in vitro or in vivo. This study aims to evaluate the individual and combined effects of dECM derived from articular cartilage and amniotic membrane on maintaining the chondrogenic phenotype of human chondrocytes. This research seeks to elucidate whether combining these biomaterials can enhance their therapeutic potential for cartilage repair and regeneration.

The aim of this work was to study the biological activity of an ECM decellularized from articular cartilage (dACM) and another from amniotic membrane (dAMM) individually and in combination on human chondrocytes. The method of decellularization used detergents and hypotonic buffers, then lyophilization and pulverization. We evaluated particle size by SEM, and the process of decellularization was validated by DAPI fluorescence staining and H and E staining. Preservation of key ECM components such as collagens and glycosaminoglycans was carried out through histological staining, and we evaluated the percentage retention of type 2 collagen and aggrecan by IH and validated it by FT-IR. The biological activity of the soluble extract of the matrices (dACM and dAMM) was evaluated individually and in combination in vitro on human chondrocytes at 24, 48, and 72 h by cell proliferation assays and expression of chondrogenic genes (SOX9, COL-II, ACAN and RUNX2).

2. Materials and Methods

2.1. *Human Cartilage Collection and Preparation of Decellularized Articular Cartilage Matrix (dACM)*

Human articular cartilage slices (thickness = 2 mm) were obtained from cadaveric donors in an aseptic environment (Research Ethical Committee approval number BI23-001). The cartilage was collected from the areas of the condyles, patella, and femoral trochlea using a No. 24 scalpel (SensiMedical, Aventura, FL, USA). The cartilage slices were washed three times with PBS (1×, Gibco, Grand Island, NY, USA) containing an Antibiotic–Antimycotic solution (100×, 10,000 units/mL of penicillin, 10,000 μg/mL of streptomycin, and 25 μg/mL of Amphotericin B, Gibco, USA). The slices were then stored at −80 °C until use.

The decellularization protocol was performed on cartilage from 20 donors as previously described by Perez-Silos [25]. Briefly, 205 g of articular cartilage was exposed to five cycles of thermal shock in liquid nitrogen for 5 min, followed by a wash in PBS (1X, Gibco, USA) for 10 min. The cartilage was then crushed with a blender. The pulverized cartilage was washed for 24 h in 10 volumes of hypotonic buffer (10 mM TRIS-HCl, 2 mM EDTA, pH 8) supplemented with 100 mM KCl (99%, SIGMA-ALDRICH, St. Louis, MO, USA) and 5 mM $MgCl_2$ (98%, SIGMA-ALDRICH, USA). Next, 10 volumes hypotonic buffer supplemented with 100 mM KCl, 5 mM $MgCl_2$, and 0.5% SDS (99%, SIGMA-ALDRICH, USA) was added for 18 h. Finally, the matrix was washed with 10 volumes of hypotonic buffer containing 0.5% SDS for 36 h. Following sterile PBS rinsing to eliminate the remaining SDS (3 times), the samples were immediately frozen at −80 °C. The samples were lyophilized for 24 h to completely remove interstitial fluid, followed by fine pulverization using a K10 pulverizer mill (Micron, Shanghai, China) and a Freezer/mill 6870 (SPEX® SamplePrep, Metuchen, NJ, USA). The decellularized articular cartilage matrix (dACM) was then sterilized with ethylene oxide and stored at −80 °C until use. The experimental strategy we followed in this study is shown in Figure 1.

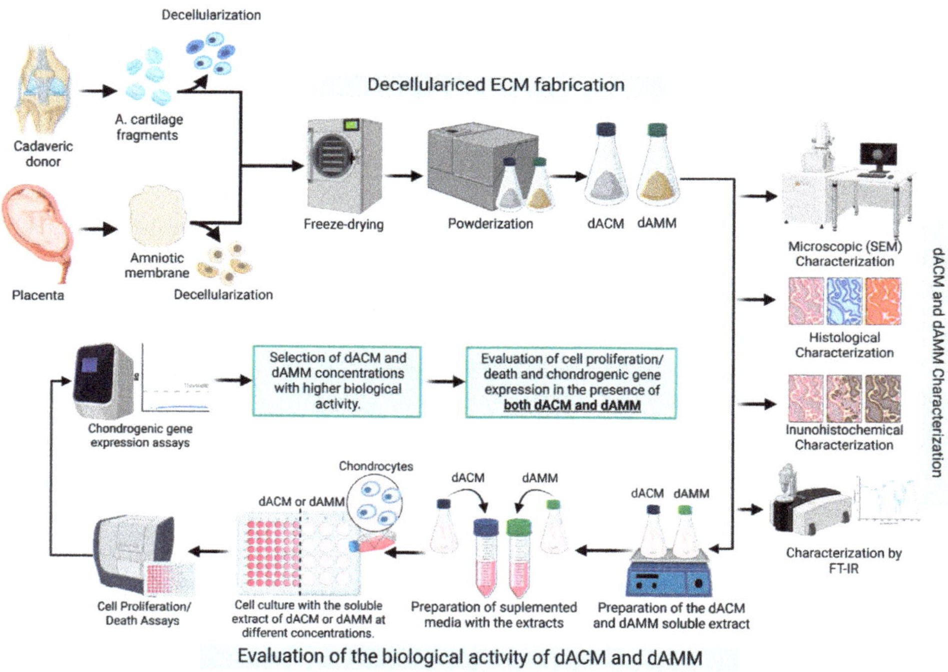

Figure 1. Experimental strategy of the study. Created with BioRender.com.

2.2. *Amniotic Membrane Collection and Preparation of Decellularized Amniotic Membrane Matrix (dAMM)*

After obtaining informed consent, the surgical intervention was performed. Human placentas were taken from the toco-surgery operating room (Research Ethical Committee approval number BI23-001). The samples were collected and transported at 4 °C in 50 mL sterile tubes containing PBS with Antibiotic–Antimycotic solution (100×, 10,000 units/mL of penicillin, 10,000 μg/mL of streptomycin, and 25 μg/mL of Amphotericin B, Gibco, USA). In a sterile environment and after three washes with PBS, the samples were cut into pieces of approximately 1 cm^2 and immediately stored at −80 °C until use. A total of 30 amniotic membranes were collected.

The decellularization protocol for the amniotic membrane was performed on the 30 collected membranes (125 g), as described previously by Villamil et al. (2019) [26]. Briefly, all amniotic membranes underwent five freezing cycles in liquid nitrogen (−196 °C) for 30 min, followed by thawing in a serological bath (Precision Scientific Inc., Chicago, IL, USA) at 37 °C for 30 min. The membranes were then treated with 0.1% Tween 80 (SIGMA-ALDRICH, USA) for 4 h, soaked in 0.1 M NaOH (97%, JALMEX, Guadalajara, México) for 1 h, and treated with 0.15% peracetic acid (PAA, 15%, Cetik, Cualiacán, México) in ethanol (96%, CTR, Monterrey, México). The membranes were bleached with NaOH 0.1 M for 1 h and PAA 0.15% for another hour. A final wash with 70% ethanol was applied for 1 h to remove residual nucleic acids and phospholipids from the tissue. Finally, all ethanol was eliminated by washing with PBS for 2 h (three times), and the membranes were stored at −80 °C. Throughout the process, the membranes were gently agitated to ensure a homogeneous wash and minimal damage to the tissue ultrastructure. The amniotic membranes were then lyophilized for 24 h to completely remove liquids, followed by fine pulverization using a K10 pulverizer mill (Micron, Shanghai, China) and a Freezer/mill 6870 (SPEX® SamplePrep, Metuchen, NJ, USA). The decellularized amniotic membrane matrix (dAMM) was sterilized with ethylene oxide and stored at −80 °C until use.

2.3. Histological and Immunohistochemical Analyses

Both the dACM and dAMM were fixed in 4% PFA for 24 h, then embedded in paraffin blocks using conventional histological methods. Subsequently, 4 μm histological sections were prepared and stained with various dyes for different analyses. To evaluate DNA, nuclear, or cell remnants on the matrices, 4′,6-diamidino-2-phenylindole (DAPI, VECTOR, Newark, CA, USA) staining was used. Hematoxylin and eosin (H and E) staining examined the general histology of the samples. Safranin O staining detected the presence of sulfated proteoglycans, while Masson's trichrome staining visualized collagen fibers in the ECM.

The conservation of ECM components, such as type II collagen fibers and aggrecans, was analyzed by immunohistochemistry (IHC). Primary antibodies, anti-col II (dilution 1:400, ab34712, ABCAM, Cambridge, UK), and anti-aggrecan (dilution 1:100, ab3778, ABCAM, Cambridge, UK) were used and incubated at 4 °C overnight. The mouse- and rabbit-specific detection system HRP/DAB (ABC) detection IHC kit (ab64264m, ABCAM, Waltham, MA, USA) was employed, with Gill's hematoxylin (SIGMA-ALDRICH, USA) as a counterstain. Negative controls for the IHC were samples of articular cartilage without the primary antibody. Images were captured with an Olympus AX70 microscope (Olympus, Tokyo, Japan). For analysis, 10 fields per slide (3 slides per tissue) were examined to calculate the respective antibody-positive area percentage using ImageJ software Version 1.54g.

2.4. dACM and dAMM Microstructure Analysis

The lyophilized dACM and dAMM were sputter-coated with gold and transferred to a scanning electron microscope (SEM) (JSM-6390LV, JEOL, Tokyo, Japan). The particle size was then measured through image analysis.

2.5. FT-IR Spectroscopy

The Fourier transform infrared spectroscopy (FT-IR) spectral profiles of dACM and dAMM were obtained using a PerkinElmer Frontier spectrometer in a spectral range of 400–4000 cm^{-1}. This analytical approach allowed for the detailed examination of the molecular composition and structural characteristics of the samples.

2.6. Chondrocytes Isolation and Cuture

Chondrocytes were isolated from the knee joints of healthy human donors and cadaveric donors (Research Ethical Committee approval number BI23-001). Briefly, the articular cartilage tissue was cut into small pieces (<1 mm) and incubated with 2.5% trypsin (Gibco, USA) at 37 °C for 30 min. Following the removal of the trypsin solution, the tissue was digested with 2 mL of 0.2% type II collagenase (Gibco, USA) at 37 °C for 2 h. Released cells were obtained by centrifugation at 4000 rpm for 10 min, and the remaining tissue was digested one more time for 90 min. The cells were cultured in Dulbecco's Modified Eagle's Medium (DMEM; Gibco, USA) supplemented with 10% FBS (Gibco, USA) containing Antibiotic–Antimycotic solution (100×, 10,000 units/mL of penicillin, 10,000 µg/mL of streptomycin, and 25 µg/mL of Amphotericin B, Gibco, USA). The medium was replaced every 2 days. Once the cells reached 80%–90% confluence, the chondrocytes were used for all subsequent experiments. Chondrocytes with more than 3 passages were not used.

2.7. In Vitro Cell Culture Assays

The soluble extract of decellularized ECM was prepared by incubating 200 mg of dACM or dAMM with 2 mL of serum-free OptiMEM (Gibco, USA) at 37 °C for 24 h. Following the incubation period, the medium was cleared by centrifugation at 4000 rpm for 20 min; then, the supernatant medium was diluted with 5% FBS OptiMEM to the concentrations of 15 mg/mL, 7.5 mg/mL, 1.5 mg/mL, and 0.15 mg/mL of dACM (Dry weigh), as described by [27], and 12 mg/mL, 6 mg/mL, 3 mg/mL, and 0.3 mg/mL of dAMM (Dry weigh), as described by [28], and stored at −20 °C or used immediately. For cellular proliferation and cytotoxicity assays, 5×10^3 chondrocytes were seeded per well in a 96-well plate with 100 µL/well OptiMEM supplemented with 5% FBS. After 24 h of incubation, the medium was removed, and the cells were treated with 100 µL/well of the medium at the different concentrations of the soluble extract of dACM or dAMM. The cells were then incubated for 24, 48, and 72 h. ATP luminescence was determined using CellTiter-Glo (Promega, Madison, WI, USA) to evaluate proliferation. A CellTiter-Glo reagent (100 µL) was added to each well, agitated for 2 min at 300–500 rpm, and incubated for 10 min at room temperature. Luminescence was quantified using a Cytation 3 plate reader (BioTek, Winooski, VT, USA). The percentage of cell death was calculated using the following formula:

$$1 - \frac{\text{treatment mean luminescence}}{\text{treatment mean luminescence}} \times 100$$

2.8. Chondrogenic Gene Expression

The chondrocytes were seeded into a 24-well plate at 1×10^5 cells/well with OptiMEM supplemented with 5% FBS. After 24 h of incubation, the cells were treated with the soluble extract at concentrations of 15 mg/mL, 7.5 mg/mL, 1.5 mg/mL, and 0.15 mg/mL for dACM, and 12 mg/mL, 6 mg/mL, 3 mg/mL, and 0.3 mg/mL for dAMM. The cells were then incubated for 24, 48, and 72 h.

RNA isolation was carried out using the RNeasy Mini Kit (Qiagen, Germantown, MD, USA) according to the manufacturer's protocol. Reverse transcription of total RNA to single-stranded cDNA was completed using the SuperScript™ III First-Strand Synthesis System kit (Invitrogen, Carlsbad, CA, USA) following the manufacturer's instructions. Gene expression was analyzed using a 7500 real-time PCR system (Applied Biosystems, Waltham, MA, USA). TaqMan probes, purchased from Applied Biosystems, included several target genes: Beta-2 microglobulin (B2M: endogenous control; ID: Hs99999907_m1), aggrecan (ACAN; ID: Hs00153936_m1), SRY (sex-determining region Y)-box 9 (SOX9; ID: Hs00165814_m1), type II collagen (COL2A1; ID: Hs00156568_m1), and RunX2. Data from 3 samples were evaluated as mRNA levels in triplicate. The $2^{-\Delta\Delta Ct}$ method was used to calculate the relative expression (RQ) of each target gene.

2.9. Statistical Analysis

All experiments were performed in triplicate and repeated three times. Statistical analysis was conducted using GraphPad Prism 9.0.2 software. First, the Shapiro–Wilk test was performed to evaluate the normal distribution of the quantitative variables including results from decellularization, morphometric analysis, relative gene expression, and cell proliferation. Afterward, an unpaired *t*-test with Welch's correction was used to evaluate significant differences in such quantitative analyses. Data are presented as mean and standard deviation (SD). Statistical significance was set at * $p < 0.05$, ** $p < 0.01$, *** $p < 0.001$.

3. Results

3.1. Particle Sizes and Assays to Evaluate the Decellularized Process

After the decellularization, lyophilization, and pulverization of the articular cartilage and amniotic membrane, fine whitish powder for dACM (Figure 2A), and slightly coarser yellowish powder for dAMM (Figure 2D) were obtained. Scanning electron microscopy analysis revealed that both matrices exhibited a porous structure with varying particle sizes. The particle size of dACM ranged from 1.6 to 600 µm (Figure 2B,C), while for dAMM, it ranged from 0.3 to 1700 µm (Figure 2E,F).

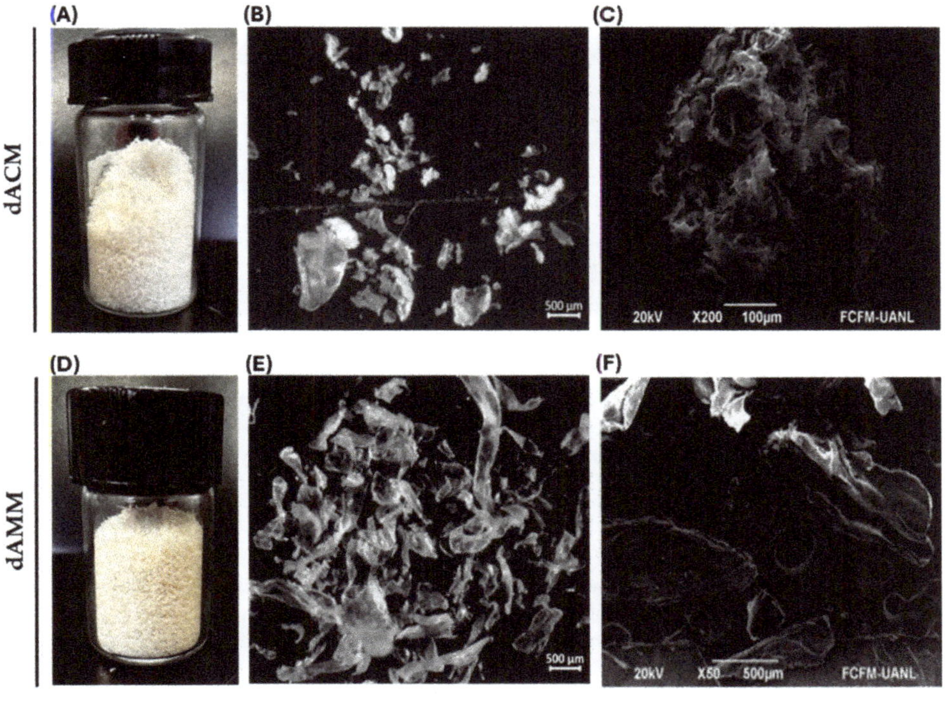

Figure 2. Macroscopic and microscopic description of dACM and dAMM: (**A**) Macroscopic photography of dACM; (**B**) scanning electron micrography image at 50× of dACM; (**C**) scanning electron micrography image at 20,000× of dACM; (**D**) macroscopic photography of dAMM; (**E**) scanning electron micrography image at 200× of dAMM; (**F**) scanning electron micrography image at 2000× of dAMM.

To assess the degree of decellularization, DAPI staining and H and E staining were conducted. The staining showed residual nuclei in the ECM of both tissues (Figure 3A,F). Image analysis indicated that dACM had less than 5% residual nuclei compared to native cartilage (Figure 3E), whereas dAMM was nearly completely decellularized, with only about 1% residual nuclei compared to native amniotic membrane (Figure 3J). H and E staining further revealed the absence of chondrocyte nuclei in decellularized cartilage

(Figure 3B) and the lack of epithelial cell nuclei in decellularized amniotic membrane (Figure 3G). The decellularization process eliminated over 90% of cells in both dACM and dAMM, as evidenced by the staining results.

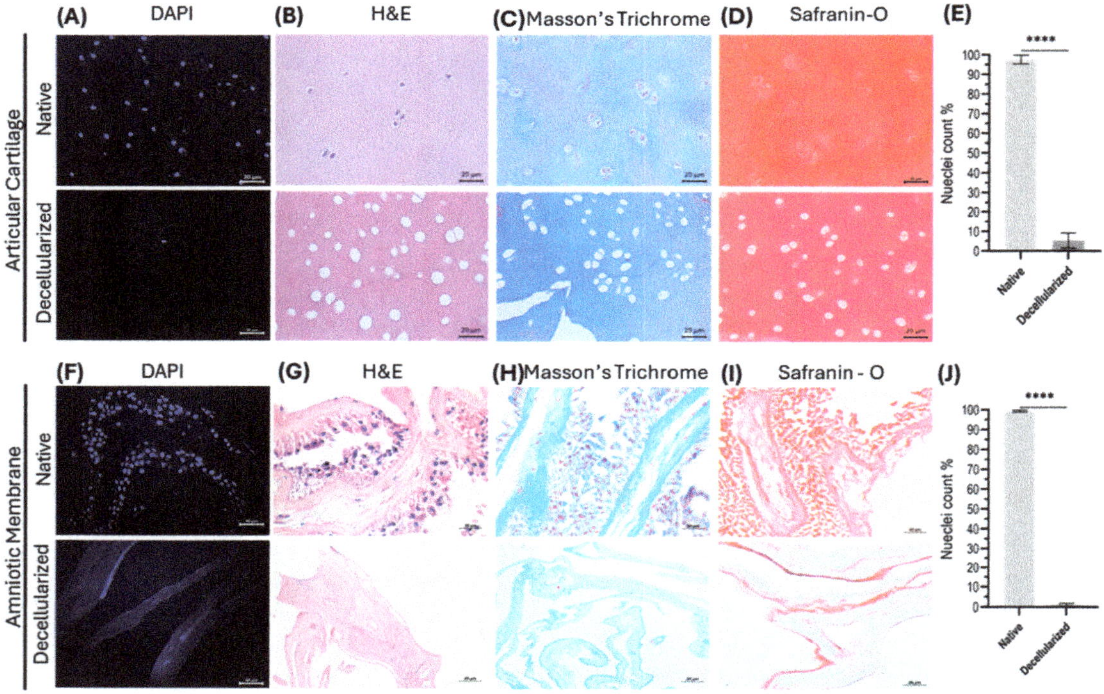

Figure 3. Histological characterization of articular cartilage (AC) and amniotic membrane (AM) before and after decellularization. (**A**) DAPI staining of AC; (**B**) hematoxylin and eosin staining of AC; (**C**) Masson's trichrome staining of AC; (**D**) Safranine O staining of AC; (**E**) quantification of AC decellularization; (**F**) DAPI staining of AM; (**G**) hematoxylin and eosin staining of AM; (**H**) Masson's trichrome staining of AM; (**I**) Safranine O staining of AM; (**J**) quantification of AM decellularization. **** = *t* test p value ≤ 0.0001.

3.2. Preservation of the Native Structure before and after Decellularized Process

To ensure that the chemical decellularization and mechanical pulverization processes did not alter the essential extracellular matrix (ECM) components of the tissues, a comprehensive histological and immunohistochemical characterization was conducted. Histological analysis using H and E staining confirmed the preservation of the ECM structure of both cartilage (Figure 3B–D) and amniotic membrane (Figure 3G–I). Eosin binding to ECM components was observed in both native and decellularized tissues. Masson's trichrome staining further demonstrated the presence of collagens in the decellularized ECMs, indicated by a blue color characteristic of aniline blue in both cases (Figure 3C for cartilage and Figure 3H for amniotic membrane).

Safranin staining was employed to assess the preservation of aggrecans. While no differences were noted in the cartilage before and after decellularization (Figure 3D), a decrease in staining intensity was observed in the decellularized amniotic membrane (Figure 3I).

Immunohistochemical staining for type 2 collagen and aggrecan was performed with semi-quantitative analysis (Figure 4A,B for cartilage and, Figure 4E,F for amniotic membrane). The results showed that dACM retained nearly 100% of the type 2 collagen found in native cartilage (Figure 4C), but there was a 30% decrease in aggrecan compared to native tissue (Figure 4D). For dAMM, approximately 89% of type IV collagen and almost 75% of aggrecan were retained (Figures 4G and 4H, respectively).

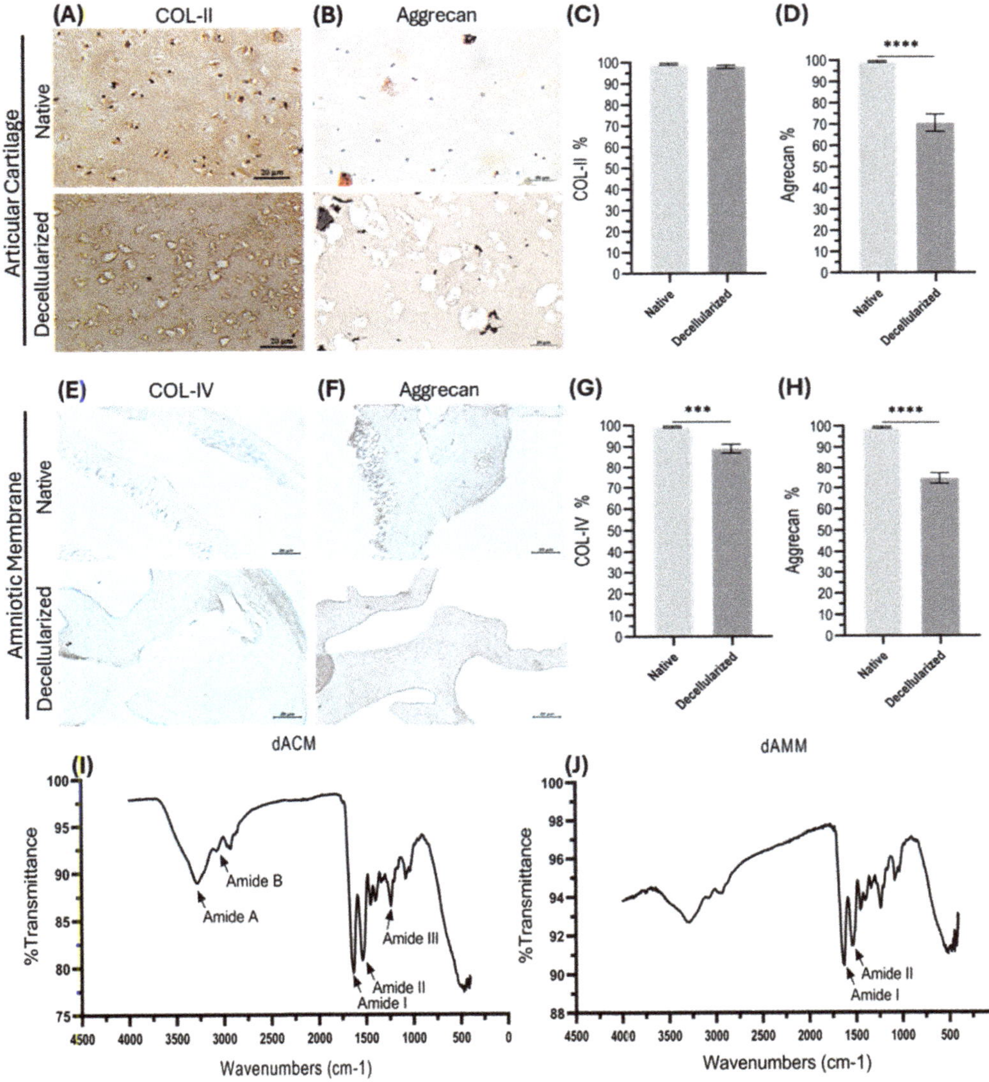

Figure 4. Characterization of articular cartilage (AC) and amniotic membrane (AM) before and after decellularization. (**A**) anti-col II IHC of AC; (**B**) anti-aggrecan IHC of AC; (**C**) quantification of collagen conservation of AC; (**D**) quantification of aggrecan conservation of AC; (**E**) anti-col IV IHC of AM; (**F**) anti-aggrecan IHC of AM; (**G**) quantification of collagen conservation of AM; (**H**) quantification of aggrecan conservation of AM; (**I**) dACM FT-IR spectrum; (**J**) dAMM FT-IR spectrum. *** = t test p value ≤ 0.001; **** = t test p value ≤ 0.0001.

To validate the preservation of the main component collagen, Fourier transform infrared (FT-IR) spectroscopy was utilized to identify the main components in the matrices.

The FT-IR spectra of dACM revealed peaks associated with structural type II collagen amides A, B, I, II, and III (Figure 4I). On the other hand, the FT-IR spectrum of dAMM exhibited peaks corresponding to type IV collagen (Figure 4J). Overall, the comprehensive characterization through histological, immunohistochemical, and FT-IR analyses confirmed the successful preservation of essential ECM components in both dACM and dAMM.

3.3. Proliferative Effect of Individually Soluble Extracts from dACM and dAMM on Human

After confirming the preservation of essential components relevant to the cartilage regeneration environment, the effects of both extracellular matrices (ECMs) on chondrocytes were analyzed to determine the concentration that elicits the greatest biological effect.

The assays revealed that the dACM and dAMM extracts can induce chondrocyte proliferation. At 24 and 48 h, no significant differences were observed compared to the negative control. However, after 72 h of culture, the dACM extract at all concentrations demonstrated the highest proliferation and viability rates in chondrocytes ($p < 0.001$), as depicted in Figure 5A. Similarly, dAMM showed no effect at 24 h, but at 48 h, treatment with dAMM extract at concentrations of 6 and 12 mg/mL increased the number of cells ($p < 0.05$). Notably, at 0.3 mg/mL, the cell count increased by almost 30%, resulting in significant differences compared to the control ($p < 0.001$). By 72 h, all evaluated concentrations showed a 40% increase in the number of cells compared to the control, as illustrated in Figure 5D ($p < 0.05$).

3.4. Evaluation of the Expression Level of Chondrogenesis-Related Markers

At 24 h, the dACM extract at concentrations of 0.15 and 1.5 mg/mL significantly increased SOX9 expression. At 48 h, a higher SOX9 expression was observed at concentrations of 1.5, 7.5, and 15 mg/mL compared to the untreated control group. Interestingly, at 72 h, the group treated with 1.5 mg/mL dACM showed the highest expression of SOX9, with levels almost three times higher than the untreated control group (Figure 5B).

Conversely, the dAMM extract increased SOX9 expression in all groups, with a significant difference observed at 6 mg/mL compared to the control. At 48 h, concentrations of 6 and 12 mg/mL significantly increased SOX9 expression by almost 6-fold. At 72 h, a decrease in SOX9 expression was further observed, although concentrations of 3 and 6 mg/mL still showed higher expression levels than the control group (Figure 5E).

Analysis of ACAN expression showed that only the dACM at 1.5 mg/mL increased the expression of this gene at 24, 48, and 72 h ($p < 0.05$) (Figure 5C). On the other hand, ACAN expression levels were increased in the presence of dAMM at 24 h, with all concentrations showing higher expression levels than the control group. Significant differences were observed at concentrations of 3 and 6 mg/mL. At 72 h, ACAN expression decreased, but concentrations of 3, 6, and 12 mg/mL still showed significantly higher expression levels compared to the control group. (Figure 5F).

Interestingly, the dACM extract did not induce RUNX2 expression in any of the treated groups. In contrast, the dAMM extract increased RUNX2 expression at 24 h, but this effect was not sustained at 48 and 72 h (Figure 5G).

3.5. Analysis the Combination of dACM and dAMM on Cell Proliferation and Chondrogenic Gene Expression

To assess the combined effect of dACM and dAMM, concentrations were selected based on previous experiments that demonstrated a higher proliferative index and increased levels of SOX9 and ACAN expression in chondrocytes. The concentrations chosen were 1.5 mg/mL of dACM and 6 mg/mL of dAMM. The mixture of these matrices significantly increased cell proliferation after 24, 48, and 72 h, showing notable differences compared to the controls ($p < 0.001$). Meanwhile, at 72 h, the dACM control exhibited a higher percentage of live cells than the mixture; the cell counts in the mixture remained consistently higher than the control (Figure 6A).

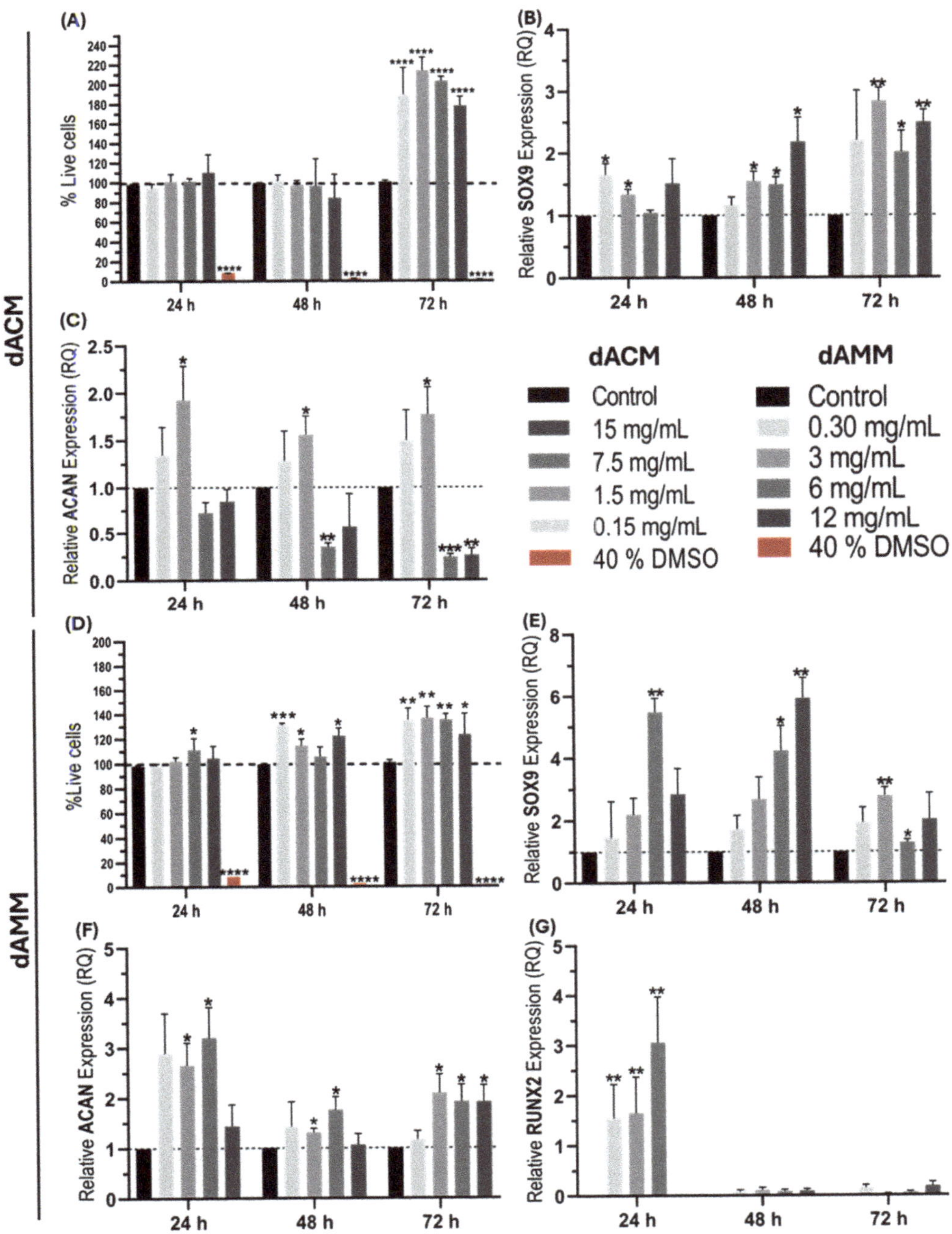

Figure 5. Effect of dACM and dAMM on in vitro chondrocytes culture: (**A**) effect of dACM on cell proliferation; (**B**) effect of dACM on SOX9 expression; (**C**) effect of dACM on ACAN expression; (**D**) effect of dAMM on cell proliferation; (**E**) effect of dAMM on SOX9 expression; (**F**) effect of dAMM on ACAN expression; (**G**) effect of dACM on RUNX expression. * = *t* test p value ≤ 0.05; ** = *t* test p value ≤ 0.005; *** = *t* test p value ≤ 0.001; **** = *t* test p value ≤ 0.0001.

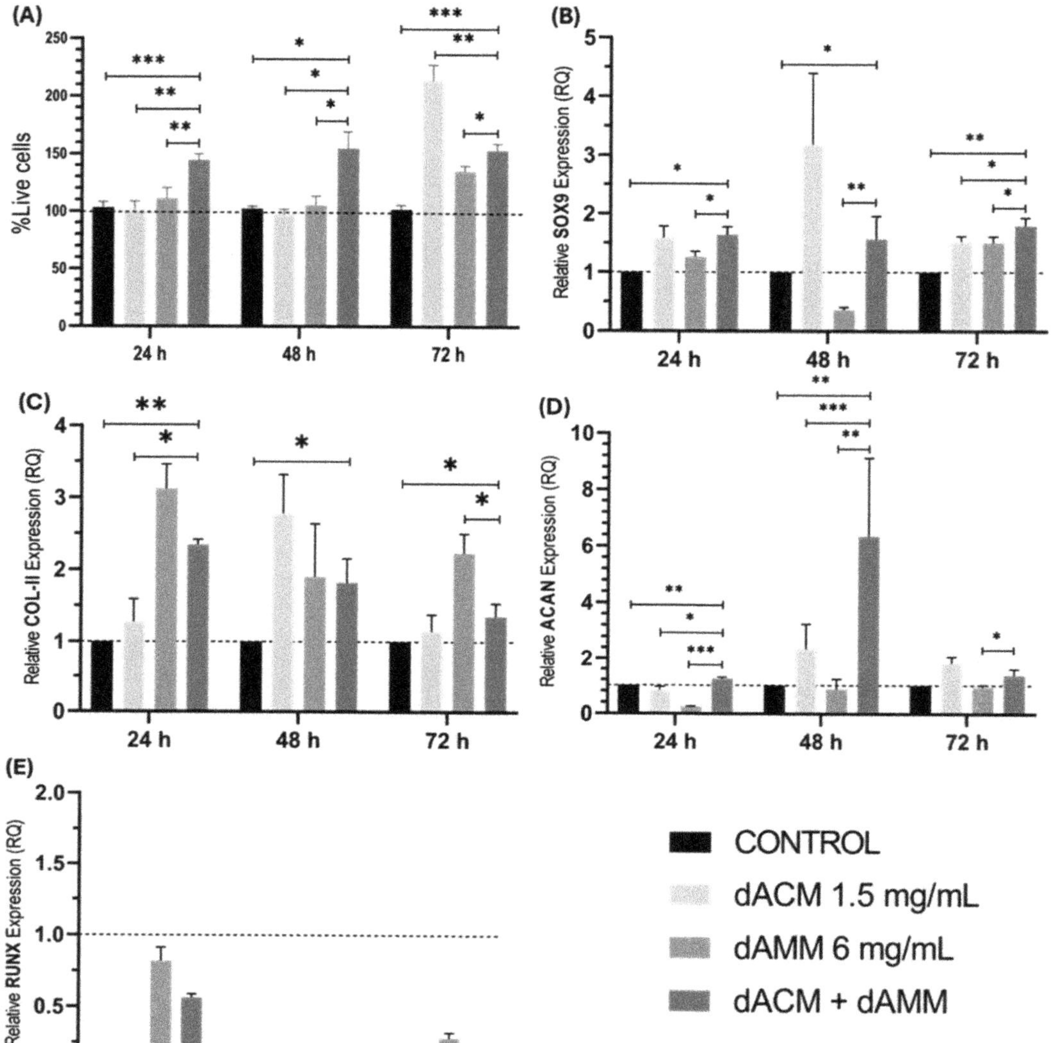

Figure 6. Effect of the mixture of dACM and dAMM on in vitro Chondrocytes culture: (**A**) effect of dACM + dAMM on cell proliferation; (**B**) effect of dACM + dAMM on SOX9 expression; (**C**) effect of dACM + dAMM on COL-II expression; (**D**) effect of dACM + dAMM on ACAN expression; (**E**) effect of dACM + dAMM on RUNX expression. * = *t* test p value ≤ 0.05; ** = *t* test p value ≤ 0.005; *** = *t* test p value ≤ 0.001.

In the presence of both matrices, SOX9 expression at 24 and 48 h surpassed that of groups treated with only dACM or dAMM, as well as the untreated control ($p < 0.05$). At 48 h, the mixed group displayed the highest SOX9 expression, significantly differing from both the control group and the group treated with dAMM alone. However, by 72 h, a higher SOX9 expression was observed after treatment with dAMM (Figure 6B). The combination of both matrices increased COL-II expression at 24, 48, and 72 h; however, this expression decreased with time, and at 72 h, the mixture showed significant differences with respect to the untreated control and dAMM group (Figure 6C). At 48 h, dAMM exhibited higher ACAN expression compared to dACM; nevertheless, by 72 h, the mixed group displayed the highest ACAN expression compared to the untreated control or either dACM or dAMM alone (Figure 6D). Notably, all experiments involving dACM and dAMM, either individually or in combination, showed reduced expression of RUNX, an osteogenic marker.

4. Discussion

In this study, we generated two decellularized biological matrices: one from articular cartilage (dACM) and the other from amniotic membrane (dAMM). Both matrices exhibited biological activity and a high content of biomolecules associated with cartilage generation.

Decellularized extracellular matrices (ECMs) have been extensively utilized in tissue engineering to treat tissue injuries by preventing scar formation and promoting regeneration [29]. They have garnered significant attention for their ability to maintain biological activity, preserve growth factors, and important proteins following chemical/physical decellularization processes. In tissue engineering, native-tissue-derived ECMs are often pulverized into powders that can be further processed into various forms, such as hydrogels, foams, nanofibers, coatings, and 3D-printed constructs [30]. This versatility in processing powdered ECM allows for the fabrication of scaffolds tailored to specific tissue-engineering needs, promoting tissue regeneration [3]. In this work, we assess the potential of blending dACM and dAMM for use in cartilage engineering.

A fundamental aspect of working with biological tissues is ensuring complete removal of cellular components. There are three essential criteria for successful tissue decellularization: (a) the decellularized ECM should contain less than 50 ng/mg dry tissue of dsDNA, (b) any remaining DNA fragments should be smaller than 200 base pairs, and (c) the decellularized ECM should show no visible nuclear material when stained with DAPI or hematoxylin and eosin [31,32]. In our study, dACM showed over 95% decellularization, with no observed nuclei in DAPI or H and E staining, consistent with other reported protocols using varying decellularization times and methods [26,27,33]. dAMM achieved almost complete removal (99.5%) of cellular components, similarly showing no visible nuclei in staining. However, to fully meet the third criterion, future studies should include DNA concentration analysis in the decellularized matrices, which was not performed in our study.

Decellularization techniques are crucial for removing cellular components while preserving ECM proteins and glycosaminoglycans [31,34], thereby maintaining the structural and functional properties essential for regenerative potential [35]. Furthermore, preservation of protein–GAG and protein–protein interactions during decellularization processes helps maintain ECM's native structure [31]. Our matrices retained major components such as type II collagen and aggrecan, as confirmed by histological staining and immunohistochemistry. While collagen preservation was nearly 100% in both matrices, a significant loss of aggrecan during decellularization was noted, likely due to its localization on hyaluronan molecules within the ECM 3D structure [36]. Nevertheless, retention levels exceeding 70% were observed in both dACM and dAMM, consistent with findings by Guo et al. [37], where collagen preservation was nearly complete, while glycosaminoglycan levels varied depending on the decellularization method [37,38]. The FT-IR analysis confirmed the presence of type II collagen in dACM and type IV collagen in dAMM.

Studies have highlighted the significant influence of ECM particle size on tissue formation and fate, independent of initial biomaterial composition [39,40]. For instance, nanometer-sized ECM particles have been proposed to enhance cellular responses under culture conditions, emphasizing particle size's role in cell proliferation, differentiation, and tissue development [41]. Our matrices exhibited heterogeneous particle sizes, beneficial for scaffold biomechanics, growth factor release, and tissue-derived protein interaction [42]. It was reported that the microarchitecture and bioactivity of porous scaffolds significantly influence cartilage-tissue-engineering outcomes, showing that the scaffold composition facilitated cartilage maturation and ECM secretion, and controllable pore sizes can induce cartilage regeneration [43].

When we analyzed the soluble extract of the decellularized matrices, we confirmed that our ECM-derived biomaterials have biological activity, as they were able to induce the proliferation of chondrocytes in culture after 48 h [16,23], which are associated with chondrocyte proliferation and maintenance of the chondrogenic lineage. This can be due to presence of soluble growth factors, such as bFGF, insulin-like growth factor-1 (IGF-1), vascular endothelial growth factor (VEGF), transforming growth factor-β1 (TGF-β1), bone morphogenic protein-2 (BMP-2), and growth differentiation factor 7 (GDF-7) in dACM or EGF, KGF, HGF, TGF-β1, TNF-α, EGF, IGF-1, PDGF–BB, HGF, VEGF, and bFGF, which can modulate the bioactivity of dAMM. According to a decellularization protocol applied to decellularized amniotic membrane, detergents, such as triton and hypotonic buffers, and NaOH do not affect the detection of structural proteins, as shown in this work [44,45]. However, the limitation of this study is the lacking characterization of soluble growth factors.

Our results demonstrated that dACM and dAMM were able to increase the expression levels of SOX9 and can maintain the chondrocyte cell lineage. The main component of hyaline cartilage is type 2 collagen; the combination of dACM and dAMM was shown to be able to improve the expression of COL-II in chondrocytes; however, although the dAMM group was shown to induce higher levels at the three measured times and dACM was shown to induce higher levels at 48 h, these show a tendency to decrease the expression with the passage of time; this could indicate that the combination of dACM and dAMM could generate more stable and maintainable results over long periods of time, but further tests are still needed to confirm this. Several authors report a significant increase in COL-II in mesenchymal stem cell cultures when exposed to decellularized matrices for 7 and 14 days [46–48]; however, this may differ from our results because we used an already differentiated cell line, which could influence the COL-II expression patterns, since chondrocytes do not go through a differentiation stage [49]; so, the bioactive components of the extracts could affect the cell line differently.

On the other hand, the expression levels of ACAN changed drastically and increased almost fourfold compared to the treatment with only one decellularized matrix (dAMM), showing an effect on ECM components synthesis. Another beneficial was used because one of the main problems with chondrocytes is that only young cell lines can be used since old cell lines tend to become undifferentiated [50]. Furthermore, as shown by the lack of RUNX2 expression, dACM alone or in combination with dAMM did not promote osteogenic differentiation. Regarding the production of cartilage components, dAMM induced higher levels of ACAN expression compared to dACM. This can be attributed to the higher amounts of TGF-β found in the amniotic membrane compared to articular cartilage [16,23]. Another important point to consider is that in several published works [27,47,48,51,52], decellularized matrices were directly added to scaffolds in 3D cultures. This can delay the release of active components of the matrices and regulate the supply of these active components; so, the effect on the cells is different. Meanwhile, in our approach, all active components are available in the 2D culture, which could explain why most of the genes that were measured (either in the presence of one or another decellularized matrix or both) presented a peak of expression at 48 h and subsequently decreased the expression.

Our results suggest that decellularized matrices derived from articular cartilage and amniotic membrane promote the proliferation and maintenance of the chondrogenic lineage, and the combination of both (dAMM and dACM) has greater potential in the engineering of cartilage tissue. Given their high collagen content, these materials could potentially serve as effective scaffolds, particularly in hydrogel formulations through simple enzymatic digestion. Future research should involve testing these materials in 3D in vitro models followed by in vivo studies

5. Conclusions

According to our results, ECM from articular cartilage and decellularized, pulverized, and freeze-dried amniotic membrane had heterogeneous particle sizes ranging from 1.6 to 600 μm for dAMM and 0.3 to 1700 μm for dACM, as well as preserved essential components of the ECM, such as collagen and ACAN. Biological activity using soluble extracts individually stimulated the proliferation of chondrocytes, and the expression of chondrogenic genes (SOX9, ACAN and COLII) at all evaluated times but were higher after 72 h. We observed the greatest effects with dACM at 1.5 mg/mL and 6 mg/mL for dAMM. The combination of dACM and dAMM showed synergistic effects, improved the expression of chondrogenic markers, and proliferated and increased the production of cartilage extracellular matrix components more than those used individually, suggesting that the combination of dACM and dAMM has potential for the cartilage regeneration.

Author Contributions: A.R.-M. contributed to designing and performing of experiments and the analysis and discussion of results. E.N.G.-T. helped to design the experiments and wrote, analyzed, and corrected the manuscript. J.L.-A. designed the experiments and performed literature analysis and discussion of results. H.L.-G. analyzed images of the SEM and FT-IR. R.F.-M. helped in the standardization of immunohistochemistry and with the software of the fluorescence microscope. N.K.M.-S. edited the text, analyzed the results, and reviewed the final manuscript. A.G.-L. helped analyze and correct the manuscript and performed the isolation of the amniotic membrane. F.V.-C. helped to write, analyze, and correct the manuscript and performed the isolation of the articular cartilage. M.S.-M. contributed to the design of experiments, analysis of the results, and the discussion and correction of the manuscript. A.R.-M., E.N.G.-T. and M.S.-M. confirm the authenticity of all the raw data. All authors have read and agreed to the published version of the manuscript.

Funding: This research received no external funding.

Institutional Review Board Statement: The study was conducted in accordance with the Declaration of Helsinki, and approved by the Ethics Committee of the School of Medicine, Universidad Autónoma de Nuevo Leon, (BI23-001, approval on 1 March 2023).

Informed Consent Statement: Written informed consent has been obtained from the patients to publish this paper.

Data Availability Statement: Data is contained within the article.

Conflicts of Interest: The authors declare no conflicts of interest.

References

1. Blache, U.; Stevens, M.M.; Gentleman, E. Harnessing the Secreted Extracellular Matrix to Engineer Tissues. *Nat. Biomed. Eng.* **2020**, *4*, 357–363. [CrossRef]
2. Seixas, M.J.; Martins, E.; Reis, R.L.; Silva, T.H. Extraction and Characterization of Collagen from Elasmobranch Byproducts for Potential Biomaterial Use. *Mar. Drugs* **2020**, *18*, 617. [CrossRef]
3. Golebiowska, A.A.; Intravaia, J.T.; Sathe, V.M.; Kumbar, S.G.; Nukavarapu, S.P. Decellularized Extracellular Matrix Biomaterials for Regenerative Therapies: Advances, Challenges and Clinical Prospects. *Bioact. Mater.* **2024**, *32*, 98–123. [CrossRef] [PubMed]
4. Luo, R.; Hu, R.; Xu, J.; Yu, P.; Wu, X.; Zhe, M.; Liu, M.; Xing, F.; Xiang, Z.; Zhou, C.; et al. Decellularized Extracellular Matrix as a Promising Biomaterial for Musculoskeletal Tissue Regeneration. *Nanotechnol. Rev.* **2023**, *12*, 20230151. [CrossRef]
5. Wang, H.; Yu, H.; Zhou, X.; Zhang, J.; Zhou, H.; Hao, H.; Ding, L.; Li, H.; Gu, Y.; Ma, J.; et al. An Overview of Extracellular Matrix-Based Bioinks for 3D Bioprinting. *Front. Bioeng. Biotechnol.* **2022**, *10*, 905438. [CrossRef] [PubMed]
6. Song, M.; Wang, W.; Ye, Q.; Bu, S.; Shen, Z.; Zhu, Y. The Repairing of Full-Thickness Skin Deficiency and Its Biological Mechanism Using Decellularized Human Amniotic Membrane as the Wound Dressing. *Mater. Sci. Eng. C* **2017**, *77*, 739–747. [CrossRef] [PubMed]
7. Nakamura, N.; Kimura, T.; Kishida, A. Overview of the Development, Applications, and Future Perspectives of Decellularized Tissues and Organs. *ACS Biomater. Sci. Eng.* **2017**, *3*, 1236–1244. [CrossRef] [PubMed]
8. Ijima, H.; Nakamura, S.; Bual, R.; Shirakigawa, N.; Tanoue, S. Physical Properties of the Extracellular Matrix of Decellularized Porcine Liver. *Gels* **2018**, *4*, 39. [CrossRef] [PubMed]
9. Hussein, K.H.; Saleh, T.; Ahmed, E.; Kwak, H.H.; Park, K.M.; Yang, S.R.; Kang, B.J.; Choi, K.Y.; Kang, K.S.; Woo, H.M. Biocompatibility and Hemocompatibility of Efficiently Decellularized Whole Porcine Kidney for Tissue Engineering. *J. Biomed. Mater. Res. A* **2018**, *106*, 2034–2047. [CrossRef]

10. Napierala, H.; Hillebrandt, K.H.; Haep, N.; Tang, P.; Tintemann, M.; Gassner, J.; Noesser, M.; Everwien, H.; Seiffert, N.; Kluge, M.; et al. Engineering an Endocrine Neo-Pancreas by Repopulation of a Decellularized Rat Pancreas with Islets of Langerhans. *Sci. Rep.* **2017**, *7*, 41777. [CrossRef]
11. Ozudogru, E.; Isik, M.; Eylem, C.C.; Nemutlu, E.; Arslan, Y.E.; Derkus, B. Decellularized Spinal Cord Meninges Extracellular Matrix Hydrogel That Supports Neurogenic Differentiation and Vascular Structure Formation. *J. Tissue Eng. Regen. Med.* **2021**, *15*, 948–963. [CrossRef] [PubMed]
12. Song, H.; Yin, Z.; Wu, T.; Li, Y.; Luo, X.; Xu, M.; Duan, L.; Li, J. Enhanced Effect of Tendon Stem/Progenitor Cells Combined with Tendon-Derived Decellularized Extracellular Matrix on Tendon Regeneration. *Cell Transpl.* **2018**, *27*, 1634–1643. [CrossRef]
13. Farag, A.; Vaquette, C.; Hutmacher, D.W.; Bartold, P.M.; Ivanovski, S. Fabrication and Characterization of Decellularized Periodontal Ligament Cell Sheet Constructs. *Methods Mol. Biol.* **2017**, *1537*, 403–412. [CrossRef]
14. Kafili, G.; Niknejad, H.; Tamjid, E.; Simchi, A. Amnion-Derived Hydrogels as a Versatile Platform for Regenerative Therapy: From Lab to Market. *Front. Bioeng. Biotechnol.* **2024**, *12*, 1358977. [CrossRef]
15. Hu, Z.; Luo, Y.; Ni, R.; Hu, Y.; Yang, F.; Du, T.; Zhu, Y. Biological Importance of Human Amniotic Membrane in Tissue Engineering and Regenerative Medicine. *Mater. Today Bio* **2023**, *22*, 100790. [CrossRef]
16. Zhang, Q.; Hu, Y.; Long, X.; Hu, L.; Wu, Y.; Wu, J.; Shi, X.; Xie, R.; Bi, Y.; Yu, F.; et al. Preparation and Application of Decellularized ECM-Based Biological Scaffolds for Articular Cartilage Repair: A Review. *Front. Bioeng. Biotechnol.* **2022**, *10*, 908082. [CrossRef] [PubMed]
17. Neishabouri, A.; Soltani Khaboushan, A.; Daghigh, F.; Kajbafzadeh, A.M.; Majidi Zolbin, M. Decellularization in Tissue Engineering and Regenerative Medicine: Evaluation, Modification, and Application Methods. *Front. Bioeng. Biotechnol.* **2022**, *10*, 805299. [CrossRef] [PubMed]
18. Li, Y.; Xu, Y.; Liu, Y.; Wang, Z.; Chen, W.; Duan, L.; Gu, D. Decellularized Cartilage Matrix Scaffolds with Laser-Machined Micropores for Cartilage Regeneration and Articular Cartilage Repair. *Mater. Sci. Eng. C Mater. Biol. Appl.* **2019**, *105*, 110139. [CrossRef]
19. Gvaramia, D.; Kern, J.; Jakob, Y.; Tritschler, H.; Brenner, R.E.; Breiter, R.; Kzhyshkowska, J.; Rotter, N. Modulation of the Inflammatory Response to Decellularized Collagen Matrix for Cartilage Regeneration. *J. Biomed. Mater. Res. A* **2022**, *110*, 1021–1035. [CrossRef]
20. Guo, W.; Zheng, X.; Zhang, W.; Chen, M.; Wang, Z.; Hao, C.; Huang, J.; Yuan, Z.; Zhang, Y.; Wang, M.; et al. Mesenchymal Stem Cells in Oriented PLGA/ACECM Composite Scaffolds Enhance Structure-Specific Regeneration of Hyaline Cartilage in a Rabbit Model. *Stem Cells Int.* **2018**, *2018*, 6542198. [CrossRef]
21. Liang, J.; Liu, P.; Yang, X.; Liu, L.; Zhang, Y.; Wang, Q.; Zhao, H. Biomaterial-Based Scaffolds in Promotion of Cartilage Regeneration: Recent Advances and Emerging Applications. *J. Orthop. Transl.* **2023**, *41*, 54–62. [CrossRef] [PubMed]
22. Fénelon, M.; Catros, S.; Meyer, C.; Fricain, J.C.; Obert, L.; Auber, F.; Louvrier, A.; Gindraux, F. Applications of Human Amniotic Membrane for Tissue Engineering. *Membranes* **2021**, *11*, 387. [CrossRef]
23. Macečková, Z.; Pergner, J.; Krbec, M.; Urban, M.; Zahradníček, M. Application of Amniotic Membrane in Osteoarthritis Management. *J. Cartil. Jt. Preserv.* **2024**, 100174. [CrossRef]
24. Cao, L.; Tong, Y.; Wang, X.; Zhang, Q.; Qi, Y.; Zhou, C.; Yu, X.; Wu, Y.; Miao, X. Effect of Amniotic Membrane/Collagen-Based Scaffolds on the Chondrogenic Differentiation of Adipose-Derived Stem Cells and Cartilage Repair. *Front. Cell Dev. Biol.* **2021**, *9*, 647166. [CrossRef] [PubMed]
25. Pérez-Silos, V.; Moncada-Saucedo, N.K.; Peña-Martínez, V.; Lara-Arias, J.; Marino-Martínez, I.A.; Camacho, A.; Romero-Díaz, V.J.; Banda, M.L.; García-Ruiz, A.; Soto-Dominguez, A.; et al. A Cellularized Biphasic Implant Based on a Bioactive Silk Fibroin Promotes Integration and Tissue Organization during Osteochondral Defect Repair in a Porcine Model. *Int. J. Mol. Sci.* **2019**, *20*, 5145. [CrossRef]
26. Ballesteros, A.C.V.; Puello, H.R.S.; Lopez-Garcia, J.A.; Bernal-Ballen, A.; Mosquera, D.L.N.; Forero, D.M.M.; Charry, J.S.S.; Bejarano, Y.A.N. Bovine Decellularized Amniotic Membrane: Extracellular Matrix as Scaffold for Mammalian Skin. *Polymers* **2020**, *12*, 590. [CrossRef]
27. Chang, C.H.; Chen, C.C.; Liao, C.H.; Lin, F.H.; Hsu, Y.M.; Fang, H.W. Human Acellular Cartilage Matrix Powders as a Biological Scaffold for Cartilage Tissue Engineering with Synovium-Derived Mesenchymal Stem Cells. *J. Biomed. Mater. Res. A* **2014**, *102*, 2248–2257. [CrossRef]
28. Adamowicz, J.; Pokrywczyńska, M.; Tworkiewicz, J.; Kowalczyk, T.; van Breda, S.V.; Tyloch, D.; Kloskowski, T.; Bodnar, M.; Skopinska-Wisniewska, J.; Marszałek, A.; et al. New Amniotic Membrane Based Biocomposite for Future Application in Reconstructive Urology. *PLoS ONE* **2016**, *11*, e0146012. [CrossRef]
29. Edgar, L.; Altamimi, A.; García Sánchez, M.; Tamburrinia, R.; Asthana, A.; Gazia, C.; Orlando, G. Utility of Extracellular Matrix Powders in Tissue Engineering. *Organogenesis* **2018**, *14*, 172. [CrossRef]
30. Xing, Q.; Qian, Z.; Jia, W.; Ghosh, A.; Tahtinen, M.; Zhao, F. Natural Extracellular Matrix for Cellular and Tissue Biomanufacturing. *ACS Biomater. Sci. Eng.* **2017**, *3*, 1462–1476. [CrossRef]
31. Crapo, P.M.; Gilbert, T.W.; Badylak, S.F. An Overview of Tissue and Whole Organ Decellularization Processes. *Biomaterials* **2011**, *32*, 3233–3243. [CrossRef]
32. Barbulescu, G.I.; Bojin, F.M.; Ordodi, V.L.; Goje, I.D.; Barbulescu, A.S.; Paunescu, V. Decellularized Extracellular Matrix Scaffolds for Cardiovascular Tissue Engineering: Current Techniques and Challenges. *Int. J. Mol. Sci.* **2022**, *23*, 13040. [CrossRef] [PubMed]

33. Chen, D.; Zhang, Y.; Lin, Q.; Chen, D.; Li, X.; Dai, J.; Sun, Y. The Effect of Cartilage Decellularized Extracellular Matrix-Chitosan Compound on Treating Knee Osteoarthritis in Rats. *PeerJ* **2021**, *9*, e12188. [CrossRef] [PubMed]
34. Shin, Y.J.; Shafranek, R.T.; Tsui, J.H.; Walcott, J.; Nelson, A.; Kim, D.H. 3D Bioprinting of Mechanically Tuned Bioinks Derived from Cardiac Decellularized Extracellular Matrix. *Acta Biomater.* **2021**, *119*, 75–88. [CrossRef]
35. Noro, J.; Vilaça-Faria, H.; Reis, R.L.; Pirraco, R.P. Extracellular Matrix-Derived Materials for Tissue Engineering and Regenerative Medicine: A Journey from Isolation to Characterization and Application. *Bioact. Mater.* **2024**, *34*, 494–519. [CrossRef]
36. Kiani, C.; Chen, L.; Wu, Y.J.; Yee, A.J.; Yang, B.B. Structure and Function of Aggrecan. *Cell Res.* **2002**, *12*, 19–32. [CrossRef]
37. Guo, P.; Jiang, N.; Mini, C.; Miklosic, G.; Zhu, S.; Vernengo, A.J.; D'Este, M.; Grad, S.; Alini, M.; Li, Z. Decellularized Extracellular Matrix Particle-Based Biomaterials for Cartilage Repair Applications. *J. Mater. Sci. Technol.* **2023**, *160*, 194–203. [CrossRef]
38. Fernández-Pérez, J.; Ahearne, M. The Impact of Decellularization Methods on Extracellular Matrix Derived Hydrogels. *Sci. Rep.* **2019**, *9*, 14933. [CrossRef]
39. Szwed-Georgiou, A.; Płociński, P.; Kupikowska-Stobba, B.; Urbaniak, M.M.; Rusek-Wala, P.; Szustakiewicz, K.; Piszko, P.; Krupa, A.; Biernat, M.; Gazińska, M.; et al. Bioactive Materials for Bone Regeneration: Biomolecules and Delivery Systems. *ACS Biomater. Sci. Eng.* **2023**, *9*, 5222–5254. [CrossRef]
40. Cramer, M.C.; Badylak, S.F. Extracellular Matrix-Based Biomaterials and Their Influence Upon Cell Behavior. *Ann. Biomed. Eng.* **2020**, *48*, 2132. [CrossRef]
41. Zahiri, S.; Masaeli, E.; Poorazizi, E.; Nasr-Esfahani, M.H. Chondrogenic Response in Presence of Cartilage Extracellular Matrix Nanoparticles. *J. Biomed. Mater. Res. A* **2018**, *106*, 2463–2471. [CrossRef] [PubMed]
42. Gresham, R.C.H.; Bahney, C.S.; Leach, J.K. Growth Factor Delivery Using Extracellular Matrix-Mimicking Substrates for Musculoskeletal Tissue Engineering and Repair. *Bioact. Mater.* **2021**, *6*, 1945. [CrossRef] [PubMed]
43. Zhang, Q.; Lu, H.; Kawazoe, N.; Chen, G. Pore Size Effect of Collagen Scaffolds on Cartilage Regeneration. *Acta Biomater.* **2014**, *10*, 2005–2013. [CrossRef]
44. Dai, C.; Liu, Y. Hepatocyte Growth Factor Antagonizes the Profibrotic Action of TGF-Beta1 in Mesangial Cells by Stabilizing Smad Transcriptional Corepressor TGIF. *J. Am. Soc. Nephrol.* **2004**, *15*, 1402–1412. [CrossRef]
45. Liu, Z.; Zhu, X.; Zhu, T.; Tang, R. Evaluation of a Biocomposite Mesh Modified with Decellularized Amniotic Membrane for Intraperitoneal Onlay Mesh Repair. *ACS Omega* **2020**, *5*, 3550. [CrossRef]
46. Hanai, H.; Jacob, G.; Nakagawa, S.; Tuan, R.S.; Nakamura, N.; Shimomura, K. Potential of Soluble Decellularized Extracellular Matrix for Musculoskeletal Tissue Engineering—Comparison of Various Mesenchymal Tissues. *Front. Cell Dev. Biol.* **2020**, *8*, 581972. [CrossRef] [PubMed]
47. Luo, L.; Eswaramoorthy, R.; Mulhall, K.J.; Kelly, D.J. Decellularization of Porcine Articular Cartilage Explants and Their Subsequent Repopulation with Human Chondroprogenitor Cells. *J. Mech. Behav. Biomed. Mater.* **2016**, *55*, 21–31. [CrossRef]
48. Cheng, N.C.; Estes, B.T.; Young, T.H.; Guilak, F. Genipin-Crosslinked Cartilage-Derived Matrix as a Scaffold for Human Adipose-Derived Stem Cell Chondrogenesis. *Tissue Eng. Part. A* **2013**, *19*, 484. [CrossRef]
49. Son, Y.B.; Jeong, Y.I.; Jeong, Y.W.; Hossein, M.S.; Olsson, P.O.; Tinson, A.; Singh, K.K.; Lee, S.Y.; Hwang, W.S. Cell Source-Dependent In Vitro Chondrogenic Differentiation Potential of Mesenchymal Stem Cell Established from Bone Marrow and Synovial Fluid of Camelus Dromedarius. *Animals* **2021**, *11*, 1918. [CrossRef]
50. Lee, J.; Lee, J.Y.; Chae, B.C.; Jang, J.; Lee, E.A.; Son, Y. Fully Dedifferentiated Chondrocytes Expanded in Specific Mesenchymal Stem Cell Growth Medium with FGF2 Obtains Mesenchymal Stem Cell Phenotype In Vitro but Retains Chondrocyte Phenotype In Vivo. *Cell Transpl.* **2017**, *26*, 1673. [CrossRef]
51. Cha, M.H.; Do, S.H.; Park, G.R.; Du, P.; Han, K.C.; Han, D.K.; Park, K. Induction of Re-Differentiation of Passaged Rat Chondrocytes Using a Naturally Obtained Extracellular Matrix Microenvironment. *Tissue Eng. Part A* **2013**, *19*, 978–988. [CrossRef] [PubMed]
52. Youngstrom, D.W.; Cakstina, I.; Jakobsons, E. Cartilage-Derived Extracellular Matrix Extract Promotes Chondrocytic Phenotype in Three-Dimensional Tissue Culture. *Artif. Cells Nanomed. Biotechnol.* **2016**, *44*, 1040–1047. [CrossRef] [PubMed]

coatings

MDPI

Article

In Vitro Evaluation of Optimized PEEK Surfaces for Enhanced Osseointegratio

Kadie Nobles [1], Amol V. Janorkar [1], Michael D. Roach [1], Lawrence Walker [2] and Randall Scott Williamson [1,*]

[1] University of Mississippi Medical Center, Department of Biomedical Materials Science, Jackson, MS 39216, USA; kadiepnobles@gmail.com (K.N.); ajanorkar@umc.edu (A.V.J.); mroach@umc.edu (M.D.R.)

[2] Zavation Medical Products, LLC, Flowood, MS 39232, USA; lawrence.walker@zavation.com

* Correspondence: rwilliamson@umc.edu

Abstract: The materials traditionally used for implant applications, such as titanium alloys, cobalt chromium, and zirconium, often require surface modifications to achieve the desired osseointegration. These materials still have the problematic stress-shielding effect. To limit stress shielding, PEEK is the superior alternative to fulfill implant needs. However, the traditional methods of modifying and functionalizing the surface of PEEK are often expensive, time consuming, and are not easily translated into commercialization. Sulfonation is a process, which is dependent on controllable factors. Thus far, no research has been performed to optimize the sulfonation process. Our data suggest that the process factors can be controlled and optimized. Cellular activity was examined on the optimized PEEK surfaces through testing with pre-osteoblast MC3T3-E1 cells through cell viability (MTT assay), cell proliferation (DNA assay), cell differentiation (ALP assay), and cell mineralization (Alizarin red assay). Overall, sulfonated and heat-treated PEEK exhibited a statistically significant increase in DNA content over the course of 21 days, indicating more cell proliferation and viability for that surface. *In vitro* testing results showed that the optimized sulfonated and heat-treated PEEK exhibited superior cell proliferation and mineralization performance over smooth PEEK and sulfonated-only PEEK.

Keywords: surface modification; PEEK; osseointegration; sulfonation; porosity

1. Introduction

The need for orthopedic implants to treat musculoskeletal disorders has been on the rise and is expected to continue increasing as the age of the population increases. There was a 113% increase in spinal fusion procedures from 1998 to 2011 and a total of 1,390,611 spinal procedures in 2011 according to the HCUP Nationwide Inpatient Sample (NIS) [1]. However, up to 10% of these implants experience failure due to causes such as poor osseointegration, infection, and failure/breakage [2–5]. Osseointegration is the direct connection between living bone and the surface of an implant and is crucial for the long-term success of an implant [6]. Successful osseointegration into an implant interface is mainly influenced by material stiffness and porosity ranging in size from 100 to 600 μm; thus, it is critical to find the appropriate material and apply a surface modification, which will be conducive to enhanced osseointegration [6–8].

Polyetheretherketone (PEEK) is a semi-crystalline non-resorbable polymer, which is radiolucent, biocompatible, and resistant to most chemicals. PEEK has gained popularity as an alternative replacement for titanium-based implants in orthopedic applications, such as spine surgery, prosthodontics surgery, and maxilla-facial surgery, over recent decades [9–13]. PEEK has an elastic modulus of 3–8 GPa, which is similar to the 7–30 GPa range of human cortical bone; therefore, the problem of stress shielding often seen with titanium implants, which have an elastic modulus of 55–110 GPa, is diminished when a PEEK-based material is utilized [13–18]. One disadvantage of utilizing PEEK as an implant material is that the material is bioinert and requires surface modifications to elicit any desirable cellular responses [9,11,19–22].

Osseointegration is influenced by surface features such as roughness and wettability and is favorable to porous surfaces because they promote cellular attachment and allow cell infiltration. Osteoblasts range in size from 10 to 50 µm, but the typical pore sizes used for orthopedic applications, which have shown propitious osseointegration, are in the size range of 100–400 µm [8,23–27]. Surface roughness on the micro-scale, typically noted as 1–50 µm, can enhance cell adhesion and osseointegration by increasing early mechanical interlocking of the osteoblast cells, and roughness on the nano-scale of 1–1000 nm can act to signal cellular attachment and differentiation, since the roughness of natural bone is around 32 nm [23–26,28–30]. It is thought that a more hydrophilic surface, which allows for increased protein adsorption and favorable protein conformations, will improve cell attachment [23,31,32]. However, studies of how wettability influences osseointegration have produced contradicting results in the literature, which shows surfaces ranging from superhydrophobic (water contact angle >150°) to superhydrophilic (water contact angle approaching 0°) to be optimal for attachment and growth of osteoblasts [31,33–36]. The literature tends to point toward wettability being a secondary influence on osseointegration, whereas features such as porosity, roughness, and coatings are the primary influences [32,35,37,38].

Sulfonation, which is the use of sulfuric acid to etch a surface, has long been performed as a method to modify aromatic polymers for the production of proton-exchange membranes for technologies such as polymer fuel cells and electrodialysis [39–41]. The porous network left behind on the surface of the sulfonated PEEK has encouraged researchers to utilize the technology for biomedical applications [42–45]. PEEK sulfonation is an electrophilic reaction in which the hydroquinone unit benzene-1,4-diol, beside the ether bridge, is sulfonated, leaving behind a sulfonic acid, the -SO_3H group, as seen in Figure 1 [39,41,44].

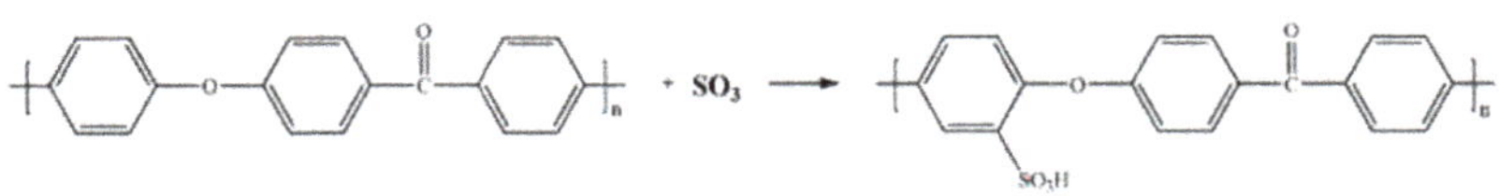

Figure 1. Chemical reaction for creating sulfonated PEEK [46].

This sulfuric acid etching creates a porous structure on the PEEK surface, which can be conducive to cellular attachment; however, the residual -SO_3H group left behind can be toxic to an *in vitro*/*in vivo* environment, thus needing to be removed through a de-sulfonation process. Hydrothermal treatment with water is a method, which can be used to remove the residual -SO_3H groups, as seen in Figure 2. In the presence of water and heat, the water molecules react with hydrogen (H) on the sulfur trioxide (SO_3) group, creating hydronium (H_3O), which reacts with the benzene ring, causing SO_3 to detach and leave as a gas when heated to its boiling point of 45 °C [47].

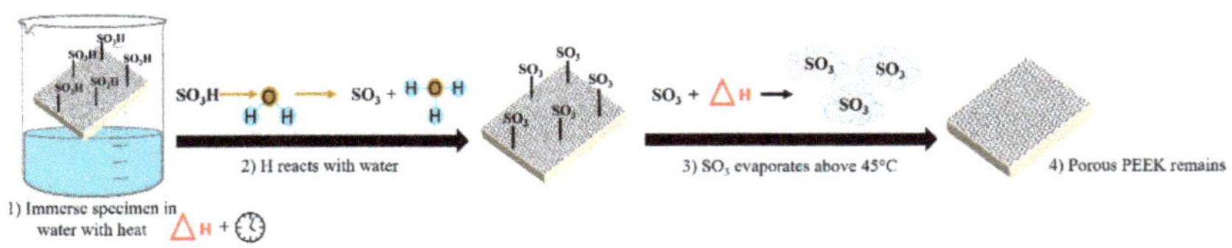

Figure 2. Schematic representation of what happens to the residual sulfur groups during hydrothermal treatment.

The purpose of this study was to investigate the potential effects sulfonation processing may have on modifying the surface of PEEK for enhanced osseointegration. Design of experiment (DOE) methods were performed to optimize surface features and surface porosity on PEEK. A hydrothermal treatment process was also utilized on the optimized sulfonated PEEK surface to remove the residual sulfocompound groups to render the surfaces advantageous to cellular attachment, proliferation, and differentiation. After successful optimization of the sulfonation process, an *in vitro* study with pre-osteoblast MC3T3-E1 cells was examined on the optimized PEEK surfaces though cell viability (MTT assay), cell proliferation (DNA assay), cell differentiation (ALP assay), and cell mineralization (Alizarin red assay).

2. Materials and Methods

2.1. Materials

Medical-grade PEEK rods with approximately ½″ diameters were provided by Zavation Medical Products, LLC (Flowood, MS, USA). Concentrated sulfuric acid (95%–98%) was purchased from Sigma-Aldrich (Burlington, MA, USA). Silicon carbide grinding paper, grit 320, was purchased from Struers (Cleveland, OH, USA) for preparation of the PEEK surfaces before sulfonation.

2.2. Specimen Preparation

Disk-shaped specimens with a thickness of approximately 4 mm and a diameter of approximately 12.5 mm were cut from the PEEK rods using a Struers Accutom-50 sectioning saw (Cleveland, OH, USA). The disk specimens were then mounted in bakelite (Struers Citopress-20, Cleveland, OH, USA) and subsequently grinded for 15 s using a 320-grit silicon carbide grinding paper, washed, and grinded again for 15 s (Struers TegraPol, Cleveland, OH, USA). The specimens were removed from the mount and underwent a 5-stage wash cycle, consisting of the following: ultrasonic clean in Alconox® (Alconox Inc., White Plains, NY, USA) for 5 min, rinse with distilled water, ultrasonic clean in distilled water for 5 min, rinse with ethanol, and rinse with distilled water. The specimens were then allowed to air dry at room temperature and stored until further use.

2.3. Experimental Design—Optimizing the Sulfonated Surface

An altered design of experiment (DOE) approach was utilized to optimize the surface porosity on the PEEK specimen. Minitab® software Version 20.1.3.0 (State College, PA, USA) was used to generate some of the experimental designs; however, because some specimens early in the optimization process did not have measurable features (i.e., pore size), a qualitative approach to narrowing down the sulfonation processing factors had to be adopted. Therefore, DOE was employed to help design the experiments when many factors were under consideration, and then, intellectual intuition was used to narrow down the sulfonation factors until measurable porosity could be attained. Hence, the authors refer to this as an altered DOE approach. It took a total of six separate designs to ultimately reach the optimized surface, and those design parameters are shown in Table 1. The following sulfonation factors were identified as the most critical to investigate for this study: pre-surface condition, soak condition, acid concentration, soak time, and soak temperature. Pre-surface condition refers to the roughness of the sulfonated surface obtained by using different grit sized silicon carbide grinding paper. There are four soak conditions mentioned in Table 1, which are stirring (S), no stirring (NS), sonication only (SO), and sonication plus stirring (SS). Acid concentration alludes to the concentration of the sulfuric acid used for the sulfonation.

After Design #1 experimentation, many of the specimens did not have measurable porosity; therefore, it was not useful to put the responses in the DOE and narrow down the significant factors in that manner. Therefore, the authors examined each specimen qualitatively and made decisions on what factors to adjust or eliminate moving forward. The results from Design #1 indicated that a rougher surface was more conducive to initiating sulfonation on the PEEK surface; therefore, the high factor of 1200-grit grinding paper was eliminated, and stirring was also eliminated as a factor due to the creation of streaks on the sulfonated surface and addition of additional variables to control for in the process.

Other factors, such as time and temperature, were also adjusted following Design #1. The soak time high of 10 min resulted in zero pores on the surface and made the surface appear melted; therefore, the authors chose to reduce the total soak time parameters significantly moving forward. Additionally, acid concentrations lower than fully concentrated sulfuric acid did not react with the PEEK surfaces to create porosity.

Table 1. Factors and related levels for each design space for optimization of surface porosity.

Design Space	Pre-Surface Condition (Grit)		Soak Condition	Acid Conc.	Soak Time (min)			Soak Temp. (°C)		
	Low	High	S/NS/SO/SS	%	Low	Center	High	Low	Center	High
Design #1	220	1200	S/NS	80–100	1	-	10	22	-	60
Design #2	220	-	NS	100	1	2.5	4	50	65	80
Design #3	220	-	NS	100	2.5	3.75	5	55	65	80
Design #4	320	-	SO	100	1	2.5	5	55	60	65
Design #5	320	-	SO/SS	100	1	1.45	2.5	55	60	65
Design #6	320	-	SO/SS	100	1	-	-	65	-	-

Design #2 and Design #3 were used to further refine the soak time and soak temperature. In between Design #3 and Design #4, the authors discovered that it would be fitting to look at a pre-surface condition, which more closely matched that of the specimen types, which would be generated by a manufacturer. A brief surface roughness side study was conducted to try and match the PEEK pre-surface condition to that of as-manufactured solid PEEK, and the authors discovered that a 320-grit grinding paper was optimal for that purpose and was therefore used going forward for Design #4–Design #6. Additionally, Minitab® was no longer used for generating designs after Design #3 was finished, and instead, designs were created based on intellectual intuition. During the design process, a beneficial method for removing the sulfuric acid and cleaning the specimen after soaking was discovered. This method involved the use of compressed air to lightly blast the acid off the surface, which enabled retention of the pores that formed during the sulfonation process. It was discovered through experimentation that the pores formed during sulfonation could be disrupted when the specimen was cleaned after soaking due to the exothermic reaction, which occurs when sulfuric acid is mixed with water.

Designs #4–#6 provided features, which were quantifiable, and thus, ImageJ software Version 1.53e (National Institutes of Health, Bethesda, MD, USA) was used to extract the pore size data for comparison between the specimens and factors.

2.4. Optimizing the Hydrothermal Treatment

Hydrothermal treatment experimentation began after optimized sulfonation factors were chosen, which provided a surface with a desirable porosity. The purpose of the hydrothermal treatment experiments was to provide an optimal range in which most of the residual sulfocompound (sulfur) groups were no longer present on the optimized sulfonated PEEK surface. The experiment was designed around a soak time low of 60 min and high of 90 min, and a soak temperature low of 45 °C and high of 80 °C, as listed in Table 2. A total of two designs were conducted for the hydrothermal treatment. The hydrothermally treated specimens were sulfonated according to the optimized parameters identified in Design #6 above and further detailed in the Results and Discussion sections. Water was continuously stirred during the hydrothermal treatment. Low and high values for the factors were initially chosen based on values originally cited in the literature [43,44]. The second design was chosen based on results from the first design. Fourier transform infrared spectroscopy (FTIR, Perkin-Elmer, Waltham, MA, USA) with a diamond/ZnSe crystal at a resolution of 4 cm^{-1} and scanned from 650 to 4000 cm^{-1} was utilized on smooth PEEK, sulfonated-only PEEK (sfPEEK), and sulfonated and heat-treated PEEK (sfPEEK-HT) to identify the functional groups of interest related to the successful removal of the residual sulfocompound groups from the optimized sulfonated PEEK surface.

Table 2. Designs and related levels for the hydrothermal treatment experiments on the optimized sulfonated PEEK surfaces.

	Specimen ID	Soak Time (min)	Soak Temp. (°C)
Design #1	H1	90	45
	H2	60	80
	H3 [a]	0	0
	H4	75	63
	H5	90	80
	H6 [b]	60	45
Design #2	sfPEEK—2 h HT	120 (2 h)	45
	sfPEEK—3 h HT	180 (3 h)	
	sfPEEK—4 h HT	240 (4 h)	
	sfPEEK—5 h HT	300 (5 h)	

[a] H3 is used as the control specimen, i.e., sulfonated PEEK without HT treatment. H3 and sfPEEK refer to the same condition when used in this paper. [b] The conditions used for H6 are the same as those used for sfPEEK—1 h HT in future sections.

2.5. Surface and Physical Characterization

2.5.1. Digital Imaging and Pore Size Measurements

All sulfonated surfaces were imaged and documented using a VHX digital microscope and its corresponding software (Keyence Corp., Osaka, Japan). Images were optimized using the contrasting and brightening features in Keyence before being processed using ImageJ software. ImageJ was used to measure the pore size and distribution data from the surfaces when applicable.

2.5.2. Atomic Force Microscopy

Atomic force microscopy (AFM, Bioscope Catalyst, Bruker, Santa Barbara, CA, USA) was performed on the optimized sulfonated and heat-treated specimens to determine the resulting surface roughness (Ra) values. The specimens were created in three batches (S1, S2, and S3) with n = 2 (A and B) for each batch. Scans with 50 μm × 50 μm area were acquired in ScanAssyst mode (0.100–0.25 Hz, and 512–256 samples/line) and further analyzed using Gwyddion software (version 2.41).

2.5.3. Contact Angle

Contact angle analysis was performed on the fully optimized surfaces from the AFM testing using 3 μL droplets of distilled water at ambient room temperature. Droplet images were captured using VHX digital microscopy (Keyence Corp., Osaka, Japan) and analyzed using Keyence software Version 1.2.0.2.

2.6. *In Vitro Characterization*

2.6.1. Cell Culture

MC3T3-E1 mouse pre-osteoblastic cells (American Type Culture Collection, Manassas, VA, USA) were maintained and expanded at 37 °C and 5% CO_2 in alpha-modified Eagle's minimum essential medium supplemented with L-glutamine, sodium pyruvate, 10% fetal bovine serum, and 1% penicillin-streptomycin, with the final pH adjusted to approximately 7.4. For experimentation, an osteogenic differentiation medium was formulated using alpha-modified Eagle's minimum essential medium supplemented with L-glutamine and sodium pyruvate, 10% fetal bovine serum, 1% penicillin-streptomycin, 0.284 mM L-ascorbic acid, and 10 mM β-glycerophosphate. Approximately 50,000 cells/specimen were seeded and acclimatized for a day; thereafter, the specimens were supplemented with 1 mL of differentiation media every 48 h for a total of 21 days. The specimen types used for all *in vitro* experimentations were smooth PEEK, sfPEEK, sfPEEK-HT, with n = 3 for each testing method. The sfPEEK-HT specimen was sulfonated according to the parameters identified during the optimization experimentation detailed earlier. The smooth PEEK specimens were used as negative control in this study because it was anticipated that cells will not proliferate or readily mineralize on the smooth surface compared to the treated surface specimens.

2.6.2. Cell Viability

To assess cell viability at Day 7 and Day 21, a CyQuant™ MTT cell proliferation assay kit (ThermoFisher, Waltham, MA, USA) was used according to the manufacturer's protocol. Each specimen was incubated with 12 mM MTT stock solution and media for 4 h, followed by solution removal and addition of DMSO. Absorbance was read at 540 nm with an ELX-800 plate reader (Winooski, VT, USA).

2.6.3. Biochemical Analysis

Cells were trypsinized and collected off each specimen at the designated time points of Days 1, 7, 14, and 21 and stored at −80 °C until use. Cells were lysed via sonication for 1 min at 10% amplitude. DNA and alkaline phosphatase (ALP) assays were performed in triplicate.

2.6.4. Cell Proliferation

A CyQuant™ DNA cell proliferation assay (ThermoFisher, Waltham, MA, USA) was used according to the manufacturer's protocol on the lysed cells. Standard cell wells were conducted in duplicate. Fluorescence was measured at an excitation wavelength of 460 nm and emission wavelength of 520 nm on a Biotek FLx800 plate reader (Winooski, VT, USA).

2.6.5. Cell Differentiation

An alkaline phosphatase (ALP) assay was performed on the lysed cells to measure cellular differentiation. A QuantiChrom ALP assay kit (BioAssay Systems, Hayward, CA, USA) was used according to the manufacturer's protocol at an absorbance of 405 nm on an ELX-800 plate reader.

2.6.6. Cell Mineralization

To verify mineralization of the pre-osteoblasts on each specimen, calcium deposition was visualized using Alizarin red staining via osteogenesis quantitation kit (EMD Millipore, Billerica, MA, USA) according to the manufacturer's protocol. A control specimen containing no cells for each specimen type was analyzed as well. A VHX digital microscope was used to image the mineralized staining.

2.7. Statistical Analysis

Welch ANOVA ($\alpha = 0.05$) was used to determine any differences among the specimen groups in terms of pore size measurement, AFM, contact angle, mechanical testing, and *in vitro* experiments. For statistical comparison among the specimen groups, a Dunnett post hoc test was performed. Normality was checked using a Shapiro–Wilk test before calculating the ANOVA. For the specimen groups with extremely large counts for pore size measurements, any slight variations in normality were ignored due to ANOVA being robust against deviations in normality, resulting in a small effect on Type I error rate. All statistical analyses were performed using GraphPad Prism Software (version 8.3.0).

3. Results

3.1. Optimized Sulfonated Surface

Measurable surface porosity was not achieved in the first few designs, leading to each experiment being scored qualitatively by visually examining the surface and determining which surface experienced changes, such as the porosity present, similar porosity distribution across the surface, and an even sulfonated surface texture. The parameters from the most visually desirable surfaces on the specimen(s) from each design were chosen, and subsequent designs were built around those factors until a fully optimized surface was attained. Contact angle analysis and surface roughness were only measured on the fully optimized surfaces, since those properties were secondary to attaining ideal porosity for this study. Each design had specimens sulfonated in random order to avoid any nuisance factors. Original non-treated PEEK has a yellowish-brown appearance in color, and once sulfonated, the surface appears white, but due to differences in lighting with the digital microscope and different surface features present on the specimens, all images were collected in the black and white color mode for pore measurements. An optimal soak time of 1 min and a soak temperature of 65 °C were determined from Design #5, but the optimal soak condition was still under investigation. Additionally, it was discovered that blasting the sulfonated surface with compressed air allowed the removal of sulfuric acid while preserving the porosity, which formed during sulfonation.

Design #6 was performed with the ideal time and temperature identified from Design #5 but with the soak condition of sonication only and sonication plus stirring under investigation. The results from this design showed similar pore surface coverage, pore count, and pore diameter, as shown in Table 3. Since similar results were obtained across the specimen types, the authors chose to move forward with *in vitro* studies using the sonication-only method, since this was more controllable, and using stirring would have required further refinement in terms of stirring speed and specimen location within the stirring vortex. Figure 3 below shows the specimens from Design #6 sulfonated with sonication only, and the images shown were enhanced in black and white for easier identification of the pores using ImageJ software. Ultimately, the optimized sulfonation parameters were a pre-surface condition using a 320-grit grinding paper, sonication-only soak method, soak time of 1 min, and soak temperature of 65 °C.

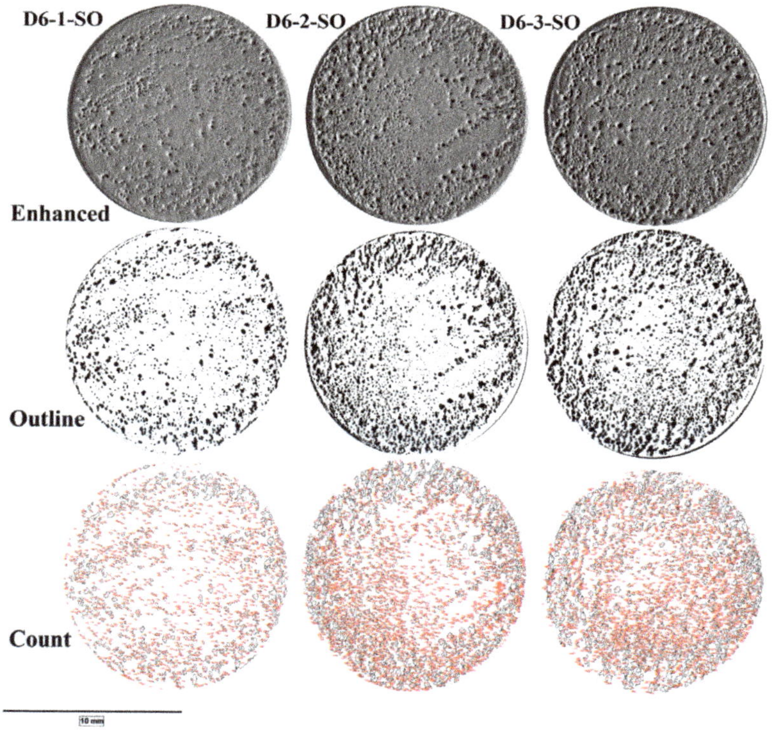

Figure 3. Digital images of the sulfonated specimens in Design #6. The red pores outlined show which ones were identified and counted by ImageJ software.

Table 3. Porosity data from specimens in Design #6. The diameter reported is the Feret diameter.

Specimen ID	Pore Coverage (% Area)	Pore Count	Pore Diameter (Avg: μm)
D6-1-SS	13	2124	103 ± 128
D6-2-SS	17	4179	88 ± 97
D6-3-SS	11	2531	94 ± 92
D6-1-SO	11	1574	120 ± 109
D6-2-SO	18	2558	115 ± 151
D6-3-SO	21	2671	118 ± 160

3.2. Hydrothermal Treatment

Specimens were prepared according to the Design #6 sample type configurations for the first hydrothermal treatment (HT) experiment (Design #1). The experiment was designed around a soak time low of 60 min and high of 90 min, and a soak temperature low of 45 °C and high of 80 °C, as listed in Table 2. Digital imaging of the specimens before and after HT showed no evidence of changing surface features, indicating that the HT only removes residual sulfocompound groups and does not change surface morphology. Fourier transform infrared spectroscopy (FTIR) was performed on each sample to identify the functional groups of interest. The spectra for samples from the hydrothermal treatment Design #1 are shown in Figure 4a. In Figure 4 graphs, sfPEEK refers to the positive control specimen, which is sulfonated PEEK without hydrothermal treatment, and H6 and sfPEEK—1 h HT both have the hydrothermal treatment parameters. The specimen labeled PEEK is the negative control, comprising untreated PEEK. In Figure 4a, there is a large, broad peak around 3400 cm^{-1} on the sfPEEK specimen, indicating the presence of the residual -SO_3H group, and another peak at approximately 1070–1100 cm^{-1}, which represents S=O stretching. The other specimen types in Figure 4a exhibit decreasing intensities of the peaks at 3400 and 1070 cm^{-1}. These data suggest that the removal of residual sulfur groups is more dependent upon time rather than temperature. Therefore, it was decided by the authors to move forward in Design #2 with 45 °C and simply adjust the time and investigate the time points of 2, 3, 4, and 5 h, as shown in Table 2. The FTIR data from the hydrothermal treatment Design #2 are shown in Figure 4b–d. The 2 h and 4 h time points were removed to better visualize the results and not overcrowd the graphs. Figure 4c shows all the results from Figure 4b on the same baseline, and Figure 4d is an enlarged portion of Figure 4b. Figure 4b–d reveal no peak in the 3400 cm^{-1} range and no peak in the 1070 cm^{-1} range on the untreated PEEK specimen, which is what is expected, given the peaks shown on the sfPEEK specimen. The data in Figure 4a–c show lower intensities of the peaks of interest, but similar values are noted for the specimen at 1, 3, and 5 h, suggesting that a hydrothermal treatment performed at 45 °C in the range of 1–5 h would be sufficient for reducing residual sulfur compound groups.

3.3. Atomic Force Microscopy

The surface roughness of the sulfonated PEEK specimens is very tortuous, with varying degrees of topography, as can be seen in the wide range of roughness values shown in Figure 5a. However, across each specimen, there was no significant difference in the mean roughness values, which were 0.298 ± 0.108 μm, 0.308 ± 0.042 μm, 0.592 ± 0.231 μm, 0.671 ± 0.089 μm, 0.538 ± 0.337 μm, and 0.440 ± 0.143 μm for specimens S1-A–S3-B. There were 15 different statistical comparisons for these specimens, and each had a $p > 0.05$, indicating no significant difference.

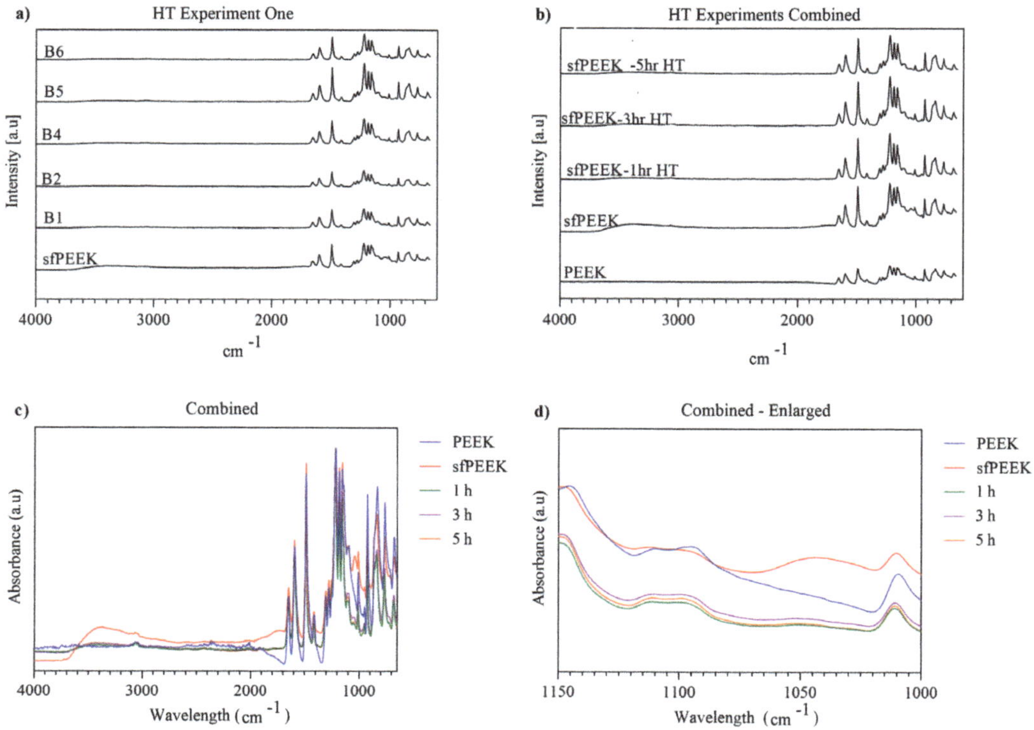

Figure 4. (**a**) FTIR spectra data for hydrothermal treatment Design #1; (**b**) FTIR spectra data for hydrothermal treatment Design #2; (**c**) FTIR spectra data for Design #2 showing wavelengths of interest in the range of 500–4000 cm^{-1} all on the same baseline; (**d**) FTIR spectra data enlarged from graph (**b**) showing Design #2 and the wavelengths of interest in the range of 1000–1150 cm^{-1}, which is the location of the S=O vibration peak.

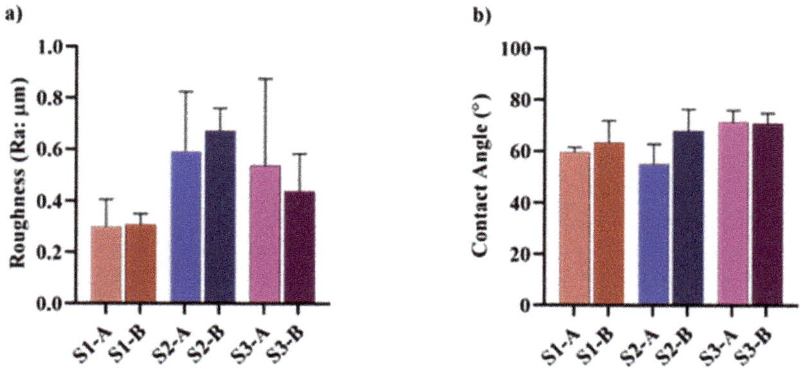

Figure 5. (**a**) Average roughness (Ra) values obtained using the atomic force microscope. (**b**) Average wetting angle obtained for each independent replicate. There were no significant differences found between any of the groups. The specimens were created in three batches (S1, S2, and S3), with n = 2 (A and B) for each batch, as further described in Section 2 earlier.

3.4. Contact Angle Analysis

Figure 5b shows that all specimens recorded hydrophilic contact angles below 90°, and there was no significant difference among the specimens. All 15 specimen comparisons had a $p > 0.05$, indicating no significance. The lowest recorded mean was 59.62°, and the highest was 71.47°.

3.5. In Vitro Evaluation

3.5.1. Cell Viability

To assess cell viability, an MTT assay was performed on the cultured specimens after 7 and 21 days. Figure 6 shows significantly higher cell viability for PEEK specimens over sfPEEK (p-value ≤ 0.0001) and sfPEEK-HT (p-value ≤ 0.0001) after Day 7, and for PEEK specimens over sfPEEK (p-value ≤ 0.0001) and sfPEEK-HT (p-value $= 0.0024$) on Day 21. The sfPEEK-HT specimen had higher viability compared to sfPEEK on Day 7 (p-value $= 0.0017$) and on Day 21 (p-value $= 0.0001$).

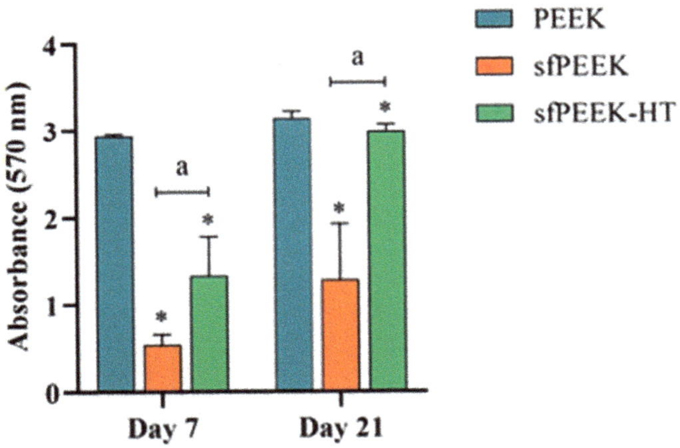

Figure 6. Absorbance readings for cells on specimen surfaces after 7 and 21 days. The MTT assay was used. The * is compared to PEEK, and a is compared to sfPEEK; significance noted at $p < 0.05$.

3.5.2. Cell Proliferation and Differentiation

Cell proliferation and differentiation on Days 1, 7, 14, and 21 were measured using a DNA and ALP assay kit, respectively. Figure 7a shows increasing cell proliferation for all specimens across each time point. Significantly more DNA content was measured for sfPEEK-HT on Day 7 (p-value ≤ 0.0001), Day 14 (p-value $= 0.0029$), and Day 21 (p-value $= 0.0301$) compared to PEEK, and significantly more DNA was measured for sfPEEK-HT on Day 7 (p-value ≤ 0.0001), Day 14 (p-value ≤ 0.0001), and Day 21 (p-value $= 0.0136$) compared to sfPEEK. Both sfPEEK and sfPEEK-HT had significantly higher cell proliferation than PEEK on Day 1 (p-value ≤ 0.0001). However, PEEK had significantly higher values on Day 14 compared to sfPEEK (p-value $= 0.0011$). Figure 7b shows low ALP activity on Day 1 for all specimens, but then, a large increase in ALP activity for all specimens occurs by Day 7. PEEK (p-value $= 0.0072$) and sfPEEK (p-value ≤ 0.0001) have significantly higher ALP activity than sfPEEK-HT on Day 7. Additionally, PEEK (p-value $= 0.0063$) and sfPEEK (p-value ≤ 0.0001) have significantly higher ALP activity on Day 14 compared to sfPEEK-HT. ALP activity by Day 21 is still significantly higher for sfPEEK compared to the other two specimen types: PEEK (p-value $= 0.0337$) and sPEEK-HT (p-value $= 0.0005$).

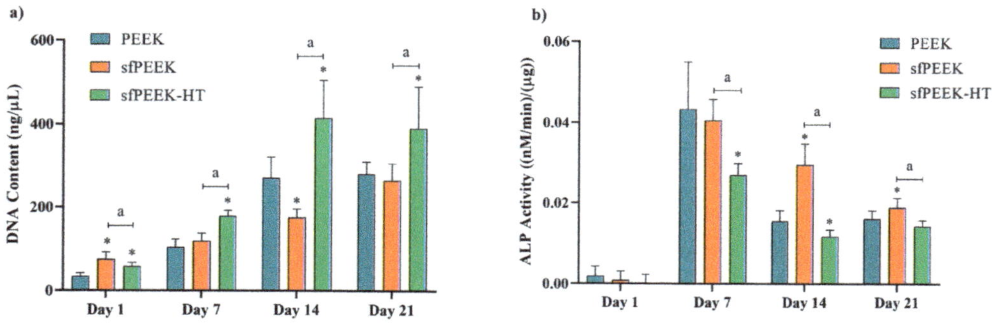

Figure 7. (**a**) DNA results for each specimen type at the designated time point. (**b**) ALP results for specimens at each time point. The * is compared to PEEK, and a is compared to sfPEEK; significance noted at $p < 0.05$.

3.5.3. Cell Mineralization

The Alizarin red staining technique and subsequent digital imaging were performed to determine whether the osteoblast cells matured and mineralized on the different specimen surfaces. Controls were used to verify that the surfaces did not uptake any of the red stain and interfere with cell identification. Alizarin red stains calcium deposits from mature osteoblasts in red. Figure 8 shows some red stained cells present on the PEEK specimens and copious amounts of red for the sfPEEK and sfPEEK-HT specimen types. This indicates more overall cell differentiation and mineralization for the sfPEEK and sfPEEK-HT specimen types compared to PEEK.

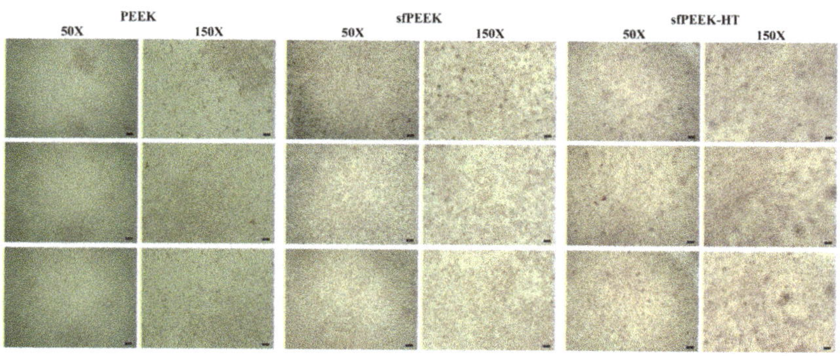

Figure 8. Mineralized cells stained with Alizarin red on each specimen type. Staining was performed on n = 3 samples of each specimen type, and the example images from each specimen are shown.

4. Discussion

The data from this study conclude that pore size and distribution may be optimized by varying the sulfonation processing factors. The most critical factors for sulfonation of PEEK are soak time, soak temperature, soak condition, and post-soak cleaning. The optimal sulfonated surface was achieved using a soak time of 1 min and a soak temperature of 65 °C. It was also determined that the best soak condition is either sonication or sonication with stirring. However, since the data of specimens with sonication only versus sonication with stirring were similar, a more efficient and controllable sulfonation process would be to use sonication only as the soak condition. Using air to blast away sulfuric acid before rinsing is the best method to clean the specimens after sulfuric acid soak.

These factors resulted in pores on the sfPEEK surface ranging from 85 to 108 µm in average size, which is in the 100–400 µm pore size range, which is conducive to osteoblast attachment and growth, which will lead to ultimately successful osseointegration [7,30]. Wan et al. studied the biomineralization ability of sulfonated PEEK with pore sizes ranging from 4.5 to 18.5 μm [46]. The research group concluded that a sulfonated PEEK with a low sulfur concentration and an average pore size of 6–14 μm had significantly higher amounts of cell adhesion and proliferation compared to untreated PEEK [46]. Another study performed by Wang et al. compared sulfonated PEEK specimen to sulfonated PEEK specimen coated with strontium and adiponectin [21]. The authors of this study showed no significant difference in cell viability for the specimen types after 5 days, no significant difference in ALP activity after 7 days, and no significant difference in the osteogenic related gene expression of Type I collagen, Runx2, or OPN [21]. However, Wang et al. did observe that the strontium and adiponectin specimen did have higher ALP activity and mineralization after Day 14, but this does not discredit the sulfonated-only PEEK, as it performed relatively well with no additional bioactive factors added [21].

The porosity distribution across the surfaces of sulfonated PEEK was visually well dispersed, as shown in Figure 3. Specimen types varied statistically when observing the pore sizes as a whole and in the 0–150 μm range. However, when the pore size range was narrowed to a range between 151 and 400 μm, there were no significant differences among the sulfonated PEEK specimens. Deviations in the total pore size averages most likely resulted from large pore size outliers or outliers caused by shadowing issues in ImageJ. Similar sizes between 151 and 400 μm are important because a pore size of at least 100 μm is considered a major influencing factor in enhancing osteogenesis [8,24]. Furthermore, the natural structure of cortical bone ranges in pore size from 10 to 500 μm, and the size of cancellous bone pores ranges from 0.2 to 1 mm [27,38]. Civantos et al. studied the osteointegration behavior of specimens with varying pore sizes of 100–200 μm, 250–355 μm, and 355–500 μm on titanium surfaces [7]. The researchers showed that there was no significant difference across the different sized porous specimens; however, all porous specimens had higher levels of ALP expression and calcium deposits compared to non-porous specimens [7]. Murphy et al. used MC3T3-E1 cells to experiment on scaffolds with mean pore sizes ranging from 85 μm to 325 μm to determine which pore size resulted in improved cell adhesion [30]. The authors showed that the initial cell attachment was approximately 62% for 325 μm pore sizes after 24 h and 48 h time points, and cell attachment for 120 μm sized pores was 45% after 24 h and 48 h [30]. No significant difference was found in cell attachment for the 85 μm, 164 μm, and 190 μm pore sized scaffolds, and the authors concluded that larger pore sizes lead to more overall cell infiltration, while smaller pore sizes have initially higher cell attachment [30].

The hydrothermal treatment experiments suggested that soaking sfPEEK in distilled water at 45 °C creates surfaces conducive to cellular growth. A study conducted by Ouyang et al. hydrothermally treated sulfonated PEEK specimens at 25 °C and 120 °C for 4 h, respectively. The EDS results showed a lower detection of sulfur content for the 120 °C treated group, and the same group displayed significantly higher amounts of cell proliferation and ALP activity compared to the 25 °C group and no hydrothermal treatment group [47]. The 25 °C hydrothermal treatment conducted by the researchers showed higher cell proliferation than the no hydrothermal treatment group; however, the sulfur content remaining was still too high to significantly outperform the 120 °C group [47]. The hydrothermal treatment experiments conducted for this study ranged in temperature from 45 °C to 80 °C, showing no difference in sulfur concentration based on temperature, but instead, the most notable difference in residual sulfur content was based on soak time. This would indicate that the temperature of 45 °C chosen is adequate for removing sulfur content from the sulfonated PEEK, which will allow proper cell viability.

FTIR was performed to observe the chemical changes, which occurred on the sulfonated PEEK after hydrothermal treatment. Figures 1 and 2 are displayed in the preceding section to show the general mechanism of the sulfonation and desulfonation of PEEK. A residual -SO_3H group is left on the sulfonated PEEK surface, and after hydrothermal treatment, this -SO_3H group is removed. On the FTIR spectrum, a broad peak representing an OH- functional group can be seen around 3400 cm^{-1}, and a small peak around 1050 cm^{-1} represents S=O vibrations [42,44,48]. Figure 4 shows higher intensities of the aforementioned peaks for specimens, which were not hydrothermally treated, compared to those, which were hydrothermally treated, which indicates the success of the treatment in removing residual sulfur groups. This is further supported by the contact angle analysis in Figure 5b displaying a significantly higher contact angle for sfPEEK-HT, which would be more hydrophobic due to no OH- group from the -SO_3H being present, compared to sfPEEK [45].

The AFM data (Figure 5a) confirm no significant differences among the replicates, with average Ra values ranging from 0.298 to 0.671 μm. This changing topography indicates the presence of micro- and submicro-structures, which are texturing the surface. Loose definitions of macro-, micro-, submicro-, and nano-scale topography are present in the literature, but typically, the scale is less than 1 mm, 100 μm, 1 μm, and 100 nm, respectively [25,28,29]. This roughening of the PEEK surface following sulfonation is conducive to osteoblast adhesion, proliferation, and mechanical interlocking. Three specimen types with average surface roughness (Ra) values of 0.6 μm, 1.5 μm, and 0.5 μm for machined (MA), micro- (MACRO), and nano- (NANO) surfaces, respectively, were studied by Salou et al. to determine osseointegration capabilities using *in vivo* animal experimentation [37]. The researchers noted that after 4 weeks of implantation, the NANO surfaces showed higher performance values for pull-out force testing and bone growth compared to the MACRO and MA surfaces [37]. The contact angle analysis for this current study shown in Figure 5b shows no significant differences among the replicates. These data observed for the independent replicates give the authors confidence that the sulfonation process used for this study is repeatable.

The MTT testing in Figure 6 revealed PEEK specimens to have overall more viability compared to the other two specimen types for Day 7 and Day 21. Additionally, the sfPEEK-HT specimen had significantly higher viability than sfPEEK over the 21 days. However, PEEK and sfPEEK-HT generally show high levels of viability, and the viability of sfPEEK increases from Day 7 to Day 21, which could indicate the possible removal of residual sulfur groups over the testing period, resulting in a more favorable cell environment. The increase in viability coincides with the increase in DNA content on all specimen surfaces seen across Days 1, 7, 14, and 21 (Figure 7a). The DNA assay is performed to represent cell proliferation, while the ALP assay is performed to observe early osteoblast differentiation. Overall, sfPEEK-HT has the most significant and highest increase in DNA content over the course of the 21 days, indicating the highest cell proliferation and presence for that surface. The lower values of DNA content seen for the sfPEEK specimens could indicate that the cells on that surface have switched from proliferation to differentiation because the ALP activity for sfPEEK on Day 7 is significantly higher than sfPEEK-HT and significantly higher than PEEK and sfPEEK-HT on Days 14 and 21.

The mineralization seen in the Alizarin red staining in Figure 8 further confirms the switch from proliferation to differentiation, which occurred on the specimen surfaces, as indicated by the ALP activity in Figure 7b. Even though the PEEK specimens have a large increase in ALP activity on Day 7, which would indicate differentiation, the DNA content is low in Figure 7a compared to the other specimen types, meaning that the overall cell count on the surface could be lower, resulting in the lower overall mature osteoblast formation seen. In contrast, the sfPEEK and sfPEEK-HT surfaces have an abundant amount of red staining. Visually, it is difficult to determine whether sfPEEK or sfPEEK-HT have more mineralization; however, the DNA and ALP data suggest that cells proliferate more on sfPEEK-HT and eventually differentiate and mineralize, while the cells on sfPEEK proliferate less but differentiate more readily and subsequently mineralize.

5. Conclusions

Porosity and porosity distribution across PEEK surfaces were able to be optimized based on varying sulfonation process factors. A hydrothermal treatment was also successfully employed to reduce residual sulfur content on the sulfonated PEEK surfaces. This study concludes that the optimal sulfonation process—which will produce a PEEK surface with the desired porosity of approximately 88 μm–120 μm, reduced sulfur content for enhanced cell viability, and cell attachment—includes an acid soak time of 1 min and a soak temperature of 65 °C. This process is followed by quick bursts of air to remove the acid and a subsequent hydrothermal treatment at 45 °C for approximately 2.5 h. After the treatments, the resulting micro-scale and nano-scale features are similar across the surfaces, as indicated by AFM roughness measurements and contact angles. FTIR and contact angle analyses were performed to confirm successful hydrothermal treatment of the sulfonated and heat-treated PEEK specimens. Successful removal of the functional groups OH- and -SO_3H, which are indicative of residual sulfur groups, was shown on the FTIR spectrum after specimens were hydrothermally treated. The contact angles went back to being more hydrophobic after hydrothermal treatment. *In vitro* experimentation proved the successful attachment and proliferation of MC3T3-E1 cells on each specimen type. Smooth PEEK had overall higher cell viability, followed by sulfonated and heat-treated PEEK. The sulfonated and heat-treated PEEK had significantly higher cell proliferation compared to smooth PEEK and sulfonated-only PEEK across all time points, indicating a favorable environment for the cells. ALP activity indicates early differentiation of the cells on smooth PEEK and sulfonated-only PEEK taking place on Day 7, while the sulfonated and heat-treated PEEK have a sustained period of differentiation due to the high amount of cell proliferation occurring. Maturation and mineralization of the osteoblast cells occurred on all specimen types, with sulfonated-only PEEK and sulfonated and heat-treated PEEK having the highest amount. Future testing should be conducted to determine sulfonated and heat-treated PEEK's ability to have enhanced osseointegration *in vivo*, but overall, it can be suggested that the optimized sulfonated and heat-treated PEEK would have superior performance over smooth PEEK and sulfonated-only PEEK due to its porosity, high cell proliferation, and mineralization performance.

Author Contributions: Conceptualization, K.N., A.V.J. and R.S.W.; methodology, K.N., A.V.J., M.D.R., L.W. and R.S.W.; formal analysis, K.N.; investigation, K.N., A.V.J. and R.S.W.; resources, L.W. and R.S.W.; data curation, K.N.; writing—original draft preparation, K.N.; writing—review and editing, A.V.J., M.D.R., L.W. and R.S.W.; visualization, K.N. and R.S.W.; supervision, R.S.W.; project administration, R.S.W.; funding acquisition, R.S.W. All authors have read and agreed to the published version of the manuscript.

Funding: This research was funded by Zavation Medical Products, LLC. Flowood, MS, USA, 39232, grant number SP14350.

Institutional Review Board Statement: Not applicable.

Informed Consent Statement: Not applicable.

Data Availability Statement: The data presented in this study are available on request from the corresponding author due to patent pending.

Conflicts of Interest: Authors Kadie Nobles, Amol V. Janorkar, and R. Scott Williamson have a patent pending with the technology (Provisional Patent #63523624, "Sulfonation Treatment of Medical and Dental Implants"; Inventors: R. Scott Williamson, PhD, UMMC, Amol V. Janorkar, PhD, UMMC, and Kadie Nobles, PhD, Institute of Defense Analysis (Washington, DC, former UMMC PhD under R. Scott Williamson); Filing Date of 27 June 2023. The specific aspects of the patent are the sulfonation process and hydrothermal process for the Poly Aryl family of polymers for medical and dental applications). Author Lawrence Walker is an employee of Zavation Medical Products, LLC, who sponsored the research for this project (Kadie Nobles, PhD, Dissertation; Author R. Scott Williamson received grant funding for this project from Zavation Medical Products, LLC. Author Michael D. Roach declares no conflicts of interest).

References

1. Watkins-Castillo, S.; Andersson, G. United States Bone and Joint Initiative: The Burden of Musculoskeletal Dieseases in the United States (BMUSA). 2014. Available online: http://www.boneandjointburden.org (accessed on 21 March 2024).

2. Afewerki, S.; Bassous, N.; Harb, S.; Palo-Nieto, C.; Ruiz-Esparza, G.U.; Marciano, F.R.; Webster, T.J.; Furtado, A.S.A.; Lobo, A.O. Advances in dual functional antimicrobial and osteoinductive biomaterials for orthopaedic applications. *Nanomed. Nanotechnol. Biol. Med.* **2020**, *24*, 102143. [CrossRef]
3. Bozic, K.J.; Kurtz, S.M.; Lau, E.; Ong, K.; Chiu, V.; Vail, T.P.; Rubash, H.E.; Berry, D.J. The epidemiology of revision total knee arthroplasty in the United States. *Clin. Orthop. Relat. Res.* **2010**, *468*, 45–51. [CrossRef]
4. Landgraeber, S.; Jäger, M.; Jacobs, J.J.; Hallab, N.J. The pathology of orthopedic implant failure is mediated by innate immune system cytokines. *Mediat. Inflamm.* **2014**, *2014*, 185150. [CrossRef]
5. Raphel, J.; Holodniy, M.; Goodman, S.B.; Heilshorn, S.C. Multifunctional coatings to simultaneously promote osseointegration and prevent infection of orthopaedic implants. *Biomaterials* **2016**, *84*, 301–314. [CrossRef] [PubMed]
6. Parithimarkalaignan, S.; Padmanabhan, T.V. Osseointegration: An update. *J. Indian Prosthodont. Soc.* **2013**, *13*, 2–6. [CrossRef] [PubMed]
7. Civantos, A.; Domínguez, C.; Pino, R.J.; Setti, G.; Pavón, J.J.; Martínez-Campos, E.; Garcia, F.J.G.; Rodríguez, J.A.; Allain, J.P.; Torres, Y. Designing bioactive porous titanium interfaces to balance mechanical properties and in vitro cells behavior towards increased osseointegration. *Surf. Coat. Technol.* **2019**, *368*, 162–174. [CrossRef]
8. Zheng, J.-P.; Chen, L.-J.; Chen, D.-Y.; Shao, C.-S.; Yi, M.-F.; Zhang, B. Effects of pore size and porosity of surface-modified porous titanium implants on bone tissue ingrowth. *Trans. Nonferr. Met. Soc. China* **2019**, *29*, 2534–2545. [CrossRef]
9. Hahn, B.-D.; Park, D.-S.; Choi, J.-J.; Ryu, J.; Yoon, W.-H.; Choi, J.-H.; Kim, J.-W.; Ahn, C.-W.; Kim, H.-E.; Yoon, B.-H.; et al. Osteoconductive hydroxyapatite coated PEEK for spinal fusion surgery. *Appl. Surf. Sci.* **2013**, *283*, 6–11. [CrossRef]
10. Kurtz, S.M.; Devine, J.N. PEEK biomaterials in trauma, orthopedic, and spinal implants. *Biomaterials* **2007**, *28*, 4845–4869. [CrossRef]
11. Mahjoubi, H.; Buck, E.; Manimunda, P.; Farivar, R.; Chromik, R.; Murshed, M.; Cerruti, M. Surface phosphonation enhances hydroxyapatite coating adhesion on polyetheretherketone and its osseointegration potential. *Acta Biomater.* **2017**, *47*, 149–158. [CrossRef]
12. Mishra, S.; Chowdhary, R. PEEK materials as an alternative to titanium in dental implants: A systematic review. *Clin. Implant. Dent. Relat. Res.* **2019**, *21*, 208–222. [CrossRef] [PubMed]
13. Zhang, J.; Tian, W.; Chen, J.; Yu, J.; Zhang, J.; Chen, J. The application of polyetheretherketone (PEEK) implants in cranioplasty. *Brain Res. Bull.* **2019**, *153*, 143–149. [CrossRef] [PubMed]
14. Evans, N.T.; Torstrick, F.B.; Lee, C.S.; Dupont, K.M.; Safranski, D.L.; Chang, W.A.; Macedo, A.E.; Lin, A.S.; Boothby, J.M.; Whittingslow, D.C.; et al. High-strength, surface-porous polyether-ether-ketone for load-bearing orthopedic implants. *Acta Biomater.* **2015**, *13*, 159–167. [CrossRef] [PubMed]
15. Kaur, M.; Singh, K. Review on titanium and titanium based alloys as biomaterials for orthopaedic applications. *Mater. Sci. Eng. C* **2019**, *102*, 844–862. [CrossRef] [PubMed]
16. Li, J.; Qin, L.; Yang, K.; Ma, Z.; Wang, Y.; Cheng, L.; Zhao, D. Materials evolution of bone plates for internal fixation of bone fractures: A review. *J. Mater. Sci. Technol.* **2020**, *36*, 190–208. [CrossRef]
17. Murr, L.E.; Gaytan, S.M.; Medina, F.; Lopez, H.; Martinez, E.; Machado, B.I.; Hernandez, D.H.; Martinez, L.; Lopez, M.I.; Wicker, R.B.; et al. Next-generation biomedical implants using additive manufacturing of complex, cellular and functional mesh arrays. *Philos. Trans. A Math Phys. Eng. Sci.* **2010**, *368*, 1999–2032. [CrossRef]
18. Wang, X.; Xu, S.; Zhou, S.; Xu, W.; Leary, M.; Choong, P.; Qian, M.; Brandt, M.; Xie, Y.M. Topological design and additive manufacturing of porous metals for bone scaffolds and orthopaedic implants: A review. *Biomterials* **2016**, *83*, 127–141. [CrossRef] [PubMed]
19. Liu, X.; Han, F.; Zhao, P.; Lin, C.; Wen, X.; Ye, X. Layer-by-layer self-assembled multilayers on PEEK implants improve osseointegration in an osteoporosis rabbit model. *Nanomedicine* **2017**, *13*, 1423–1433. [CrossRef]
20. Sun, Y.; Liu, X.; Tan, J.; Lv, D.; Song, W.; Su, R.; Li, L.; Liu, X.; Ouyang, L.; Liao, Y. Strontium ranelate incorporated 3D porous sulfonated PEEK simulating MC3T3-E1 cell differentiation. *Regen. Biomater.* **2021**, *8*, rbaa043. [CrossRef]
21. Wang, S.; Yang, Y.; Li, Y.; Shi, J.; Zhou, J.; Zhang, L.; Deng, Y.; Yang, W. Strontium/adiponectin co-decoration modulates the osteogenic activity of nano-morphologic polyetheretherketone implant. *Colloids Surf. B Biointerfaces* **2019**, *176*, 38–46. [CrossRef]
22. Gao, C.; Wang, Z.; Jiao, Z.; Wu, Z.; Guo, M.; Wang, Y.; Liu, J.; Zhang, P. Enhancing antibacterial capability and osseointegration of polyetheretherketone (PEEK) implants by dual-functional surface modification. *Mater. Des.* **2021**, *205*, 109733. [CrossRef]
23. Chen, S.; Guo, Y.; Liu, R.; Wu, S.; Fang, J.; Huang, B.; Li, Z.; Chen, Z.; Chen, Z. Tuning surface properties of bone biomaterials to manipulate osteoblastic cell adhesion and the signaling pathways for the enhancement of early osseointegration. *Colloids Surf. B Biointerfaces* **2018**, *164*, 58–69. [CrossRef]
24. Necula, B.; Apachitei, I.; Fratila-Apachitei, L.; van Langelaan, E.; Duszczyk, J. Titanium bone implants with superimposed micro/nano-scale porosity and antibacterial capability. *Appl. Surf. Sci.* **2013**, *273*, 310–314. [CrossRef]
25. Perez, R.A.; Mestres, G. Role of pore size and morphology in musculo-skeletal tissue regeneration. *Mater. Sci. Eng. C Mater. Biol. Appl.* **2016**, *61*, 922–939. [CrossRef] [PubMed]
26. Torstrick, F.B.; Lin, A.S.; Potter, D.; Safranski, D.L.; Sulchek, T.A.; Gall, K.; Guldberg, R.E. Porous PEEK improves the bone-implant interface compared to plasma-sprayed titanium coating on PEEK. *Biomaterials* **2018**, *185*, 106–116. [CrossRef] [PubMed]
27. Souza, J.C.M.; Sordi, M.B.; Kanazawa, M.; Ravindran, S.; Henriques, B.; Silva, F.S.; Aparicio, C.; Cooper, L.F. Nano-scale modification of titanium implant surfaces to enhance osseointegration. *Acta Biomater.* **2019**, *94*, 112–131. [CrossRef]

28. Coelho, P.G.; Jimbo, R.; Tovar, N.; Bonfante, E.A. Osseointegration: Hierarchical designing encompassing the macrometer, micrometer, and nanometer length scales. *Dent. Mater.* **2015**, *31*, 37–52. [CrossRef]
29. Gittens, R.A.; Olivares-Navarrete, R.; Schwartz, Z.; Boyan, B.D. Implant osseointegration and the role of microroughness and nanostructures: Lessons for spine implants. *Acta Biomater.* **2014**, *10*, 3363–3371. [CrossRef] [PubMed]
30. Murphy, C.M.; Haugh, M.G.; O'Brien, F.J. The effect of mean pore size on cell attachment, proliferation and migration in collagen-glycosaminoglycan scaffolds for bone tissue engineering. *Biomaterials* **2010**, *31*, 461–466. [CrossRef]
31. Wei, J.; Igarashi, T.; Okumori, N.; Igarashi, T.; Maetani, T.; Liu, B.; Yoshinari, M. Influence of surface wettability on competitive protein adsorption and initial attachment of osteoblasts. *Biomed. Mater.* **2009**, *4*, 045002. [CrossRef]
32. Zhao, G.; Raines, A.; Wieland, M.; Schwartz, Z.; Boyan, B. Requirement for both micron- and submicron scale structure for synergistic responses of osteoblasts to substrate surface energy and topography. *Biomaterials* **2007**, *28*, 2821–2829. [CrossRef]
33. Eriksson, C.; Nygren, H.; Ohlson, K. Implantation of hydrophilic and hydrophobic titanium discs in rat tibia: Cellular reactions on the surfaces during the first 3 weeks in bone. *Biomaterials* **2004**, *25*, 4759–4766. [CrossRef] [PubMed]
34. Kennedy, S.B.; Washburn, N.R.; Simon, C.G., Jr.; Amis, E.J. Combinatorial screen of the effect of surface energy on fibronectin-mediated osteoblast adhesion, spreading and proliferation. *Biomaterials* **2006**, *27*, 3817–3824. [CrossRef]
35. Rodriguez-Contreras, A.; Guadarrama Bello, D.; Nanci, A. Surface nanoporosity has a greater influence on osteogenic and bacterial cell adhesion than crystallinity and wettability. *Appl. Surf. Sci.* **2018**, *445*, 255–261. [CrossRef]
36. Buck, E.; Li, H.; Cerruti, M. Surface Modification Strategies to Improve the Osseointegration of Poly(etheretherketone) and Its Composites. *Macromol. Biosci.* **2020**, *20*, e1900271. [CrossRef] [PubMed]
37. Salou, L.; Hoornaert, A.; Louarn, G.; Layrolle, P. Enhanced osseointegration of titanium implants with nanostructured surfaces: An experimental study in rabbits. *Acta Biomater.* **2015**, *11*, 494–502. [CrossRef] [PubMed]
38. Wang, F.; Shi, L.; He, W.-X.; Han, D.; Yan, Y.; Niu, Z.-Y.; Shi, S.-G. Bioinspired micro/nano fabrication on dental implant–bone interface. *Appl. Surf. Sci.* **2013**, *265*, 480–488. [CrossRef]
39. Huang, R.Y.M.; Shao, P.; Burns, C.M.; Feng, X. Sulfonation of poly(ether ether ketone)(PEEK): Kinetic study and characterization. *J. Appl. Polym. Sci.* **2001**, *82*, 2651–2660. [CrossRef]
40. Khomein, P.; Ketelaars, W.; Lap, T.; Liu, G. Sulfonated aromatic polymer as a future proton exchange membrane: A review of sulfonation and crosslinking methods. *Renew. Sustain. Energy Rev.* **2021**, *137*, 110471. [CrossRef]
41. Zaidi, S.M.J. Polymer sulfonation—A versatile route to prepare proton-conducting membrane material for advanced technologies. *Arab. J. Sci. Eng. Sect. B Eng.* **2003**, *28*, 183–194.
42. He, M.; Huang, Y.; Xu, H.; Feng, G.; Liu, L.; Li, Y.; Sun, D.; Zhang, L. Modification of polyetheretherketone implants: From enhancing bone integration to enabling multi-modal therapeutics. *Acta Biomater.* **2021**, *129*, 18–32. [CrossRef] [PubMed]
43. Sun, Z.; Ouyang, L.; Ma, X.; Qiao, Y.; Liu, X. Controllable and durable release of BMP-2-loaded 3D porous sulfonated polyetheretherketone (PEEK) for osteogenic activity enhancement. *Colloids Surf. B Biointerfaces* **2018**, *171*, 668–674. [CrossRef] [PubMed]
44. Wang, W.; Luo, C.J.; Huang, J.; Edirisinghe, M. PEEK surface modification by fast ambient-temperature sulfonation for bone implant applications. *J. R. Soc. Interface* **2019**, *16*, 20180955. [CrossRef]
45. Zhao, Y.; Wong, H.M.; Wang, W.; Li, P.; Xu, Z.; Chong, E.Y.; Yan, C.H.; Yeung, K.W.; Chu, P.K. Cytocompatibility, osseointegration, and bioactivity of three-dimensional porous and nanostructured network on polyetheretherketone. *Biomaterials* **2013**, *34*, 9264–9277. [CrossRef] [PubMed]
46. Wan, T.; Jiao, Z.; Guo, M.; Wang, Z.; Wan, Y.; Lin, K.; Liu, Q.; Zhang, P. Gaseous sulfur trioxide induced controllable sulfonation promoting biomineralization and osseointegration of polyetheretherketone implants. *Bioact. Mater.* **2020**, *5*, 1004–1017. [CrossRef] [PubMed]
47. Ouyang, L.; Zhao, Y.; Jin, G.; Lu, T.; Li, J.; Qiao, Y.; Ning, C.; Zhang, X.; Chu, P.K.; Liu, X. Influence of sulfur content on bone formation and antibacterial ability of sulfonated PEEK. *Biomaterials* **2016**, *83*, 115–126. [CrossRef]
48. Qin, W.; Li, Y.; Ma, J.; Liang, Q.; Cui, X.; Jia, H.; Tang, B. Osseointegration anvd biosafety of graphene oxide wrapped porous CF/PEEK composites as implantable materials: The role of surface structure and chemistry. *Dent. Mater.* **2020**, *36*, 1289–1302. [CrossRef]

coatings

MDPI

Article

Coralline Hydroxyapatite Coated with a Layer Biomimetic Calcium Phosphate Containing BMP-2 Induces Dose-Related Ectopic Bone Formation in Wistar Rats

Henri J. J. Uijlenbroek [1,†], Xingnan Lin [2,†], Xin Zhang [3], Liquan Deng [2], Daniel Wismeijer [4], Mingjie Wang [1], Lingfei Wei [1,5], Yuanna Zheng [2] and Yuelian Liu [1,*]

1 Department of Oral Cell Biology, Academic Centre for Dentistry Amsterdam (ACTA), Gustav Mahlerlaan 3004, 1081LA Amsterdam, The Netherlands; hjju@shre.nl (H.J.J.U.); m.wang@acta.nl (M.W.); weilingfei@hotmail.com (L.W.)
2 School/Hospital of Stomatology, Zhejiang Chinese Medical University, Binwen Road 548, Hangzhou 310053, China; xingnanphd@gmail.com (X.L.); totti.12345@163.com (L.D.); zyn218@126.com (Y.Z.)
3 Stomatology Hospital, School of Stomatology, Zhejiang University School of Medicine, Clinical Research Centre for Oral Diseases of Zhejiang Province, Key Laboratory of Oral Biomedical Research of Zhejiang Province, Cancer Centre of Zhejiang University, Hangzhou 310006, China; 7315053@zju.edu.cn
4 Private Practice, Zutphensestraatweg 26, 6955AH Ellecom, The Netherlands; Danwismeijer@gmail.com
5 Department of Oral Implantology, Yantai Stomatological Hospital, No.142 Beida Street, Yantai 264001, China
* Correspondence: y.liu@acta.nl
† These authors contributed equally.

Abstract: In order to evaluate loading methods and the dose dependency of bone morphogenetic protein 2 (BMP-2) in ectopic bone formation, an osteoinductive material consisting of commercially available coralline hydroxyapatite (CHA) was coated with a layer of biomimetic calcium phosphate (BioCaP) containing BMP-2 in different ways. Eight groups—each containing samples of 0.25 g CHA—were formed and coated with, respectively, BioCaP with internally incorporated BMP-2 in concentrations of 1, 5, 10, 20, 40 and 60 μg per sample, and the two control groups with BioCaP only and BioCaP with 20 μg of adsorbed BMP-2 per sample. The samples were implanted subcutaneously in 27 male Wistar rats. The histological results show that there is no bone formation in the group in which no BMP-2 was included. All samples with BioCaP containing BMP-2 show bone formation. The group with 20 μg of adsorbed BMP-2 per sample shows the least bone formation. Coating-incorporated BMP-2 is more efficient in inducing bone formation than adsorbed BMP-2. The group with 5 μg of coating-incorporated BMP-2 per sample shows the most bone formation. Increasing the amount of coating-incorporated BMP-2 up to 60 μg does not improve ectopic bone formation.

Keywords: osteoinduction; coralline hydroxyapatite; CHA; bone morphogenetic protein-2; BMP-2; ectopic bone formation; biomimetic calcium phosphate; BioCaP

1. Introduction

Bone defects develop due to different causes, such as congenital disorders, infections, cysts, trauma and tumors. A bone defect that is too large to heal spontaneously is called a critical-sized bone defect [1]. The repair of a critical-sized bone defect often results in fibrous connective tissue instead of bone [1]. To prevent this unwanted effect, the use of autologous graft materials are used when treating critical-sized bone defects, which was first described in 1875 [2]. However, the use of autologous graft material implies a donor site, which, when harvesting bone from that site, is sometimes accompanied by both complications at the donor site itself and inconvenience for the patient [3,4]. To avoid these undesirable effects, other graft materials have been developed, such as allograft, xenograft and synthetic graft materials [5]. Although autologous graft material is often described as the gold standard [6], research shows that no graft biomaterial is predominant with regard to its healing capacity [7].

Marine coral is used as a natural bone graft material. Through a hydrothermal process, its calcium carbonate skeleton converses from coral to hydroxyapatite [8], thus forming coralline hydroxyapatite (CHA). CHA has been used since 1979 [9] as a bone graft material. CHA is osteoconductive and, as such, provides a biocompatible lattice for the passage and assembly of vascular, fibroblastic and osteoblastic tissues. It also provides support for surrounding osseous structures [10]. The benefits of CHA as a bone graft are predominantly its safety, biocompatibility and osteoconduction [8].

The best material for bone healing has the following properties: osteoconduction, osteoinduction and osteogenesis [6]. For surgery, primary tension-free closure, angiogenesis, space creation/maintenance and stability are prerequisites [11]. CHA has no osteoinductive properties [8]. This can, however, be achieved by incorporating an osteoinductive agent [12]. By coating CHA with our biomimetic coating and incorporating bone morphogenetic protein-2 (BMP-2), CHA could develop both osteoconductive and osteoinductive properties [13].

BMP-2 is a pleiotropic morphogen that induces a sequential cascade of events in the bone healing process, including chemotaxis, the regulation of osteoblasts, differentiation, angiogenesis and osteoclasts apoptosis [14]. In vitro and in vivo studies have demonstrated that BMPs are capable of enhancing the osteoinduction of mesenchymal stem cells (MSCs), regulating their proliferation and differentiation into osteoblasts [14,15]. BMP-2 incorporated into biomimetic calcium phosphate (BioCaP) coatings are capable of inducing ectopic bone formation in vivo [12]. BMP-2 can be bound to BioCaP by (i) adsorption, (ii) coating incorporation or (iii) internal incorporation [15].

At (i) adsorption, the BMP-2 protein is absorbed by immersing the BioCaP in a solution with BMP-2. The BMP-2 is mainly absorbed on the surface of the outside of the material. Adsorbed BMP-2 is released with a diffusion-controlled burst release [15,16]. During the production of the (ii) BioCaP coating-incorporated BMP-2, BMP-2 and BioCaP precipitate simultaneously on a calcium-based support. A BioCaP latticework with incorporated BMP-2 grows onto the support material: the BioCaP BMP-2 coating. The BMP-2 can only be released after BioCaP is dissolved [15]. If BMP-2 and BioCaP simultaneously precipitate on amorphous CaP (ACaP) to form a unit consisting exclusively of BioCaP and BMP-2, we refer to it as (iii) BioCaP with internally incorporated BMP-2 [12].

Our earlier studies [12,16] show that BMP-2 can be incorporated in BioCaP and is an effective bone formation inducer. Previous studies also show that the coating of CHA with BioCaP and BMP-2 induces bone formation [13]. However, we do not know which concentrations of BMP-2 are effective in ectopic bone formation when CHA is coated with BioCaP containing BMP-2. An animal model is often used as a method to study the efficacy of biomaterials [17]. We made samples consisting of coated CHA with BioCaP coating-incorporated BMP-2 in different concentrations, as well as BioCaP-adsorbed BMP-2. The samples were placed subcutaneously in Wistar rats to study the efficacy of the BMP-2 concentrations at ectopic bone formation.

2. Material and Methods

2.1. Material Preparation

2.1.1. Coating of CHA Granules with BioCaP

CHA granules (YHJ Bio-Osteon®, Beijing, China) with a granule size of 0.4–1.0 mm and a total weight of 0.25 g were biomimetically coated with a layer of BioCaP according to the established protocols [12,15,18]. The biomaterial is first immersed—whilst stirring at 150 r.p.m—in a fivefold concentrated simulated body fluid (684 mM NaCl; 12.5 mM $CaCl_2 \cdot 2H_2O$; 21 mM $NaHCO_3$; 5 mM $Na_2HPO_4 \cdot 2H_2O$) for 24 h at 37 °C under high-nucleation conditions (in the presence of 7.5 mM $MgCl_2 \cdot 2H_2O$) to inhibit crystal growth. A fine, dense layer of amorphous calcium phosphate (ACaP) is formed on the surface of the CHA [15,19]. This layer serves as a seeding substratum for the deposition of a more substantial crystalline layer. The crystalline layer is produced by immersing the ACaP biomaterials in a supersaturated calcium phosphate (CaP) solution [40 mM HCl; 2 mM $Na_2HPO_4 \cdot 2H_2O$; 4 mM $CaCl_2 \cdot 2H_2O$; 50 mM TRIS base (pH 7.4)] for 48 h at 37 °C, 60 r.p.m. whilst shaking, buffered with tris(hydroxymethyl)aminomethane (3.76 mM; pH 7.4). Through this procedure, CHA granules with an outer layer of octa calcium phosphate (OCP) are prepared [13,19]. The hydrated and carbonated calcium phosphate is now called BioCaP.

2.1.2. Preparation of BioCaP with and without Coating-Incorporated BMP-2

Adding BMP-2 (INFUSE®, Medtronic, Minneapolis, MN, USA) in the different concentrations of none, 1, 5, 10, 20, 40 and 60 µg/sample to this supersaturated CaP solution will result in co-precipitation of BMP-2 into the OCP coatings of the CHA granules, thus creating the group in which CHA was coated with BioCaP coating-incorporated BMP-2. All of the samples were freeze-dried at a temperature of around minus 30 °C. The entire procedure was performed under sterile conditions.

2.1.3. Preparation of CHA Coated with BioCaP and Adsorbed BMP-2

The samples with 0.25 g CHA coated with BioCaP and adsorbed 20 µg of BMP-2 per sample were made by merging 0.2 mL BMP-2 solution (100 µg/mL) with 0.25 g of CHA granules, after which followed centrifuge for 2 s with a relative centrifugal force (RCF) of 1019 and vortex. All of the samples were freeze-dried. The entire procedure was performed under sterile conditions.

2.1.4. Control of the Amount of Adsorbed or Incorporated BMP-2

To determine the BMP-2 incorporation, the BMP-2 release is determined, as very well described by Wu [18]. By monitoring the release kinetics of a coating-incorporated depot of protein, BMP-2 (10 µg/mL) was introduced into the supersaturated solution of calcium phosphate. Six samples were used to determine the total amount of incorporated BMP-2. These samples were immersed in 1 mL of 0.5% EDTA (pH 8.0) and vortexed twice for 5 min to ensure complete dissolution of the coatings. The supernatants were withdrawn for analysis of total loading of BMP-2. To monitor the release kinetics, six samples of CHA bearing a coating-incorporated depot of BMP-2 and six samples of CHA bearing an equivalent amount of adsorbed BMP-2 (included for the purpose of comparison) were incubated in sealed 10-mL glass tubes containing 2 mL of phosphate-buffered 0.9% saline (pH 7.4). The tubes were incubated for up to 35 days in a shaking water bath (60 agitations/min), which was maintained at 37 °C. Before each sampling, a mild centrifuge step was performed to collect all of the solution (part of which may stick to the cap and wall of the tubes due to evaporation in the 37 °C water bath). Triplicate 200-µL aliquots of the medium (containing released BMP-2) were withdrawn for analysis after 3 h, 6 h, 9 h, 1 day, 2 days, 3 days, 5 days, 7 days, 10 days, 14 days, 18 days, 23 days, 28 days and 35 days. The fluorescence density was measured in a spectrophotometer (Biotek Instruments Inc., Winooski, VT, USA; excitation wavelength: 485 nm; emission wavelength: 519 nm). The fluorescence readings were converted to amounts of protein using a standard curve, which was generated by preparing a dilution series of BMP-2 in 5 mL of phosphate-buffered 0.9% saline. The temporal release of BMP-2 was expressed as a percentage of the total amount that had been incorporated into the crystalline layer of the calcium phosphate coating or that had been adsorbed directly onto the CHA granules. The absolute amount of 20 µg of BMP-2 was directly added on the surface of 0.25 g CHA BioCaP-coated samples, which were the absorbed ones.

2.1.5. Surgical Procedure

Twenty-seven adult male Wistar rats, each weighing between 200 and 220 g and approximately 3–4 months old, were used in this study. We used male Wistar rats to avoid any differential effects due to animal gender. Every surgical intervention took place under anesthesia by administering pentobarbital sodium (50 mg/kg IP). A systematic random sampling protocol was used for the distribution of the samples over the rats. They were numbered from 1 to 27, after which, the samples out of one group were divided at random over 6 rats, so each rat received 2 different samples from different groups. The animals were fed a standard diet and had unlimited access to water. After shaving, the skin was disinfected. An incision was made, after which, two subcutaneous pockets were created in each rat: one on the right dorsal side and one on the left. According to the distribution in Table 1, one sample of 0.25 g CHA granules was directly placed into the surgically created pocket on each side in 27 rats, following a systematic random sampling protocol. The incisions were closed by suturing and thus entrapping the samples.

Table 1. Sample distribution over the rats.

Group	N	CHA	BMP-2	BMP-2 Loaded to CHA
1	6	0.25 g	-	-
2	8	0.25 g	1 μg	coating-incorporated
3	8	0.25 g	5 μg	coating-incorporated
4	8	0.25 g	10 μg	coating-incorporated
5	6	0.25 g	20 μg	coating-incorporated
6	6	0.25 g	20 μg	adsorbed
7	6	0.25 g	40 μg	coating-incorporated
8	6	0.25 g	60 μg	coating-incorporated

Group 1; n = 6: 0.25 g CHA containing no BMP-2 per sample. Group 2: n = 8: 0.25 g CHA containing 1 μg of coating-incorporated BMP-2 per sample. Group 3: n = 8: 0.25 g CHA containing 5 μg of coating-incorporated BMP-2 per sample. Group 4: n = 8: 0.25 g CHA containing 10 μg of coating-incorporated BMP-2 per sample. Group 5: n = 6: 0.25 g CHA containing 20 μg of coating-incorporated BMP-2 per sample. Group 6: n = 6: 0.25 g CHA containing 20 μg of adsorbed BMP-2 per sample. Group 7: n = 6: 0.25 g CHA containing 40 μg of coating-incorporated BMP-2 per sample. Group 8: n = 6: 0.25 g CHA containing 60 μg of coating-incorporated BMP-2 per sample.

2.1.6. Study Design

All of the animal experiments were carried out according to the ethics laws and regulations of People's Republic of China. The experiment was approved by the committee of the Board of Animal Experiments, traditional Chinese Medicine University, Hangzhou, China (Accreditation number: ZSLL-2016-46). Fifty-four samples, each containing 0.25 g of coralline hydroxyapatite (CHA) granules, were made. BioCaP with concentrations of 1, 5, 10, 20, 40 and 60 μg/sample of incorporated BMP-2 were made, as well as 20 μg/sample of adsorbed BMP-2. The CHA samples were coated with BioCaP, thus creating five groups containing six samples per group and three groups containing eight samples per group. Thus, eight groups were formed, as displayed in Table 1.

2.1.7. Histological Observation

Five weeks after implantation of the samples, the animals were sacrificed by an overdose of gaseous carbon dioxide. The samples with their surrounding tissues were obtained. These were dehydrated in serial steps in alcohol and xylol and embedded in methylmethacrylate (MMA, Technovit® 7200 VLC Exact®, Heraeus Kulzer Wehrheim, Germany) [20,21]. The samples were cut using a systematic random sampling protocol with a diamond saw parallel to their longitudinal axis. The distance between the slices was set at 1.0 mm. Six slices per sample were obtained, and a total of 324 slices were evaluated. Every slice was mounted separately on a plexiglass holder and polished to approximately a thickness of 100 μm. The slices were then stained with McNeal's tetrachrome, basic fuchsine and toluidine blue [20].

2.1.8. Histomorphometric Analysis

With a Nikon-Eclipse light microscope (Nikon, Tokyo, Japan), photos were taken of all of the slices. The photos were printed in color in A4 format. The total area of interest was bounded by the external boundary of the connective tissue surrounding the samples. The volumes of bone, CHA and total area were determined per slice. The histomorphometric analysis was performed by randomly placing an appropriate grid over the prints and manually point counting according to Cruz-Orive [22].

2.1.9. Statistical Analysis

The hits that were achieved by manual point counting of newly formed bone, remaining CHA and total area were determined for each slice. Out of these hits, the means of the hits for newly formed bone, remaining CHA and total area were calculated per sample. The percentage of newly formed bone in relation to the total area, as well as the remaining CHA in relation to the total area, were calculated. These percentages were used as input for the statistical analysis, using IBM SPPS statistical software version 26.0. for Windows 10 (IBM corporation, Armonk, NY, USA). The total p-value of the one-way analysis of variance (ANOVA) was <0.01 for all three percentages, meaning that the means are statistically significant for the different groups. An LSD test was then used for pairwise comparisons between two groups within one percentage, setting $p < 0.05$ as statistically significant. All numerical data are presented as the average of one sample—obtained out of all the slices of that sample - per group with means and outliers.

3. Results

3.1. In Vitro Results

3.1.1. Material Preparation

The scanning electron microscope (SEM) images (Figure 1) showed that the structure of CHA is well interconnected. After coating the CHA with OCP containing BMP-2, an interwoven network of needle-like and plate-like crystals was formed. Our previous study [13] showed that the pore size of the obtained CHA varied from approximately 100 till 600 μm and that the OCP coating thickness was approximately 20 to 30 μm, which is approximately 3% to 30% of the CHA pore size. The trabecular structure is not changed by coating the CHA [13].

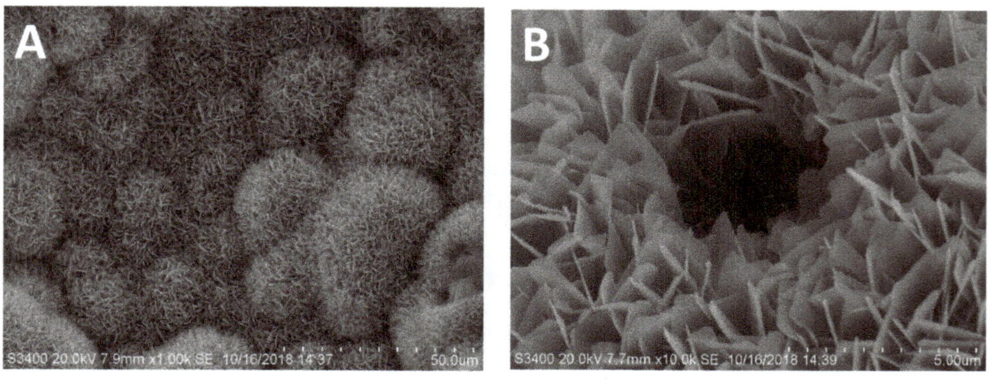

Figure 1. Scanning electron microscope (SEM) images showing an overview of BioCaP-coated and BMP-2-functionalized coralline hydroxyapatite (CHA) granules (**A**). High magnification of a crystalline layer of OCP coating with an incorporated depot of BMP-2 (**B**).

3.1.2. Amount of Encapsulated BMP-2

Using an enzyme-linked immunosorbent assay (Elisa, PreproTech EC, London, UK), as described in our earlier studies [12,23], showed that the amount of BMP-2 encapsulated in the samples was approximately 0, 1, 5, 10, 20, 40 and 60 μg of BMP-2 per sample.

3.2. Observations

After the implantation, the wounds healed well in all rats, and there were no wound infections clinically visible. We experienced no losses of rats and/or samples during the clinical phase.

3.3. Histological Findings

All samples were completely surrounded by soft tissues, which is to be expected. In the group of 0.25 g CHA coated with BioCaP containing no BMP-2 per sample, no bone formation was found (Figure 2A). The CHA granules were surrounded by fibrous tissue. In all other groups, which all contained BMP-2, newly formed bone was present in direct contact with the BioCaP coating of the CHA particles, which is visible in Figure 2B–G. Occasionally, we saw an island of solitary newly formed bone with no direct contact to the BioCaP coating of the CHA particles (Figure 3). It seemed that newly formed bone was more located in the center of the samples than in their periphery.

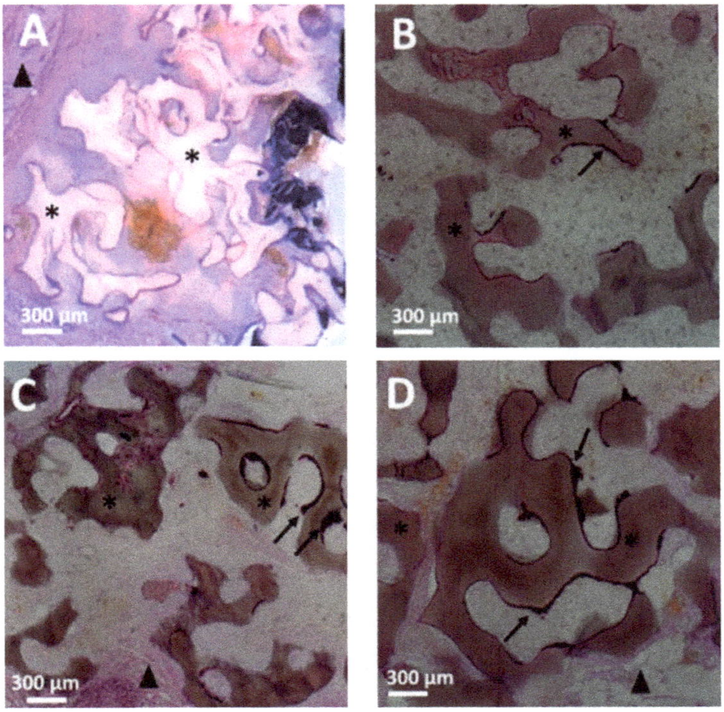

Figure 2. *Cont.*

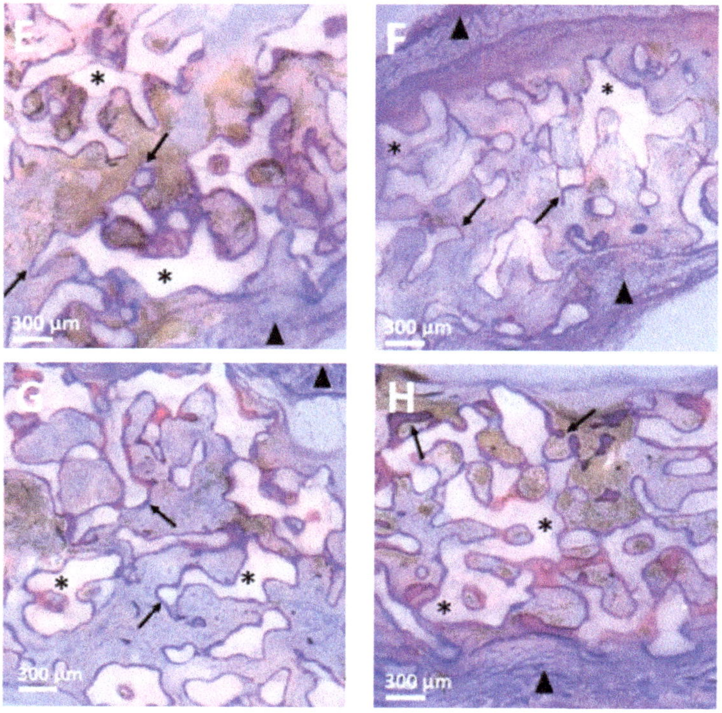

Figure 2. Bone formation in the groups 1 to 8. Pictures showing group 1 as A, group 2 as B, group 3 as C etcetera. All groups containing 0.25 g CHA; with no BMP-2 (**A**); respectively, 1, 5, 10, 20 µg coating-incorporated BMP-2 (**B–E**); 20 µg adsorbed BMP-2 (**F**) and, respectively, 40, 60 µg of coating-incorporated BMP-2 (**G,H**). ▲—fibrous connective tissue; ↗—newly formed bone; *—CHA.

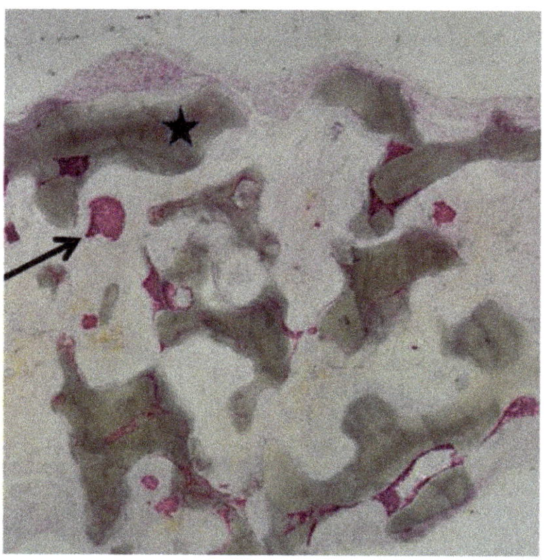

Figure 3. Newly formed bone (↗) not in contact with CHA (★).

3.4. Histomorphometric Analysis

The bone/total area percentage was zero in the group of 0.25 g CHA coated with BioCaP and containing no BMP-2 per sample, indicating that there was no bone formed at all. The bone/total area percentage of the group with 0.25 g CHA coated with BioCaP containing 20 μg of adsorbed BMP-2 per sample was significantly less than all other groups with BMP-2, with the exception of the group of 0.25 g CHA coated with BioCaP containing 20 μg of coating-incorporated BMP-2 per sample.

The CHA/total area percentage shows that the groups with 0.25 g CHA containing 1, 5 and 10 μg of coating-incorporated BMP-2 per sample behave differently compared to groups with 0.25 g CHA containing 0, 20, 40, 60 μg of coating-incorporated BMP-2 per sample, as well as with the group of 0.25 g CHA containing 20 μg of adsorbed BMP-2 per sample. There seems to be more remaining CHA in groups with 0.25 g CHA containing 1, 5 and 10 μg of coating-incorporated BMP-2 per sample compared to the other groups.

3.5. Statistical Findings CHA/Total Area Percentage

In Figures 4–6, the whisker extending down represents the minimum value, excluding any outliers. The whisker extending up represents the maximum value, excluding any outliers. An outlier is shown as a dot (Figure 6). The bottom of the rectangle represents quartile one, and the top of the rectangle represents quartile three. The median is shown by the horizonal line in the rectangle. The mean is marked with X in the box. The tests performed, with their confidence intervals, are described in Section 2.1.9.

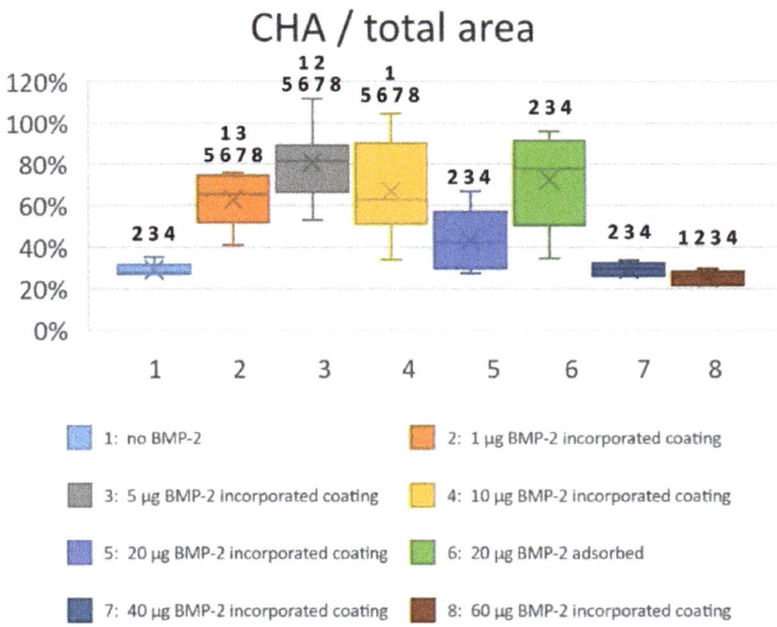

Figure 4. The CHA/total area percentage presented as a box plot with average, median and outliers. Vertical axis: percentage; horizontal axis: groups. There are statistically significant differences between the group represented by the bar and the groups represented by a number above the bar. The group of 0.25 g CHA containing 5 μg of coating-incorporated BMP-2 per sample shows the highest CHA/total area percentage.

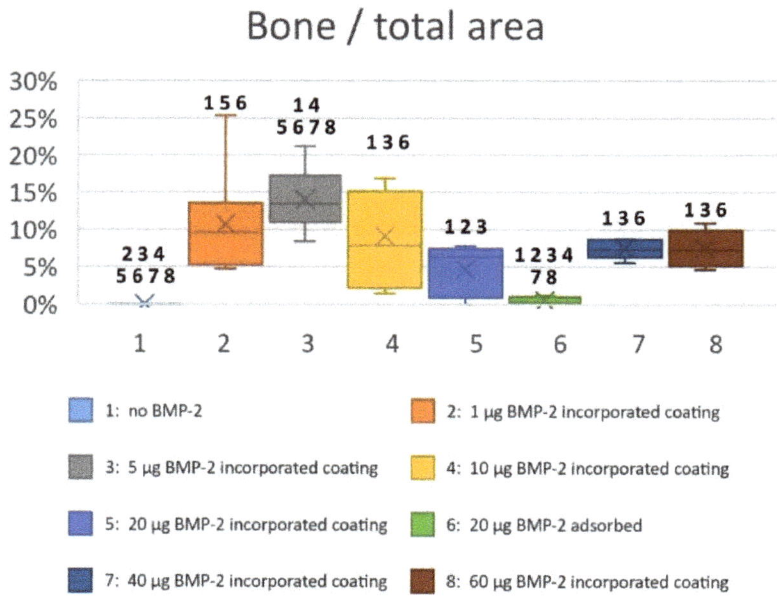

Figure 5. The bone/total area percentage presented as a box plot with average, median and outliers. Vertical axis: percentage; horizontal axis: groups. There are statistically significant differences between the group represented by the bar and the groups represented by a number above the bar. The group with 0.25 g CHA containing 5 µg of coating-incorporated BMP-2 shows the highest average bone/total area percentage.

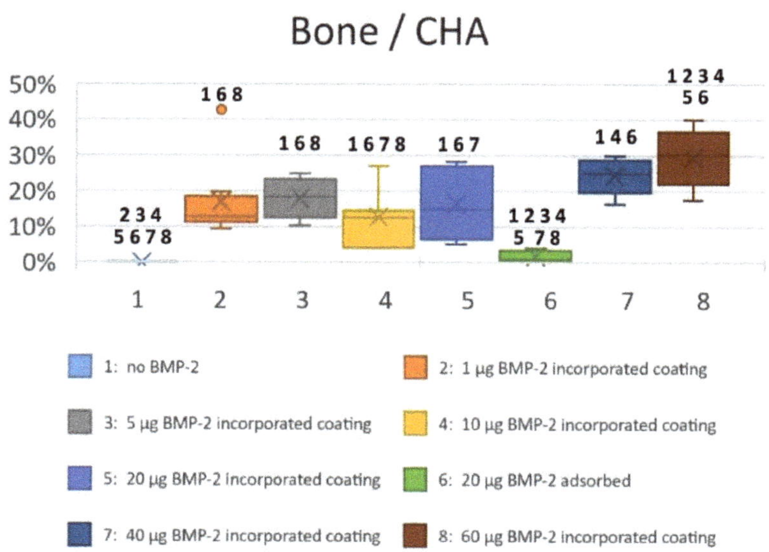

Figure 6. The bone/CHA percentage presented as a box plot with average, median and outliers. Vertical axis: percentage; horizontal axis: groups. There are statistically significant differences between the group represented by the bar and the groups represented by a number above the bar.

Figure 4 illustrates that groups of 0.25 g CHA containing 1, 5, 10 µg/sample of coating-incorporated BMP-2 and 20 µg/sample of adsorbed BMP-2 contain the most non-degraded CHA within all groups, whereby the group of 5 µg/sample of coating-incorporated BMP-2 within these groups shows the least spread in CHA/total area percentage. Two outliers show a percentage > 100, which can be caused by the cutting of the slices and the position of the grid at point counting.

3.6. Statistical Findings Bone/Total Area Percentage

The group containing 20 μg/sample of adsorbed BMP-2 induces less ectopic bone formation compared to all other groups with coating-incorporated BMP-2. In addition, the box-plot in Figure 5 shows that the group with 0.25 g CHA containing 20 μg of adsorbed BMP-2 per sample is less efficient in bone formation than the group with 0.25 g CHA containing 20 μg of coating-incorporated BMP-2 per sample; however, there is no statistically significant difference (Figure 5). Increasing the amount of incorporated BMP-2 from 5 up to 60 μg per sample does not result in more ectopic bone formation. There are statistically significant differences between all groups, whereby the group of 5 μg of coating-incorporated BMP-2 per sample shows the highest bone/total area percentage within the range of 1 to 60 μg/sample of coating-incorporated BMP-2, when the outlier of the group of 1 μg/sample of coating-incorporated BMP-2 is excluded.

3.7. Statistical Findings Bone/CHA Percentage

The group without BMP-2 has a bone/CHA percentage of 0, whereas the group of 0.25 g CHA containing 20 μg of adsorbed BMP-2 per sample shows the lowest bone/CHA percentage (Figure 6). The groups of 0.25 g CHA containing 40 and 60 μg of coating-incorporated BMP-2 per sample show the highest bone/CHA percentage. There are no statistical differences (Figure 6) between the groups of 0.25 g CHA containing 1, 5, 10 and 20 μg of coating-incorporated BMP-2 per sample.

4. Discussion

To evaluate the efficacy in bone formation of different BMP-2 concentrations in coating-incorporated CHA, in vivo research is necessary. Choosing the right animal model depends on various factors, such as significant physiological and pathophysiological analogies when compared to humans, its manageability to operate, after which it is easily possible to observe a multiplicity of study objects over time, its post-surgical period, which is preferably relatively short, the costs for animal acquisition and care, animal availability, tolerance to captivity, ease of housing and acceptability to society [17]. Male Wistar rats meet the described requirements well. However, rodent models do not show Haversian-type remodeling in the bone cortex, whereas larger animals do [17]. We opted for an ectopic ossification model because we wanted to exclude influences of any possible spontaneous bone healing.

Studies have shown that a burst release of BMP-2 alone leads to insufficient bone formation [6,24], whilst coating-incorporated BMP-2 is more effective compared to the same amount of adsorbed BMP-2 [16]. This is due to the long, slow and stable release of the coating-incorporated BMP-2 compared to the burst release of adsorbed BMP-2 [24]. However, we did not know if this is also the case with coated CHA [16]. We also did not know which concentration of BMP-2 was the most effective using coated CHA [13]. Our initial research was with five groups and a total of 15 animals. Due to the fact that we did not know which coating-incorporated amount of BMP-2 would be effective in ectopic bone formation, we started with 0 μg of BMP-2 in the control group, increasing in every next group with 20 up to 60 μg/sample per group. To minimize animal sacrifice, we made only one control group with absorbed BMP-2. This group (6) contained 20 μg of adsorbed BMP-2 per sample. For bone/total area percentages, no statistically significant differences were found (Figure 5) between groups 5, 7 and 8 (20, 40 and 60 μg of coating-incorporated BMP-2 per sample). However, the concentrations of the coating-incorporated BMP-2 per sample differ from 20 to 60 μg. A concentration above 20 μg of coating-incorporated BMP-2 per sample did not result in more ectopic bone formation. This is why we started our second research. We created a greater sample size of three groups with a total of 12 animals and a coating-incorporated BMP-2 of 1, 5 and 10 μg per sample.

Ideal bone graft material should provide an osteoconductive matrix, which allows for vascular invasion and cellular infiltration, as well as osteoinductive factors, which recruit and induce mesenchymal stem cells to differentiate [6]. CHA alone has osteoconductive characteristics, but no osteoinductive characteristics [8]. BMP-2 does have osteoinductive characteristics [12,16]. In normal male Wistar rats, no bone is present in the dorsal subcutaneous pocket we created. Therefore, this study proves, like many others [12,16,25], that BMP-2 is capable of inducing bone formation.

In our histological slices, bone was visible in direct contact with the graft material, as well as solitary, and not in direct contact with the BioCaP coating of the CHA (Figure 3). This is due to two reasons. Firstly, research shows that BMP-2 can induce bone formation at a distance of up to 340 μm from the biomimetic calcium phosphate as a carrier on a titanium disc implanted in an ectopic model [12]. Secondly, it is due to the fabrication method of the histological sections [26]. Regarding the slice level of the samples, it shows that the angle at which the sample is cut into slices, from three-dimensional objects to two-dimensional slices, determines the display image of the two-dimensional slice. For this reason, we can find bone in a slice that is not connected with the CHA surface. The bridges of newly formed bone between the two granules of CHA are based on the same phenomenon, displaying a three-dimensional structure in two dimensions.

When we exclude the control groups (1 and 6: only CHA without BMP-2 and 20 μg of adsorbed BMP-2) and look at the CHA/total area percentages (Figure 4) of our first experiment, with the groups 5, 7 and 8 (20, 40 and 60 μg of coating-incorporated BMP-2 per sample), it is striking that there are no statistically significant differences between these groups. There are also no statistically significant differences for CHA/total area percentages between the groups of our second experiment (groups 2, 3 and 4: containing 1, 5 and 10 μg of coating-incorporated BMP-2 per sample). There are, however, statistically significant differences between the groups for the CHA/total area percentage of our first and second experiment. There may have been minor differences in the material composition between the groups from our first and second experiments, although the same protocols were followed. Less CHA means a faster degradation of CHA. The amount of BMP-2 above 5 μg/sample could be considered as an ineffective burst release, causing no bone formation. However, it does cause extra blood supply [6], which can be responsible for the CHA resorption. Our previous study [27] has proven that BMP-2 can stimulate the degradation as well as the formation of bone. A higher concentration of BMP-2 does not always lead to more bone formation; it can cause a reduced bone volume, due to the fact that BMP-2 stimulates osteoblasts as well as osteoclasts. Mesenchymal stem cells (MSCs) are a prerequisite for heterotopic bone formation because they can differentiate into osteoblasts [28]. A certain concentration of BMP-2 could be enough for the differentiation of all the MSCs present, thus explaining why increasing the concentration of coating-incorporated BMP-2 above a critical value does not generate additional bone.

There are no statistically significant differences between groups 2 and 3 (containing 1 and 5 μg of coating-incorporated BMP-2 per sample) for the bone/total area percentage (Figure 5). There are statistically significant differences between groups 3 (containing 5 μg of coating-incorporated BMP-2 per sample) and all other groups, containing a higher dose of BMP-2 per sample for the bone/total area percentage (Figure 5). Group 3 (containing 5 μg of coating-incorporated BMP-2 per sample) induces more bone formation than the other groups (Figure 5). This finding would indicate that there is a maximum concentration, above which, an increase in concentration is of no use, and in this study, there is an optimal concentration of 5 μg of coating-incorporated BMP-2 per sample.

It is to be expected that there is a statistically significant difference between groups 5 and 6, whereby the coating-incorporated BMP-2 behaves as superior in bone formation compared to the same amount of adsorbed BMP-2. The spreads (Figures 4 and 5), of the CHA/total area and bone/total area percentage indicate this. However, we found only a statistical difference in the bone/CHA percentage (Figure 6). This may be caused by the relatively high doses of BMP-2, as we now know that increasing the amount of incorporated BMP-2 from 5 up to 60 μg/sample does not result in more ectopic bone formation.

5. Conclusions

CHA, with BioCaP alone, is not able to generate ectopic bone formation. CHA coated with BioCaP and coating-incorporated BMP-2 is effective in ectopic bone formation and more effective than CHA coated with BioCaP and adsorbed BMP-2. In this study CHA coated with BioCaP and coating-incorporated BMP-2 in different concentrations above 5 μg per sample does not result in more ectopic bone formation.

Author Contributions: X.L. and H.J.J.U. equally contributed in study design and data analysis, completing the manuscript. Writing by H.J.J.U., L.W., X.Z. and L.D. took care of the animals, surgery, data collections and revision of the manuscript. Y.Z. and M.W. assisted with the animal care, histological preparation and data analysis. D.W.

gave input the manuscript writing and its revision. Y.L., as a correspondence author, was involved in the study design, animal care, data collection and analysis, finalizing the manuscript. All authors have read and agreed to the published version of the manuscript.

Funding: This research was funded by Uijlenbroek & Partners, dentists. This project was practically supported by the Zhejiang Provincial Natural Science Foundation of China (LQ19H280008) and Shandong Taishan Scholar Program to Y.L.

Institutional Review Board Statement: The experiment was approved by the committee of the Board of Animal Experiments, traditional Chinese Medicine University, Hangzhou, China, according to Chinese law (Accreditation number: ZSLL-2016-46).

Informed Consent Statement: Not applicable.

Data Availability Statement: Not applicable.

Acknowledgments: The authors would like to thank Naichuan Su for the assistance on the statistical analysis.

Conflicts of Interest: The authors declare no conflict of interest.

References

1. Schmitz, J.P.; Hollinger, J.O. The critical size defect as an experimental model for craniomandibulofacial nonunions. *Clin. Orthop. Relat. Res.* **1986**, *205*, 299–308. [CrossRef]
2. Van Heest, A.; Swiontkowski, M. Bone-graft substitutes. *Lancet* **1999**, *353* (Suppl. S1), S28–S29. [CrossRef]
3. Goulet, J.A.; Senunas, L.E.; DeSilva, G.L.; Greenfield, M.L. Autogenous iliac crest bone graft: Complications and functional assessment. *Clin. Orthop. Relat. Res.* **1997**, *339*, 76–81. [CrossRef] [PubMed]
4. Dimitriou, R.; Mataliotakis, G.I.; Angoules, A.G.; Kanakaris, N.K.; Giannoudis, P.V. Complications following autologous bone graft harvesting from the iliac crest and using the RIA: A systematic review. *Injury* **2011**, *42* (Suppl. S2), S3–S15. [CrossRef] [PubMed]
5. Chavda, S.; Levin, L. Human studies of vertical and horizontal alveolar ridge augmentation comparing different types of bone graft materials: A systematic review. *J. Oral Implantol.* **2018**, *44*, 74–84. [CrossRef]
6. Miron, R.J.; Zhang, Y.F. Osteoinduction: A review of old concepts with new standards. *J. Dent. Res.* **2012**, *91*, 736–744. [CrossRef]
7. Winkler, T.; Sass, F.A.; Duda, G.N.; Schmidt-Bleek, K. A review of biomaterials in bone defect healing, remaining shortcomings and future opportunities for bone tissue engineering: The unsolved challenge. *Bone Joint Res.* **2018**, *7*, 232–243. [CrossRef]
8. Damien, E.; Revell, P.A. Coralline hydroxyapatite bone graft substitute: A review of experimental studies and biomedical applications. *J. Appl. Biomater. Biomech.* **2004**, *2*, 65–73.
9. Holmes, R.E. Bone regeneration within a coralline hydroxyapatite implant. *Plast. Reconstr. Surg.* **1979**, *63*, 626–633. [CrossRef]
10. Rahimi, F.; Maurer, B.T.; Enzweiler, M.G. Coralline hydroxyapatite: A bone graft alternative in foot and ankle surgery. *J. Foot Ankle Surg.* **1997**, *36*, 192–203. [CrossRef]
11. Wang, H.L.; Boyapati, L. "PASS" principles for predictable bone regeneration. *Implant Dent.* **2006**, *15*, 8–17. [CrossRef]
12. Liu, Y.; de Groot, K.; Hunziker, E.B. BMP-2 liberated from biomimetic implant coatings induces and sustains direct ossification in an ectopic rat model. *Bone* **2005**, *36*, 745–757. [CrossRef]
13. Lin, X.; Hunziker, E.B.; Liu, T.; Hu, Q.; Liu, Y. Enhanced biocompatibility and improved osteogenesis of coralline hydroxyapatite modified by bone morphogenetic protein 2 incorporated into a biomimetic coating. *Mater. Sci. Eng. C* **2019**, *96*, 329–336. [CrossRef] [PubMed]
14. Lissenberg-Thunnissen, S.N.; de Gorter, D.J.; Sier, C.F.; Schipper, I.B. Use and efficacy of bone morphogenetic proteins in fracture healing. *Int. Orthop.* **2011**, *35*, 1271. [CrossRef]
15. Liu, Y.; Wu, G.; de Groot, K. Biomimetic coatings for bone tissue engineering of critical-sized defects. *J. R. Soc. Interface* **2010**, *7* (Suppl. S5), S631–S647. [CrossRef]
16. Liu, Y.; Huse, R.O.; de Groot, K.; Buser, D.; Hunziker, E.B. Delivery mode and efficacy of BMP-2 in association with implants. *J. Dent. Res.* **2007**, *86*, 84–89. [CrossRef] [PubMed]
17. Li, Y.; Chen, S.K.; Li, L.; Qin, L.; Wang, X.L.; Lai, Y.X. Bone defect animal models for testing efficacy of bone substitute biomaterials. *J. Orthop. Translat.* **2015**, *3*, 95–104. [CrossRef] [PubMed]
18. Wu, G.; Hunziker, E.B.; Zheng, Y.; Wismeijer, D.; Liu, Y. Functionalization of deproteinized bovine bone with a coating-incorporated depot of BMP-2 renders the material efficiently osteoinductive and suppresses foreign-body reactivity. *Bone* **2011**, *49*, 1323–1330. [CrossRef] [PubMed]
19. Liu, Y.; Layrolle, P.; de Bruijn, J.; van Blitterswijk, C.; de Groot, K. Biomimetic coprecipitation of calcium phosphate and bovine serum albumin on titanium alloy. *J. Biomed. Mater. Res. Off. J. Soc. Biomater. Jpn. Soc. Biomater. Aust. Soc. Biomater. Korean Soc. Biomater.* **2001**, *57*, 327–335. [CrossRef]
20. Schenk, R.K.; Olah, A.J.; Herrmann, W. *Preparation of Calcified Tissues for Light Microscopy*; Dickson, G.R., Ed.; Methods Calcif. Tissue Prep; Elsevier: Amsterdam, The Netherlands, 1984; pp. 1–56.

21. Maniatopoulos, C.; Rodriguez, A.; Deporter, D.A.; Melcher, A.H. An improved method for preparing histological sections of metallic implants. *Int. J. Oral Maxillofac. Implant.* **1986**, *1*, 31–37.
22. CRUZ-ORIVE, L.M. Precision of Cavalieri sections and slices with local errors. *J. Microsc.* **1999**, *193*, 182–198. [CrossRef]
23. Wu, G.; Liu, Y.; Iizuka, T.; Hunziker, E.B. The effect of a slow mode of BMP-2 delivery on the inflammatory response provoked by bone-defect-filling polymeric scaffolds. *Biomaterials* **2010**, *31*, 7485–7493. [CrossRef] [PubMed]
24. Liu, Y.; Schouten, C.; Boerman, O.; Wu, G.; Jansen, J.A.; Hunziker, E.B. The kinetics and mechanism of bone morphogenetic protein 2 release from calcium phosphate-based implant-coatings. *J. Biomed. Mater. Res. A* **2018**, *106*, 2363–2371. [CrossRef] [PubMed]
25. Liu, T.; Wu, G.; Zheng, Y.; Wismeijer, D.; Everts, V.; Liu, Y. Cell-mediated BMP-2 release from a novel dual-drug delivery system promotes bone formation. *Clin. Oral Implant. Res.* **2014**, *25*, 1412–1421. [CrossRef] [PubMed]
26. Johansson, C.; Morberg, P. Cutting directions of bone with biomaterials in situ does influence the outcome of histomorphometrical quantifications. *Biomaterials* **1995**, *16*, 1037–1039. [CrossRef]
27. Liu, Y.; Enggist, L.; Kuffer, A.F.; Buser, D.; Hunziker, E.B. The influence of BMP-2 and its mode of delivery on the osteoconductivity of implant surfaces during the early phase of osseointegration. *Biomaterials* **2007**, *28*, 2677–2686. [CrossRef]
28. Kenkre, J.; Bassett, J. The bone remodelling cycle. *Ann. Clin. Biochem.* **2018**, *55*, 308–327. [CrossRef]

coatings MDPI

Article

Galvanic Deposition of Calcium Phosphate/Bioglass Composite Coating on AISI 316L

Claudio Zanca [1], Alessandro Milazzo [1], Simona Campora [2], Elisa Capuana [1], Francesco Carfì Pavia [1,3], Bernardo Patella [1], Francesco Lopresti [1], Valerio Brucato [1], Vincenzo La Carrubba [1,4] and Rosalinda Inguanta [1,*]

1 Department of Engineering, University of Palermo, Viale delle Scienze, 90128 Palermo, Italy
2 Department of Biological, Chemical and Pharmaceutical Science and Technologies, University of Palermo, Viale delle Scienze, Ed. 16, 90128 Palermo, Italy; simona.campora@unipa.it
3 Consorzio Universitario di Caltanissetta, Corso Vittorio Emanuele 92, 93100 Caltanissetta, Italy
4 ATeN Center, University of Palermo, Viale delle Scienze, 90133 Palermo, Italy
* Correspondence: rosalinda.inguanta@unipa.it

Abstract: Calcium phosphate/Bioglass composite coatings on AISI 316L were investigated with regard to their potential role as a beneficial coating for orthopedic implants. These coatings were realized by the galvanic co-deposition of calcium phosphate compounds and Bioglass particles. A different amount of Bioglass 45S5 was used to study its effect on the performance of the composite coatings. The morphology and chemical composition of the coatings were investigated before and after their aging in simulated body fluid. The coatings uniformly covered the AISI 316L substrate and consisted of a brushite and hydroxyapatite mixture. Both phases were detected using X-ray diffraction and Raman spectroscopy. Additionally, both analyses revealed that brushite is the primary phase. The presence of Bioglass was verified through energy-dispersive X-ray spectroscopy, which showed the presence of a silicon peak. During aging in simulated body fluid, the coating was subject to a dynamic equilibrium of dissolution/reprecipitation with total conversion in only the hydroxyapatite phase. Corrosion tests performed in simulated body fluid at different aging times revealed that the coatings made with 1 g/L of Bioglass performed best. These samples have a corrosion potential of −0.068V vs. Ag/AgCl and a corrosion current density of 8.87×10^{-7} A/cm^2. These values are better than those measured for bare AISI 316L (−0.187 V vs. Ag/AgCl and 2.52×10^{-6} A/cm^2, respectively) and remained superior to pure steel for all 21 days of aging. This behavior indicated the good protection of the coating against corrosion phenomena, which was further confirmed by the very low concentration of Ni ions (0.076 ppm) released in the aging solution after 21 days of immersion. Furthermore, the absence of cytotoxicity, verified through cell viability assays with MC3T3-E1 osteoblastic cells, proves the biocompatibility of the coatings.

Keywords: AISI 316L; Bioglass 45S5; coating; corrosion; cytotoxicity; galvanic deposition; hydroxyapatite; orthopedic implant

1. Introduction

Ceramic materials are inorganic compounds that have been gradually incorporated into our everyday lives [1,2]. Thanks to their thermal and electrical insulating properties, ceramic materials can be applied as technological solutions in the fields of telecommunications, energy, and the environment [3]. Ceramic materials are popularly used in the field of biomedical applications, and those useful in this field are usually referred to as bioceramics [4–7].

Bioceramics play a crucial role in tissue engineering and orthopedics because of their chemical inertia and high wettability, favoring proteins and cell adhesion [8–11]. Therefore, different types of ceramic materials have been studied [12].

Bioceramics can be subdivided into three main groups depending on the interactions between the used materials and the human body. Bioinert ceramics do not interact with the human environment and, therefore, can be incorporated into implanted living bone without presenting any sign of toxicity [13,14]. In contrast, bioactive ceramics possess the ability to initiate a biological response after implantation into the body, such as cell stimulation and stem cell proliferation for the regeneration of damaged tissue or whole organs [15–17]. Finally, when in contact with body fluids, bioresorbable ceramics can be fully metabolized, easily dissolved, and subsequently excreted [18,19].

Among the ceramic materials used in biomedical applications, Bioglass (BG) and calcium phosphate (CaP) compounds have found great applications in repairing and reinforcing bone tissue in orthopedics and dentistry [20–24].

Bioglasses are silica-based bioactive ceramics developed in the 1970s by Hench through mixing different oxides using the melt-quenching method [25]. The bioactivity of Bioglass is linked to the chemical composition [23], which, in this case, refers to how easily surface chemical reactions occur, as it can generate a hydroxyapatite layer and promote excellent osseointegration [26]. Specifically, Bioglass with 45–52 wt% SiO_2 ensures good adhesion to bone and soft tissue, whereas a content of 52–60 wt% SiO_2 only allows for bone adhesion [27,28]. As soon as its SiO_2 concentration increases above 60 wt%, Bioglass loses its bioactivity, with a dramatic decrease in terms of osseointegration [29]. Approved by the Food and Drug Administration in 1985, Bioglass® 45S5 (45% SiO_2, 24.5% CaO, 24.5% Na_2O, and 6% P_2O_5) has been widely used to repair bone and dental defects [30]. Bioglasses have remarkable osteoconductivity and osteoinductivity properties in that they induce primitive cells to differentiate, forming new bone tissue [30,31]. Baino et al. [32] shed light on the potential of Bioglass, which has not only been used in the field of orthopedics (e.g., as scaffolds for bone regeneration or coatings for orthopedic implants) but also for soft tissue regeneration, ocular implants, wound healing, and percutaneous catheters [33]. Bioglass–polymer composites have also been proposed [34–36].

Calcium phosphates are another class of ceramic materials that have been successfully applied in the biomedical field [37,38]. In contrast to Bioglass, calcium phosphate compounds allow for the osteoconduction of pre-differentiated bone cells [21,39–41]. A comprehensive overview of calcium-phosphate-based biomaterials in orthopedics was presented by Hou et al. [42], who summarized the advantages and disadvantages of these materials. Since the early 1970s, hydroxyapatite ($Ca_{10}(PO_4)_6(OH)_2$) has increasingly been utilized in orthopedics since it represents the mineral part of bone tissue [43,44]. Thanks to its great stability [45], it is suitable for the fabrication of scaffolds, bone void fillers, and maxillofacial and dental surgery [42,46]. Despite its weak mechanical strength, hydroxyapatite has largely been utilized as the main biomaterial for the fabrication of coatings in orthopedics [47]. Many studies have demonstrated the improvement of cellular response thanks to this material's high biocompatibility and enhanced protection against corrosion phenomena [48–50]. Moreover, it has been demonstrated that the biological and physicochemical properties of hydroxyapatite can be increased by doping it with various ions such as iron [51], magnesium [52], europium [53], samarium [54], silver [55], and cerium [56].

The properties of pure calcium phosphate compounds and Bioglass are synergically increased in their composites [57]. From a biological point of view, the synergistic action of hydroxyapatite and Bioglass improves cell adhesion and proliferation [58] and promotes osteogenic activity [59]. Moreover, in vitro tests conducted on hydroxyapatite/Bioglass-coated substrates have shown that the presence of the coating remarkably increases cell viability and proliferation compared to no coated metallic surfaces [60,61].

Regarding the deposition methods used for composite coatings, a comprehensive, state-of-the-art technique was proposed by Maximov et al. [62]. These authors reported different kinds of coating deposition techniques. Most of the used methods are characterized by high-vacuum and/or high-temperature steps, such as enameling [63], plasma spray deposition [64], air plasma spraying [65], suspension plasma spraying [66], radio frequency magnetron sputtering [67], laser-pulsed deposition [68], thermal spraying [69], and laser cladding [70]. In addition, these methods are difficult to manage and require the use of specific equipment and specialized personnel. Electrophoretic deposition [71], electrodeposition [72,73], and sol–gel deposition [74,75] represent other methods of deposition that are more user-friendly to perform, but they require the use of specific equipment. Furthermore, typically after electrophoresis, a high-temperature treatment step is required to improve the adhesion strength between the Bioglass coating and the substrate [76].

In this work, galvanic deposition was used to achieve the co-deposition of a calcium phosphate/Bioglass composite coating on 316L stainless steel, constituting the first-ever use of this method for such an application. This substrate was selected from among those usable for orthopedic implants [77] because of its excellent biocompatibility and good mechanical properties [78]. The calcium phosphate/Bioglass composite coating is designed to enhance the biocompatibility of this substrate [79].

Unlike the methods of deposition mentioned earlier, galvanic deposition is a simple technique that does not require a power supply [80]. The galvanic coupling between two metals (a working electrode and a sacrificial anode) plays a crucial role in the process. The metals must be characterized by different values of standard redox potential. In addition, another positive point to highlight is the controllability of the galvanic process since the deposition rate depends on the ratio between anodic and cathodic areas. Using this method, we have obtained different types of coatings on stainless steel [81–88]. The galvanic deposition of brushite on Ti alloy, which was converted into hydroxyapatite through a subsequent hydrothermal treatment, was also reported by Chen et al. [89]. Moreover, this technique can be useful for the fabrication of nanowire arrays via the template-assisted method [90–94]. Galvanic replacement was also used to prepare bimetallic and ternary electrocatalysts for applications in fuel cells and the electrolysis and electrosynthesis reactions [95]. Since it could be considered a spontaneous process, galvanic deposition does not require sophisticated equipment or an experienced operator. In addition, it can be considered an eco-friendly process since a sacrificial anode (e.g. aluminum) can be procured from end-life materials to provide added value to waste. A limitation of this method is that it can only be used for cathodic deposits; moreover, if the deposition parameters are not suitably optimized, the coatings obtained may present numerous defects [96].

In this work, we will attempt to demonstrate that the galvanic method is also useful for achieving the co-deposition of calcium phosphate compounds and Bioglass and that the coating obtained has excellent performance in terms of corrosion resistance and biocompatibility and does not require further heat treatments. To achieve this aim, physical–chemical and electrochemical characterizations were carried out to investigate the coatings' morphologies, chemical compositions, and performance. In addition, biocompatibility and the release of metal ions from the steel substrate during aging in Simulated Body Fluid (SBF) were evaluated. The results demonstrate that composite coatings increase the resistance of the analyzed substrate against corrosion phenomena without inducing any cytotoxic effects.

2. Materials and Methods

Calcium phosphate/Bioglass composite coatings were obtained through galvanic coupling between commercial AISI 316L (working electrode) and zinc (sacrificial anode). AISI 316L bars (1.5 cm × 7 cm × 0.2 cm, UNS S31603, 0.022% wt. C, 16.61% wt. Cr, 10.02% wt. Ni, 1.244% wt. Mn, 0.321% wt. Si, 0.029% wt. P, 0.001% wt. S, 2.010% wt. Mo, Fe at balance) were mechanically pretreated with abrasive paper (#150, #320, #800, and #1200, LECO). Afterward, a degreasing step applied to the metallic surface was carried out in an ultrasonic bath, first in acetone and then in water, for 10 min for each sample. The zinc sheets (3 cm × 7 cm) were pretreated with the same procedure. After cleaning, the working electrode and sacrificial anode were delimited with an insulator lacquer to expose an electroactive area of 1.13 cm^2 and 27 cm^2, respectively.

Galvanic deposition was carried out in a two-compartment cell (Figure S1 in Supplementary Materials). The electrolyte solution for the cathodic compartment consisted of 0.061M $Ca(NO_3)_2$ $4H_2O$, 0.036M $NH_4H_2PO_4$, and 1 M $NaNO_3$. Different amounts of Bioglass 45S5 were added to the solution (0.5 gL^{-1}, 1 gL^{-1}) to observe the effects related to corrosion behavior. All reagents were purchased from Sigma Aldrich. Deposition was conducted while stirring at 400 rpm so that the Bioglass particles remained suspended.

A 1 M sodium chloride solution was used in the anode compartment. The corresponding deposition was carried out for 24 h at room temperature. After the deposition, the samples were air-dried.

Coating morphology was scrutinized using a field emission scanning microscope (QUANTA 200, FEI) equipped with an energy-dispersive X-ray spectroscopy (EDS, Ametek, Berwyn, PA, USA) probe. The spatial resolution of EDS measurements is about 1 μm, and the relative error is lower than 3%.

The crystallographic structures were studied via X-ray diffraction using a RIGAKU instrument (D-MAX 25600 HK). The analyses were carried out in the 2-theta range from 10° to 60° using copper Kα radiation (λ = 1.54 Å) with the following setup conditions: tube voltage of 40 kV, current of 100 mA, a scan speed of $4°min^{-1}$, and sampling of 0.01°. The X-Ray diffraction patterns were studied and compared with the International Centre of Data Diffraction database [97].

Raman spectroscopy was performed using a Renishaw (inVia Raman Microscope) spectrometer. The resolution was 0.5 cm^{-1}. Excitation originated from the 532 nm line of an Nd:YAG laser calibrated by the Raman peak of polycrystalline Si (520 cm^{-1}). For this laser with a 100× objective, the theoretical spatial resolution is about 360 nm. The Raman spectra were analyzed via comparison with the RHUFF database [98].

To quantify the metal ion concentrations released from the sample after 21 days of aging in SBF at 37 ± 1 °C, Inductively Coupled Plasma Optical Emission Spectroscopy (ICP-OES, PerkinElmer Optima 2100 DV) was also executed. A calibration line was obtained for each ion (Fe, Ni, Cr, Mo, Ca, and P) using standard calibration solutions. The evaluation of corrosion behavior was studied by aging the samples in a simulated body fluid prepared according to the procedure reported in [86]. The samples were kept in SBF at 37 °C for 21 days and were periodically extracted to perform electrochemical tests. Corrosion tests consisted of the monitoring of open-circuit potential (OCP), Potentiodynamic polarization (PP), and Electrochemical Impedance Spectroscopy (EIS). These tests were performed in a conventional three-electrode cell with a Pt wire used as the counter electrode and 3.0 M Ag/AgCl as the reference electrode. Corrosion potential (E_{corr}) and corrosion current density (i_{corr}) were determined by extrapolation of Tafel's curves. Polarization measurements were performed with a scan rate of 0.166 mVs^{-1} in a potential range of ±150 mV to open-circuit potential. Electrochemical Impedance Spectroscopy was carried out in the frequency range from 100 kHz to 0.1 Hz, with 0.010 V of AC perturbation. The impedance data were fitted using ZSimpWin software with an equivalent circuit (EC).

Before performing biocompatibility tests, the AISI 316L and CaP—1 gL^{-1} BG samples (1.5 cm × 3 cm × 0.2 cm) were sterilized by soaking them in a 70% (*v*/*v*) ethanol bath for 24 h followed by UV treatment for 2 h (1 h on each side) under a laminar fume hood.

Each sample was incubated with Dulbecco's modified Eagle's high-glucose medium (D-MEM, Sigma Aldrich, Saint Louis, MO, USA) supplemented with 10% (*v*/*v*) fetal bovine serum (Euroclone, Celbar), 100 units per ml of penicillin G, 100 μgmL^{-1} of streptomycin (Euroclone, Celbar, Pero, Italy), and 2 mM L-glutamine (Euroclone, Celbar) at 37 °C. The incubations were performed in a humidified atmosphere of 5% CO_2 for 24 h with a surface-to-volume area ratio of 3 cm^2/mL. Subsequently, each treated medium was collected in a 50 mL Falcon tube to carry out biocompatibility tests (named AISI 316L and CaP—1 gL^{-1} BG D-MEM, respectively).

Mouse pre-osteoblastic MC3T3-E1 cell lines at passage 7 (ECACC, European Collection of Cells Cultures) were cultured in complete D-MEM. The culture consisted of DMEM supplemented with 10% (*v*/*v*) fetal bovine serum, 100 units per ml of penicillin G, 100 μg/mL of streptomycin, and 2 mM L-glutamine. Cell growth transpired at 37 °C and in a 5% CO_2 atmosphere. A total of 1×10^4 cells/well were seeded on a twenty-four-well culture plate (with a single-well diameter of 14 mm) and incubated with complete D-MEM at 37 °C and 5% CO_2. After 24 h, the medium was replaced with fresh complete D-MEM (control), AISI 316L, or CaP—1 gL^{-1} BG D-MEM-treated media. At different times (2, 5, and 7 days) following cell seeding, the viability and cell proliferation rate of each sample were evaluated using the cell-counting Kit-8 (CCK-8) colorimetric assay (Sigma-Aldrich) according to the manufacturer's recommendations. Cells grown with each type of medium (complete D-MEM, AISI 316L D-MEM, or the CaP—1 gL^{-1} BG D-MEM) were incubated with CCK-8 reagent solution (10% in culture medium) for 2.5 h in a humidified incubator (37 °C; 5% CO_2).

Absorbance intensity (450 nm), which changes according to the degree of cell viability, was evaluated using a microplate reader (Synergy HT, Biotek, Winooski, VT, USA). The assay was performed ten times for each condition. The obtained data were compared using Student's t-test. A p value < 0.05 was considered significant.

Optical microscopy of MC3T3-E1 cells grown with complete D-MEM (control), AISI 316L D-MEM, and CaP—1 gL^{-1} BG D-MEM was performed at different times (2, 5, and 7 days).

A workflow of the fabrication and characterization process is reported in Scheme 1.

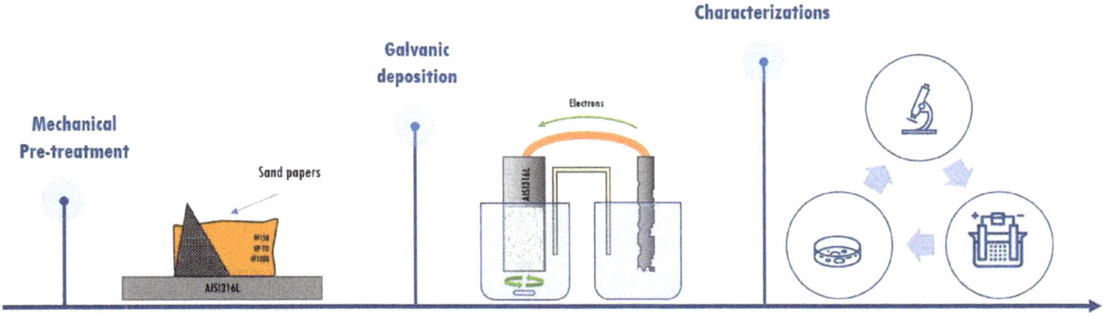

Scheme 1. Workflow of the fabrication and characterization process.

3. Results and Discussion

Galvanic deposition was carried out in a two-compartment cell shown in Figure 1. Several reactions occurred once the working electrode and sacrificial anode were short-circuited and placed in their corresponding solutions.

Zinc dissolution occurred in the anodic compart and has been represented in Equation (1):

$$Zn \rightarrow Zn^{2+} + 2e^- \qquad (E^\circ = -0.76\ V/NHE) \qquad (1)$$

Regarding the cathodic compartment, base electrogeneration reactions [99,100] occurred thanks to the electrons stemming from Reaction (1). At the cathode, surface nitrate ions, water molecules, and dissolved oxygen react in the following reactions (Equations (2)–(4)):

$$NO_3^- + H_2O + 2e^- \rightarrow NO_2^- + 2OH^- \qquad (E_{eq} = 0.0835 - 0.059\ pH\ V/NHE) \qquad (2)$$

$$2H_2O + 2e^- \rightarrow 2OH^- + H_2 \qquad (E_{eq} = 0.0 - 0.059\ pH\ V/NHE) \qquad (3)$$

$$O_2 + 2H_2O + 4e^- \rightarrow 4OH^- \qquad (E_{eq} = 1.23 - 0.059\ pH\ V/NHE) \qquad (4)$$

Due to the electrogeneration of hydroxyl ions, the deposition of calcium phosphate compounds is possible. The increase in the local pH at the cathode's surface drives the mechanism of deposition, shifting the dissociation equilibrium of $H_2PO_4^-$ toward HPO_4^{2-}, as shown in Equation (5):

$$H_2PO_4^- + OH^- \leftrightarrow HPO_4^{2-} + H_2O \qquad (pH = 7.19) \qquad (5)$$

Hydrogen phosphate ions formation induces the precipitation of brushite ($CaHPO_4 \cdot 2H_2O$) according to Equation (6):

$$Ca^{2+} + HPO_4^{2-} + 2H_2O \rightarrow CaHPO_4 \cdot 2H_2O \quad (6)$$

Since a non-uniform, non-conductive, and porous layer of brushite is formed, the reactions of base electrogeneration are not hindered. This ensures a continuous increase in the interface pH; therefore, the HPO_4^{2-} ion dissociates in the PO_4^{3-} ion according to equilibrium Equation (7):

$$HPO_4^{\ 2-} + OH^- \leftrightarrow PO_4^{3-} + H_2O\ (pH = 12.03) \quad (7)$$

As soon as the pH reaches a value above 12, hydroxyapatite precipitation takes place, as shown in Equation (8):

$$10Ca^{2+} + 6PO_4^{3-} + 2OH^- \rightarrow Ca_{10}(PO_4)_6(OH)_2 \downarrow \quad (pH > 12) \quad (8)$$

The composite coating is formed purely through physical incorporation during the deposition of calcium phosphates. The Bioglass present acts as a nucleation point for the formation of new crystals in the coating.

The mechanism of calcium phosphate formation in the presence of Bioglass has been extensively investigated in the literature [24]. In the first stage, there is ion exchange between H^+ from the solution with the Na^+ and Ca^{2+} from the Bioglass, which leads to the formation of silanol groups due to the hydrolysis of the silica groups. The reaction mechanism begins with the following reaction (Equation (9)):

$$Si\text{-}O\text{-}Ca^{2+} + H^+ \rightarrow Si\text{-}O\text{-}H + Ca^{2+} \quad (9)$$

Afterward, the dissolution of silica leads to the formation of $Si(OH)_4$ on the surface of the Bioglass, as shown in Equation (10):

$$Si\text{-}O\text{-}Si + H_2O \rightarrow 2Si\text{-}OH \quad (10)$$

The dissolution of the lattice leads to the realization of an insoluble form of silica that will serve as a nucleation center for the formation of calcium phosphates. A silica gel layer with a thickness ranging from 1 to 2 μm is formed on the surface of the Bioglass particles. The precipitation and migration of calcium ions on the surface occur, followed by the incorporation of hydroxide anions and phosphates.

To obtain a composite coating, the Bioglass must be kept in suspension in the deposition solution. Therefore, a preliminary study was carried out to evaluate the effect of the stirring of the solution during the galvanic process. Stirring (shown in Figure S2 in the Supplementary Materials) leads to the formation of a calcium phosphate coating with a typical needle-like shape and a smaller crystal size compared with the product obtained under unstirred conditions. This effect was studied by Azar et al. [101], who observed the electro-crystallization of hydroxyapatite coatings on Nitinol while changing the fluid-dynamic conditions of the solution. The growth of hydroxyapatite crystals is a step that is favored over nucleation in unstirred conditions. A concentration gradient of the precursors is produced within the diffusive layer close to the electrode. Therefore, the ions involved are incorporated within the crystal lattice, and crystal growth occurs perpendicularly to the surface. Under stirred conditions, the diffusive layer is thinner, and the bulk concentration is almost the same as that at the electrode/electrolyte interface. So, in this case, crystal nucleation is favored with respect to growth. However, the precursor concentration is high enough to allow crystal growth parallel to the surface.

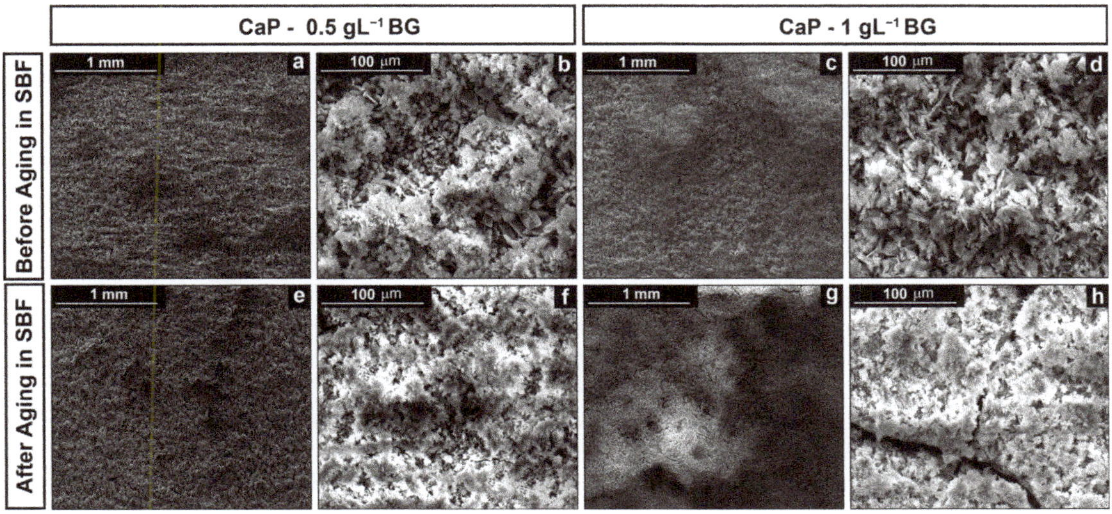

Figure 1. Micrographs of the coatings: (**a**,**b**) CaP—0.5 gL^{-1} BG and (**c**,**d**) CaP—1 gL^{-1} BG post-deposition; (**e**,**f**) CaP—0.5 gL^{-1} BG; and (**g**,**h**) CaP—1 gL^{-1} BG after aging in simulated body fluid.

A comparison between the energy-dispersive X-ray spectra post-deposition and after aging is shown in Figure 2. These characterizations provide semi-qualitative information regarding the chemical composition of the coating. In particular, the atomic ratio Ca/P accounts for the composition of the coating. On the other hand, Ca/Fe provides semi-qualitative information about thickness.

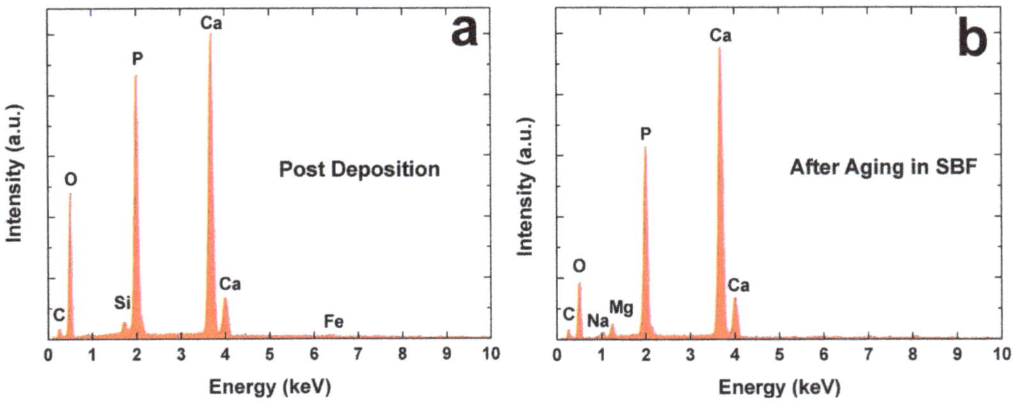

Figure 2. Energy-dispersive X-ray spectra of calcium phosphate/Bioglass coating: (**a**) post-deposition and (**b**) after aging in simulated body fluid.

From the results reported in Table 1, it can be concluded that the thickness of the coating increased with the increase in the amount of Bioglass in the deposition solution. This is due to the Bioglass acting as a nucleation center for the formation of calcium phosphates. The results shown in Table 1 also demonstrate that the coatings are composed of a mixture of brushite (Ca/P = 1) and hydroxyapatite (Ca/P = 1.59~1.86). Although brushite could be the main crystalline phase, the Ca/P ratio increases to 1.66 after 21 days of aging in simulated body fluid.

Table 1. Ca/P and Ca/Fe ratios, obtained using energy dispersive X-ray spectroscopy, of the coatings (post-deposition and after aging in simulated body fluid). The mean deviation is 1.2%.

	Post Deposition		After Aging in SBF	
	Ca/P	Ca/Fe	Ca/P	Ca/Fe
CaP—0.5 gL^{-1} BG	1.05	57	1.78	No Fe
CaP—1 gL^{-1} BG	1.03	92	1.66	No Fe

This result can be explained by referring to Reaction (11) proposed by Nur et al. [102], which leads to the total conversion of the brushite in the hydroxyapatite during aging in simulated body fluid:

$$10CaHPO_4 + 2OH^- \leftrightarrow Ca_{10}(PO_4)_6(OH)_2 + 4PO_4{}^{3-} + 10H^+ \ (pH < 12) \quad (11)$$

Although Fe atoms could be detected after galvanic deposition, no Fe atoms were found after aging in simulated body fluid for 21 days. Thus, due to dissolution/reprecipitation phenomena, an increase in the thickness of the coating occurs with the aging time. The Si peak observed corresponds to the Bioglass particles incorporated in the coating.

It is interesting to notice how the Si peak is not identified after aging due to the bioactivity of Bioglass to produce hydroxyapatite. The absence of a Si peak is due to the coating's dissolution in SBS. As reported in [103], the dissolution processes of composite coatings are quicker than pure calcium phosphate and Bioglass coatings. These high dissolution rates lead to rapid supersaturation at the coating/SBF interface and thus to the rapid precipitation of the hydroxyapatite phase [104]. Therefore, after immersion in SBF, the hydroxyapatite phase must be the only one present in the coating. To verify this phenomenon, the post-aged samples were characterized via XRD and Raman spectroscopy, and the results were compared with the pre-aged samples.

Figure 3 shows X-ray diffraction patterns of the samples' varying Bioglass concentration. In Figure 3a, the characteristic brushite peaks were identified, with values of 2-theta equal to 11.65°, 20.95°, 29.3°, and 30.54°, as well as some hydroxyapatite peaks with lower intensity. In addition, some peaks belonging to the substrate were present. Nevertheless, the intensity of the peaks was shielded by the presence of the coating. No diffraction peaks could be detected for Bioglass due to its amorphous nature (see Figure S3 in the Supplementary Materials; the very broad band at about 32°–33° is typical of an amorphous structure).

After 21 days of aging in simulated body fluid (Figure 3b), the brushite peaks disappeared, and hydroxyapatite peaks are the only ones present, which is a finding that is in line with the results of energy-dispersive X-ray spectroscopy.

Grain size was evaluated using Scherrer's equation, considering the main peaks were located at around 2-theta 11.65° and 20.95°. The results (Figure S4 in the Supplementary Materials) demonstrate that the galvanic deposition of calcium phosphate in an unstirred solution leads to the formation of coatings with larger crystalline grain sizes. In a stirred solution, since the phenomenon of crystal nucleation is the predominant process compared to crystal growth, as reported earlier, a lower grain dimension was measured.

The addition of Bioglass in the solution causes a further decrease in the main grain sizes. This behavior is due to the incorporation of Bioglass in the calcium phosphate deposit, which continuously stimulates the formation of new nucleation sites and limits crystal growth [103]. The low crystallinity of the composite coating (which corresponds to a high level of reactivity and, therefore, a high dissolution/reprecipitation rate) is another reason for its high bioactivity in SBF [105]. Confirmation of these hypotheses was achieved using RAMAN spectroscopy. The spectra of the coatings following the deposition process (Figure 4a) show that the brushite modes were identified. In particular, the modes at 985 cm^{-1} and 878 cm^{-1} are related to the stretching (ν_1 P-O) of the phosphate groups. The stretching ($\nu 3$ P-O) of the PO_4 group was observed at 1081 cm^{-1} and 1059 cm^{-1}. The bending of the HPO_4 group (ν_2 OPO) corresponds to the vibrational modes at 379 cm^{-1} and 415 cm^{-1}. Furthermore, the stretching (ν_3 P-O) of the $HPO_4{}^{2-}$ group was observed at 1121 cm^{-1} and is characterized by low intensity. Finally, the stretching (ν_4 O-P-O: 530 cm^{-1}) of the $HPO_4{}^{2-}$ group and the bending (ν_4 P-O: 593 cm^{-1}) of the PO_4 group were observed. In addition, the typical hydroxyapatite stretching mode (ν_1: 960 cm^{-1}) was detected, although it had low intensity compared to the brushite main mode.

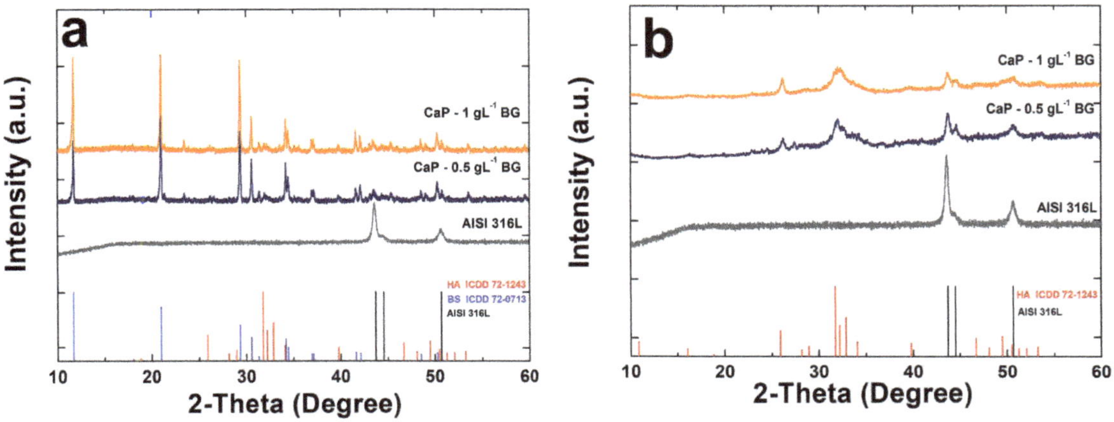

Figure 3. X−ray diffraction patterns of calcium phosphate/Bioglass coatings: (**a**) before and (**b**) post-aging in simulated body fluid. The pattern of uncoated AISI 316L was also reported for comparison.

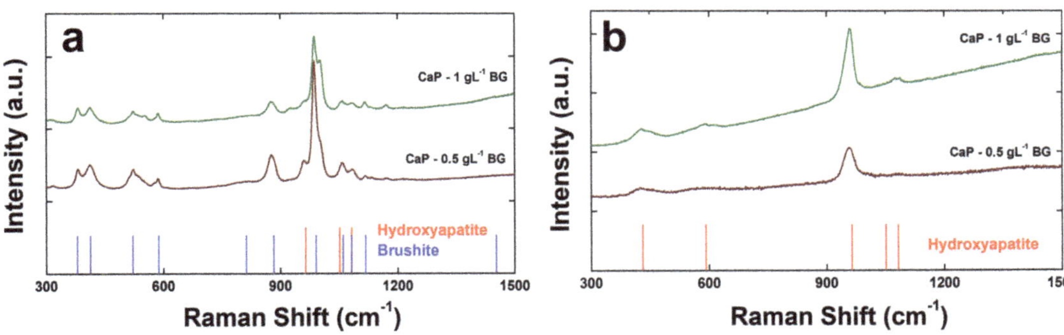

Figure 4. RAMAN spectra of calcium phosphate/Bioglass coatings: (**a**) post-deposition and (**b**) post−aging in simulated body fluid.

As shown in Figure 4b, in both samples after aging, only the hydroxyapatite phase was detected, confirming the total conversion of brushite, which is in line with the results observed via energy-dispersive X-ray spectroscopy and X-ray diffraction.

Corrosion tests were carried out in simulated body fluid for 21 days to evaluate the performance of the coatings in terms of their resistance to corrosion phenomena. Figure S5 (in the Supplementary Materials) shows the open-circuit potential (for 30 min) in simulated body fluid. The potential remained almost constant, and no unexpected changes were found, which could, in this case, occur due to the solubilization/precipitation of calcium phosphate. In addition, the value of the open-circuit potential shifts to more noble values than that of bare steel due to increased corrosion resistance. The sample CaP/BG 0.5 gL^{-1} (Figure S5a) was characterized by a constant potential. In contrast, the largest changes were observed during the first few days of immersion. A decrease in open-circuit potential value is observed from day 14. This phenomenon is attributable to the phase change induced by the brushite/hydroxyapatite equilibrium. Despite this phenomenon, nobler open-circuit potential values were recorded compared to bare steel.

The sample CaP-BG 1 gL^{-1} (Figure S5b) shows a similar trend. There was a significant increase in open-circuit potential value after the first day of immersion; then, a lower variation was noted under the dynamic equilibrium of calcium phosphate. A maximum open-circuit potential value of 0.081 V was recorded on day 14. Compared with the previous sample, higher values were recorded. This behavior is due to the greater presence of Bioglass, leading to the formation of a thicker coating.

Tafel curves have been provided in Figure 5. The values of corrosion potential (E_{corr}) and corrosion current density (i_{corr}) were evaluated through the extrapolation of the Tafel curves and have been reported in Table 2. The samples were characterized by a higher value of E_{corr} than bare steel. In agreement with the results regarding open-circuit potential monitoring, the potential changes after day 14 were due to the equilibrium established between calcium phosphate compounds. Regarding i_{corr}, the values were lower than the uncoated substrate.

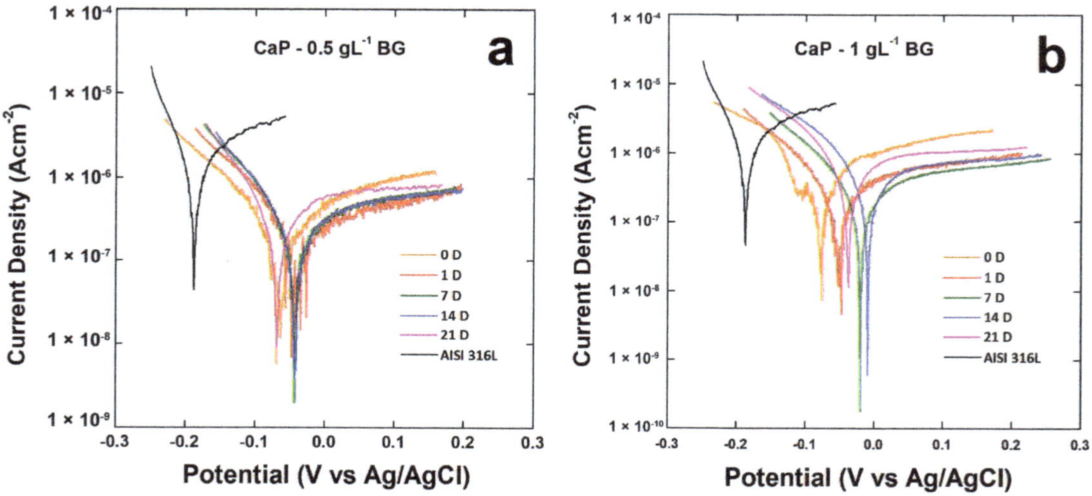

Figure 5. Tafel plots of (**a**) CaP—0.5 gL^{-1} BG coating and (**b**) CaP—1 gL^{-1} BG coating during 21 days of aging in the simulated body fluid solution.

The crystallinity of the coating can be related to its corrosion resistance. The CaP-BG 1gL^{-1} sample (with the smallest grain size and thus the lowest crystallinity) has the best E_{corr} and i_{corr} values due to the high coating dissolution/reprecipitation rate in simulated body fluid. The increased coating thickness and high reprecipitation rate due to the increased presence of Bioglass lead to the formation of a thicker deposit during aging, which results in better protective action against corrosive phenomena.

Electrochemical impedance spectroscopy measurements are reported in Figure 6. The best fitting was obtained using the equivalent circuit ($R_s(CPE_1(R_1(CPE_2(R_2W))))$) schematized in Figure S6 (presented in the Supplementary Materials). This equivalent circuit was proposed by Kathavate et al. [106] and simulated the behavior of inhomogeneous coatings modeled by Jüttner [107] to describe coatings with defects. In particular, the proposed model considers the presence of a porous coating in which diffusive processes occur. Rs represents the resistance of the simulated body fluid. CPE1 and R1 describe the behavior of the outer layer of the coating; alternatively, R2 corresponds to the pore resistance. CPE2 and R3 refer to the double-layer capacitance and charge transfer resistance, respectively. The Warburg element (W) was included to simulate diffusive processes within the porous structure. The CPE (Constant Phase Element) was used by Hinderliter [108] to model inhomogeneous coatings, the presence of defects, and composition variation.

Table 2. E_{corr} and i_{corr} of calcium phosphate/Bioglass coatings obtained by extrapolation of Tafel's curves in Figure 5. For comparison, E_{corr} and i_{corr} of uncoated AISI 316L were also calculated. The mean standard deviation was ±1.7%.

	Time					
	0	**1**	**7**	**14**	**21**	**AISI 316L**
CaP—0.5 gL^{-1} BG						
E_{corr} [V]	−0.088	−0.036	−0.057	−0.054	−0.088	−0.187
i_{corr} [Acm^{-2}]	4.83×10^{-7}	2.43×10^{-7}	4.31×10^{-7}	3.91×10^{-7}	6.14×10^{-7}	2.52×10^{-6}
CaP—1 gL^{-1} BG						
E_{corr} [V]	−0.068	0.056	−0.029	0.009	−0.048	−0.187
i_{corr} [Acm^{-2}]	8.85×10^{-7}	4.13×10^{-7}	4.07×10^{-7}	5.84×10^{-7}	8.16×10^{-7}	2.52×10^{-6}

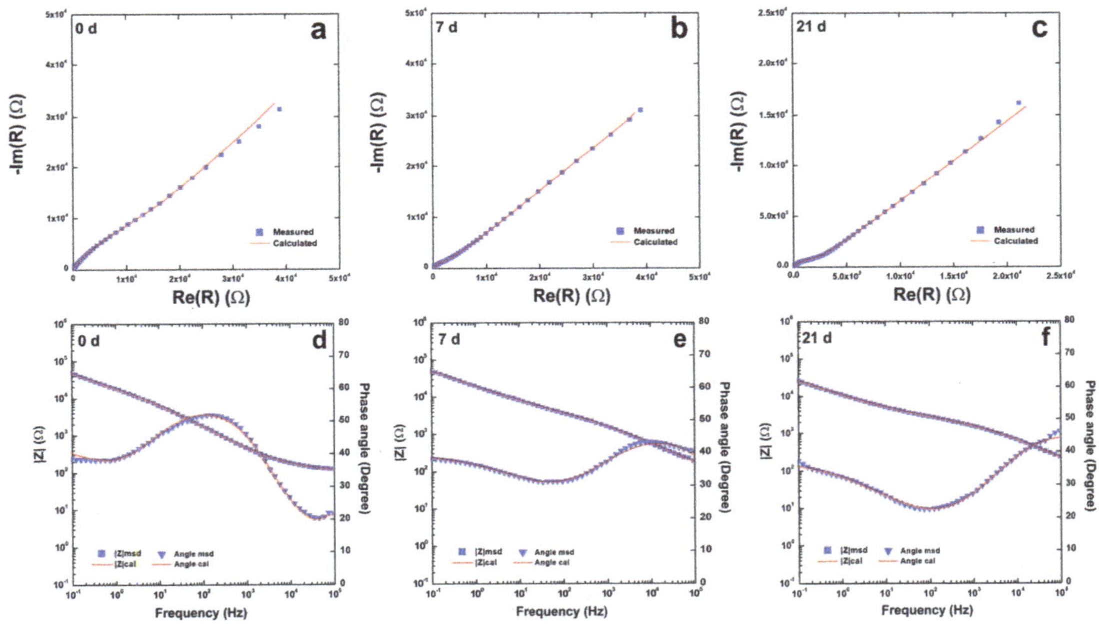

Figure 6. Impedance spectra of calcium phosphate/Bioglass coating during 21 days of aging in simulated body fluid.

The obtained fitting (reported in Table S1) is characterized by a χ^2 value of the magnitude of 10^{-4}, and the relative error of each parameter is less than 10%. The evolution of the coating due to the equilibrium between brushite/hydroxyapatite was noted during aging. The values of the constant phase element exponents, n_1 and n_2, are lower than those of uncoated steel since the coating has a more complex morphology than a passive film on stainless steel. In any case, after 21 days, the overall impedance was higher than 10^4 Ω.

These results show that the galvanic method yields coatings with good corrosion resistance properties, which are better than those obtained through electrodeposition [109] and electrophoretic deposition [110]. However, it is worth noting that a direct comparison with other results is difficult to achieve. The performance of the coating is linked to its morphology, composition, and adhesion to the substrate, which strongly depend on the substrate itself, the deposition method, and the deposition conditions (i.e., temperature, agitation, and pH). However, we can compare these results with those concerning other coatings we obtained using the galvanic deposition method [82,84,86–88]. According to these results, it can be concluded that the composite with Bioglass has excellent performance, which is only surpassed by the composite with biopolymers such as chitosan and collagen.

Inductive plasma spectroscopy confirmed the ability of the coatings to limit corrosion phenomena. In particular, the concentration of metal ions (Fe, Ni, Cr, and Mo) released in the solution from the substrate after 21 days of aging in simulated body fluid was quantified. As can be observed in Table 3, the concentration of metal ions in the SBF used to age the samples is very low. In the case of iron, which is the most abundant element in steel, a concentration ten times lower than that of bare steel was found. A very low value was also detected in the case of nickel, which was well below the threshold limits for human health [111]. Furthermore, after 21 days, the concentration of Ca ions decreased in the simulated body fluid solution due to the incorporation of Ca ions within the calcium phosphate crystal structure [112] that occurs during aging. Contemporaneously, the number of P ions increased due to the equilibrium proposed by Nur et al. [102], wherein phosphate ions are released (Equation (11)).

Table 3. Ion concentrations in the simulated body fluid solution after 3 weeks of aging. For comparison, the concentrations of Ca and P ions in the as-prepared simulated body fluid solution were reported (for both measured and calculated values). The mean standard deviation was 0.7%.

	Ions Concentration (ppm)					
	Fe	**Cr**	**Ni**	**Mo**	**Ca**	**P**
SBF Mis.	0	0	0	0	103.75	31.74
SBF Cal.	0	0	0	0	105	31
AISI 316L	0.088	0	0.197	0	89.16	28.82
CaP—BG 1gL^{-1}	0.001	0	0.076	0	39.26	66.74

The biocompatibility of calcium phosphate/Bioglass-coated (CaP$-$1gL^{-1}BG) or uncoated steel (AISI 316L) samples was investigated with respect to MC3T3-E1 pre-osteoblastic cells grown in a medium previously incubated with AISI 316L or CaP—1 gL^{-1} BG as reported in the Materials and Methods. Cells grown in a D-MEM medium were used as a control. The viability assay was performed after 2, 5, and 7 days of treatment using the CCK-8 assay. As reported in Figure 7, both samples (AISI 316L and CaP—1 gL^{-1} BG) presented viability values comparable to those of the control at each time point, suggesting the good biocompatibility of the devices. Therefore, the percentage of cell viability of AISI 316L and CaP—1 gL^{-1} BG (considering the untreated control cells' percentage to be 100%) was more than 91% at each timepoint (reported in Table 4): after 2 days, 97.37% and 96.90%; after 5 days, 91.07% and 98.88%; and after 7 days, 91.82% and 92.26% for AISI 316L and CaP—1 gL^{-1} BG, respectively. The degree of cell viability, which was well over 70% compared to the control, confirms the non-cytotoxicity of the tested materials as determined by the ISO standard.

These data were confirmed through an analysis of cell morphology (Figure 8), which showed an increase in cell density after 5 and 7 days at the same level as the control. Moreover, cell morphology observations showed discrete intra-cytoplasmatic granules, well-spread cells, and less than 20% round cells, indicating an absent or very low cytotoxic effect (less than 10%) and suggesting good biocompatibility. Therefore, these tests show that the calcium phosphate/Bioglass-coated samples are biocompatible and suitable for biomedical applications since they do not release cytotoxic products that alter cell growth.

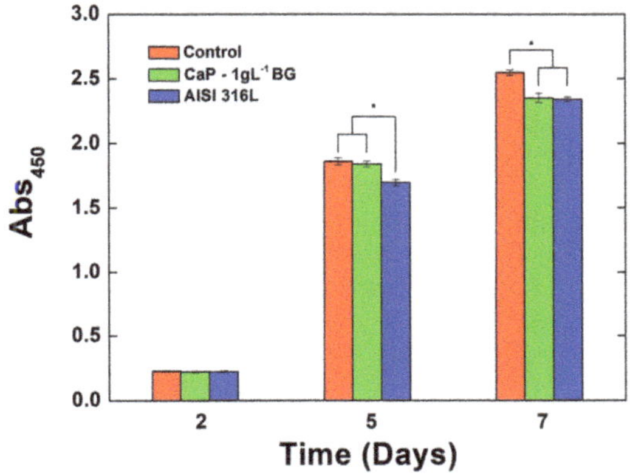

Figure 7. Cell viability assay (CCK-8) of MC3T3-E1 cells grown for different times (2, 5, and 7 days) with complete D-MEM (red, positive control); the CaP—1 gL^{-1} BG (green) or AISI 316L D-MEM (blue). Viability was expressed as absorbance at 450 nm (Abs), which is directly proportional to cell viability. Error bars represent means ± SD for n = 10 (* $p < 0.05$).

Table 4. Cell viability percentages normalized with respect to the control (100% viability).

Time (Days)	CaP—1gL^{-1} BG	AISI 316L
2	96.90 ± 0.57	97.39 ± 0.48
5	98.88 ± 2.33	91.07 ± 2.44
7	92.26 ± 3.56	91.82 ± 1.85

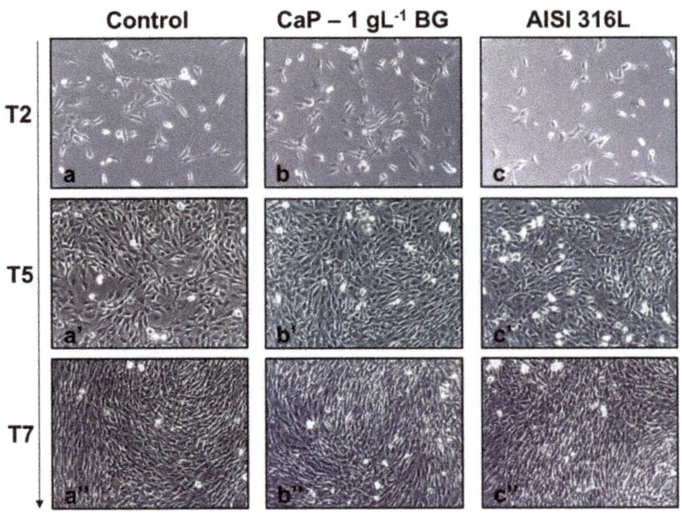

Figure 8. Optical microscopy of MC3T3-E1 cells grown for different times (2, 5 (′), and 7 (″) days) with: (**a**,**a′**,**a″**) complete D-MEM (control); (**b**,**b′**,**b″**) AISI 316L D-MEM; (**c**,**c′**,**c″**) the CaP—1 gL^{-1} BG D-MEM. Magnificence 10X.

4. Conclusions

The galvanic deposition method was used to obtain composite coatings of calcium phosphate/Bioglass on AISI 316L. the coatings were achieved with different amounts of Bioglass 45S5, which was suspended in the deposition solution so that it could be incorporated during the deposition of the calcium phosphate compounds.

The obtained micrographs showed the formation of a uniform deposit characterized by small crystals, whose formation was probably due to the physical incorporation of Bioglass during deposition, which acts as nucleation points for new calcium phosphate crystals.

Through energy-dispersive X-ray spectroscopy, in addition to detecting the presence of Si originating from the Bioglass, the Ca/P ratio was calculated, and the values obtained were attributable to the deposition of both brushite and hydroxyapatite phases. The presence of these two phases was verified through both Raman spectroscopy and X-ray diffraction. Both techniques also revealed that the deposit mainly consists of brushite. Given its amorphous nature, no diffraction peaks of the Bioglass were detected. From the X-ray diffraction patterns and Scherrer's equation, it was found that the presence of Bioglass in the bath leads to the formation of deposits with a smaller average grain size than deposits without Bioglass.

The coatings were aged for 21 days in simulated body fluid at 37 °C. After aging, the morphology of the deposit changed slightly, while its composition was completely different. Both Raman spectroscopy and X-ray diffraction showed the presence of only hydroxyapatite, and energy-dispersive X-ray spectroscopy also revealed the complete disappearance of Si. This behavior is typical of calcium phosphate compounds, which naturally tend to transform into hydroxyapatite when immersed in simulated body fluid. The disappearance of Si was also expected and is a confirmation of the Bioglass's bioactivity, which, in simulated body fluid, further stimulates the dissolution of the coating and its reprecipitation in solely the hydroxyapatite phase.

Of great interest are the results of electrochemical characterization, which showed a nobler corrosion potential and lower corrosion current density for all samples compared to those of bare steel from the first day of aging in the simulated body fluid. The coating with the best performance in terms of resistance to corrosive phenomena was obtained with a concentration of Bioglass equal to 1 gL^{-1}. Under these conditions, the coating with the lowest crystallinity and the highest thickness was obtained, which was characterized by a high dissolution/reprecipitation rate in simulated body fluid and thus better protective action against corrosive phenomena. The slowdown of corrosive phenomena in the presence of the coating was also confirmed through the inductive plasma spectroscopy analysis of the SBF used for aging. Lower concentrations of Fe (10 times lower) and Ni (more than half) were found in these solutions compared to that measured in the solution used to treat bare steel.

Finally, biological tests carried out with MCT3-E1 pre-osteoblastic cells revealed that the coated samples do not influence typical cellular growth and, consequently, can be considered non-cytotoxic and suitable for biomedical purposes. Further research is underway to evaluate the scalability of the method using areas much larger than those studied herein and deposition on nonplanar substrates to simulate the actual shape of a prosthesis. Experimentation is also underway using the Ti alloy (Ti-6Al-4V), which we are coating with a CaP-Polymer-Drug composite, again using the galvanic method.

Supplementary Materials: The following supporting information can be downloaded at: https://www.mdpi.com/article/10.3390/coatings13061006/s1, Figure S1: Cell layout; Figure S2: SEM images of CaP coatings obtained via galvanic deposition under different conditions: (a) unstirring (b) and stirring; Figure S3: XRD pattern of Bioglass 45S5 powder; Figure S4: Grain size calculated via Scherrer's equation at 2-theta degree equal to 11.65 and 20.95; Figure S5: OCP curves of (a) CaP—BG 0.5 gL^{-1} and (b) CaP—BG 1 gL^{-1}; Figure S6: Equivalent circuit used to fit EIS data; Table S1: Composition of SBF; Table S2: Fitting parameters of EIS data obtained for CaP—BG 1gL^{-1}.

Author Contributions: Conceptualization, C.Z., A.M., B.P., F.L. and F.C.P.; methodology, C.Z., A.M., B.P., F.L., E.C., S.C. and F.C.P.; validation, C.Z., B.P., F.L., E.C., S.C. and F.C.P.; formal analysis, C.Z.; investigation, C.Z., A.M., B.P., F.L., E.C., S.C. and F.C.P.; data curation, C.Z.; writing—original draft preparation, C.Z., B.P., E.C., F.C.P. and R.I.; writing—review and editing, R.I., V.B., and V.L.C.; supervision, R.I., V.B. and V.L.C. All authors have read and agreed to the published version of the manuscript.

Funding: This research received no external funding.

Institutional Review Board Statement: Not applicable.

Informed Consent Statement: Not applicable.

Data Availability Statement: Not applicable.

Conflicts of Interest: The authors declare no conflict of interest.

References

1. Carter, C.B.; Norton, M.G. *Ceramic Materials: Science and Engineering*; Springer Science+Business Media, LLC Springer e-books: New York, NY, USA, 2007.
2. Pomeroy, M.; Cambier, F. (Eds.) *Encyclopedia of Materials: Technical Ceramics and Glasses*; Elsevier: Amsterdam, The Netherlands, 2021.
3. Hojo, J. (Ed.) *Materials Chemistry of Ceramics*; Springer: Singapore, 2019.
4. Huang, J.; Best, S. Ceramic biomaterials for tissue engineering. In *Tissue Engineering Using Ceramics and Polymers*; Elsevier: Amsterdam, The Netherlands, 2014; pp. 3–34. [CrossRef]
5. Osaka, A.; Narayan, R. (Eds.) *Bioceramics: From Macro to Nanoscale*; Elsevier: Amsterdam, The Netherlands; Oxford, UK; Cambridge, MA, USA, 2021.
6. Farid, S.B.H. *Bioceramics: For Materials Science and Engineering*; Elsevier/Woodhead Publishing: Duxford, UK; Cambridge, MA, USA, 2019.
7. Ravaglioli, A.; Krajewski, A. *Bioceramics*; Springer: Dordrecht, The Netherlands, 1992. [CrossRef]
8. Abbas, Z.; Dapporto, M.; Tampieri, A.; Sprio, S. Toughening of Bioceramic Composites for Bone Regeneration. *J. Compos. Sci.* **2021**, *5*, 259. [CrossRef]
9. Wang, K.; Zhou, C.; Hong, Y.; Zhang, X. A review of protein adsorption on bioceramics. *Interface Focus.* **2012**, *2*, 259–277. [CrossRef] [PubMed]
10. Pawelec, K.M.; Planell, J.A. (Eds.) *Bone Repair Biomaterials: Regeneration and Clinical Applications*, 2nd ed.; Elsevier/Woodhead Publishing: Duxford, UK; Cambridge, MA, USA, 2019.
11. Mistry, S.; Kundu, D.; Datta, S.; Basu, D. Comparison of bioactive glass coated and hydroxyapatite coated titanium dental implants in the human jaw bone: Comparison of bioactive glass coated and hydroxyapatite coated implants. *Aust. Dent. J.* **2011**, *56*, 68–75. [CrossRef] [PubMed]
12. Ammarullah, M.; Santoso, G.; Sugiharto, S.; Supriyono, T.; Wibowo, D.; Kurdi, O.; Tauviqirrahman, M.; Jamari, J. Minimizing Risk of Failure from Ceramic-on-Ceramic Total Hip Prosthesis by Selecting Ceramic Materials Based on Tresca Stress. *Sustainability* **2022**, *14*, 13413. [CrossRef]
13. Buj-Corral, I.; Tejo-Otero, A. 3D Printing of Bioinert Oxide Ceramics for Medical Applications. *J. Funct. Biomater.* **2022**, *13*, 155. [CrossRef]
14. Piconi, C. Bioinert Ceramics: State-of-the-Art. *Key Eng. Mater.* **2017**, *758*, 3–13. [CrossRef]
15. Hench, L.L. Bioactive Ceramics: Theory and Clinical Applications. In *Bioceramics*; Elsevier: Amsterdam, The Netherlands, 1994; pp. 3–14.
16. Ohtsuki, C.; Kamitakahara, M.; Miyazaki, T. Bioactive ceramic-based materials with designed reactivity for bone tissue regeneration. *J. R. Soc. Interface.* **2009**, *6*, S349–S360. [CrossRef]
17. Salinas, A.J.; Vallet-Regí, M. Bioactive ceramics: From bone grafts to tissue engineering. *RSC Adv.* **2013**, *3*, 11116. [CrossRef]
18. Poitout, D.G. (Ed.) *Biomechanics and Biomaterials in Orthopedics*, 2nd ed.; Springer: London, UK, 2016. [CrossRef]
19. Bohner, M. Bioresorbable ceramics. In *Degradation Rate of Bioresorbable Materials*; Elsevier: Amsterdam, The Netherlands, 2008; pp. 95–114. [CrossRef]
20. Dehghanghadikolaei, A.; Fotovvati, B. Coating Techniques for Functional Enhancement of Metal Implants for Bone Replacement: A Review. *Materials* **2019**, *12*, 1795. [CrossRef]
21. Harun, W.S.W.; Asri, R.I.M.; Alias, J.; Zulkifli, F.H.; Kadirgama, K.; Ghani, S.A.C.; Shariffuddin, J.H.M. A comprehensive review of hydroxyapatite-based coatings adhesion on metallic biomaterials. *Ceram. Int.* **2018**, *44*, 1250–1268. [CrossRef]
22. Skallevold, H.E.; Rokaya, D.; Khurshid, Z.; Zafar, M.S. Bioactive Glass Applications in Dentistry. *Int. J. Mol. Sci.* **2019**, *20*, 5960. [CrossRef]
23. Ylänen, H.O. (Ed.) *Bioactive Glasses: Materials, Properties and Applications*, 2nd ed.; Elsevier/Woodhead Publishing: Duxford, UK, 2018.
24. Karadjian, M.; Essers, C.; Tsitlakidis, S.; Reible, B.; Moghaddam, A.; Boccaccini, A.; Westhauser, F. Biological Properties of Calcium Phosphate Bioactive Glass Composite Bone Substitutes: Current Experimental Evidence. *Int. J. Mol. Sci.* **2019**, *20*, 305. [CrossRef]
25. Kaur, G. *Bioactive Glasses: Potential Biomaterials for Future Therapy*, 1st ed.; Springer International Publishing: Cham, Swizerland, 2017. [CrossRef]
26. Varila, L.; Fagerlund, S.; Lehtonen, T.; Tuominen, J.; Hupa, L. Surface reactions of bioactive glasses in buffered solutions. *J. Eur. Ceram. Soc.* **2012**, *32*, 2757–2763. [CrossRef]

27. Fernandes, H.R.; Gaddam, A.; Rebelo, A.; Brazete, D.; Stan, G.E.; Ferreira, J.M.F. Bioactive Glasses and Glass-Ceramics for Healthcare Applications in Bone Regeneration and Tissue Engineering. *Materials* **2018**, *11*, 2530. [CrossRef]
28. Drago, L.; Toscano, M.; Bottagisio, M. Recent Evidence on Bioactive Glass Antimicrobial and Antibiofilm Activity: A Mini-Review. *Materials* **2018**, *11*, 326. [CrossRef]
29. Hench, L.L.; Jones, J.R. Bioactive Glasses: Frontiers and Challenges. *Front. Bioeng. Biotechnol.* **2015**, *3*, 194. [CrossRef]
30. Baino, F.; Hamzehlou, S.; Kargozar, S. Bioactive Glasses: Where Are We and Where Are We Going? *J. Funct. Biomater.* **2018**, *9*, 25. [CrossRef]
31. Fiume, E.; Barberi, J.; Verné, E.; Baino, F. Bioactive Glasses: From Parent 45S5 Composition to Scaffold-Assisted Tissue-Healing Therapies. *J. Funct. Biomater.* **2018**, *9*, 24. [CrossRef]
32. Baino, F.; Verné, E. Glass-based coatings on biomedical implants: A state-of-the-art review. *Biomed. Glas.* **2017**, *3*, 1–17. [CrossRef]
33. Oliver, J.N.; Su, Y.; Lu, X.; Kuo, P.-H.; Du, J.; Zhu, D. Bioactive glass coatings on metallic implants for biomedical applications. *Bioact. Mater.* **2019**, *4*, 261–270. [CrossRef]
34. Camponogara, F.; Zanotti, F.; Trentini, M.; Tiengo, E.; Zanolla, I.; Pishavar, E.; Soliani, E.; Scatto, M.; Gargiulo, P.; Zambito, Y.; et al. Biomaterials for Regenerative Medicine in Italy: Brief State of the Art of the Principal Research Centers. *Int. J. Mol. Sci.* **2022**, *23*, 8245. [CrossRef] [PubMed]
35. Vaez, S.; Emadi, R.; Sadeghzade, S.; Salimijazi, H.; Kharaziha, M. Electrophoretic deposition of chitosan reinforced baghdadite ceramic nano-particles on the stainless steel 316L substrate to improve biological and physical characteristics. *Mater. Chem. Phys.* **2022**, *282*, 125991. [CrossRef]
36. Smith, J.R.; Lamprou, D.A.; Larson, C.; Upson, S.J. Biomedical applications of polymer and ceramic coatings: A review of recent developments. *Trans. IMF* **2022**, *100*, 25–35. [CrossRef]
37. Wang, Y.; Wu, B.; Ai, S.; Wan, D. Electroplating of HAp-brushite coating on metallic bioimplants with advanced hemocompatibility and osteocompatibility properties. *J. Appl. Biomater. Funct. Mater.* **2022**, *20*, 228080002211039. [CrossRef]
38. Jiménez-García, F.N.; Giraldo-Torres, L.R.; Restrepo-Parra, E. Electrochemically Deposited Calcium Phosphate Coatings Using a Potentiostat of In-house Design and Implementation. *Mat. Res.* **2021**, *24*, e20210098. [CrossRef]
39. Ghiasi, B.; Sefidbakht, Y.; Rezaei, M. Hydroxyapatite for Biomedicine and Drug Delivery. In *Nanomaterials for Advanced Biological Applications*; Rahmandoust, M., Ayatollahi, M.R., Eds.; Springer International Publishing: Cham, Swizerland, 2019; Volume 104, pp. 85–120. [CrossRef]
40. LeGeros, R.Z. Properties of Osteoconductive Biomaterials: Calcium Phosphates. *Clin. Orthop. Relat. Res.* **2002**, *395*, 81–98. [CrossRef]
41. Eliaz, N.; Metoki, N. Calcium Phosphate Bioceramics: A Review of Their History, Structure, Properties, Coating Technologies and Biomedical Applications. *Materials* **2017**, *10*, 334. [CrossRef]
42. Hou, X.; Zhang, L.; Zhou, Z.; Luo, X.; Wang, T.; Zhao, X.; Lu, B.; Chen, F.; Zheng, L. Calcium Phosphate-Based Biomaterials for Bone Repair. *J. Funct. Biomater.* **2022**, *13*, 187. [CrossRef]
43. Ielo, I.; Calabrese, G.; De Luca, G.; Conoci, S. Recent Advances in Hydroxyapatite-Based Biocomposites for Bone Tissue Regeneration in Orthopedics. *Int. J. Mol. Sci.* **2022**, *23*, 9721. [CrossRef]
44. Shi, H.; Zhou, Z.; Li, W.; Fan, Y.; Li, Z.; Wei, J. Hydroxyapatite Based Materials for Bone Tissue Engineering: A Brief and Comprehensive Introduction. *Crystals* **2021**, *11*, 149. [CrossRef]
45. Layrolle, P.; Daculsi, G. Physicochemistry of Apatite and Its Related Calcium Phosphates. In *Thin Calcium Phosphate Coatings for Medical Implants*; León, B., Jansen, J., Eds.; Springer: New York, NY, USA, 2009; pp. 19–21. [CrossRef]
46. Dutta, S.R.; Passi, D.; Singh, P.; Bhuibhar, A. Ceramic and non-ceramic hydroxyapatite as a bone graft material: A brief review. *Ir. J. Med. Sci.* **2015**, *184*, 101–106. [CrossRef]
47. Awasthi, S.; Pandey, S.K.; Arunan, E.; Srivastava, C. A review on hydroxyapatite coatings for the biomedical applications: Experimental and theoretical perspectives. *J. Mater. Chem. B* **2021**, *9*, 228–249. [CrossRef]
48. El Hadad, A.; Peón, E.; García-Galván, F.; Barranco, V.; Parra, J.; Jiménez-Morales, A.; Galván, J. Biocompatibility and Corrosion Protection Behaviour of Hydroxyapatite Sol-Gel-Derived Coatings on Ti6Al4V Alloy. *Materials* **2017**, *10*, 94. [CrossRef]
49. Grebņevs, V.; Leśniak-Ziółkowska, K.; Wala, M.; Dulski, M.; Altundal, Ş.; Dutovs, A.; Avotiņa, L.; Erts, D.; Viter, R.; Vīksna, A.; et al. Modification of physicochemical properties and bioactivity of oxide coatings formed on Ti substrates via plasma electrolytic oxidation in crystalline and amorphous calcium phosphate particle suspensions. *Appl. Surf. Sci.* **2022**, *598*, 153793. [CrossRef]
50. Zhang, G.; Xu, Y.; Zeng, Z.; Cao, B. Enhanced the corrosion resistance and biocompatibility of magnesium alloy by hydroxyapatite composite coating of AZ31/Ti/PDA/HA. *Surf. Topogr. Metrol. Prop.* **2021**, *9*, 025042. [CrossRef]
51. Predoi, D.; Iconaru, S.L.; Ciobanu, S.C.; Predoi, S.-A.; Buton, N.; Megier, C.; Beuran, M. Development of Iron-Doped Hydroxyapatite Coatings. *Coatings* **2021**, *11*, 186. [CrossRef]
52. Predoi, D.; Ciobanu, S.C.; Iconaru, S.L.; Predoi, M.V. Influence of the Biological Medium on the Properties of Magnesium Doped Hydroxyapatite Composite Coatings. *Coatings* **2023**, *13*, 409. [CrossRef]
53. Ciobanu, C.S.; Predoi, M.V.; Buton, N.; Megier, C.; Iconaru, S.L.; Predoi, D. Physicochemical Characterization of Europium-Doped Hydroxyapatite Thin Films with Antifungal Activity. *Coatings* **2022**, *12*, 306. [CrossRef]
54. Iconaru, S.L.; Groza, A.; Gaiaschi, S.; Rokosz, K.; Raaen, S.; Ciobanu, S.C.; Chapon, P.; Predoi, D. Antimicrobial Properties of Samarium Doped Hydroxyapatite Suspensions and Coatings. *Coatings* **2020**, *10*, 1124. [CrossRef]

55. Iconaru, S.L.; Predoi, D.; Ciobanu, C.S.; Motelica-Heino, M.; Guegan, R.; Bleotu, C. Development of Silver Doped Hydroxyapatite Thin Films for Biomedical Applications. *Coatings* **2022**, *12*, 341. [CrossRef]
56. Predoi, D.; Iconaru, S.L.; Predoi, M.V.; Groza, A.; Gaiaschi, S.; Rokosz, K.; Raaen, S.; Negrila, C.C.; Prodan, A.-M.; Costescu, A.; et al. Development of Cerium-Doped Hydroxyapatite Coatings with Antimicrobial Properties for Biomedical Applications. *Coatings* **2020**, *10*, 516. [CrossRef]
57. Filip, D.G.; Surdu, V.-A.; Paduraru, A.V.; Andronescu, E. Current Development in Biomaterials—Hydroxyapatite and Bioglass for Applications in Biomedical Field: A Review. *J. Funct. Biomater.* **2022**, *13*, 248. [CrossRef] [PubMed]
58. Ryu, J.-H.; Kwon, J.-S.; Kim, K.-M.; Hong, H.J.; Koh, W.-G.; Lee, J.; Lee, H.-J.; Choi, H.-J.; Yi, S.; Shin, H.; et al. Synergistic Effect of Porous Hydroxyapatite Scaffolds Combined with Bioactive Glass/Poly(lactic-*co*-glycolic acid) Composite Fibers Promotes Osteogenic Activity and Bioactivity. *ACS Omega* **2019**, *4*, 2302–2310. [CrossRef]
59. Ebrahimi, S.; Hanim, Y.U.; Sipaut, C.S.; Jan, N.B.A.; Arshad, S.E.; How, S.E. Fabrication of Hydroxyapatite with Bioglass Nanocomposite for Human Wharton's-Jelly-Derived Mesenchymal Stem Cell Growing Substrate. *Int. J. Mol. Sci.* **2021**, *22*, 9637. [CrossRef]
60. Mesquita-Guimarães, J.; Detsch, R.; Souza, A.C.; Henriques, B.; Silva, F.S.; Boccaccini, A.R.; Carvalho, O. Cell adhesion evaluation of laser-sintered HAp and 45S5 bioactive glass coatings on micro-textured zirconia surfaces using MC3T3-E1 osteoblast-like cells. *Mater. Sci. Eng. C* **2020**, *109*, 110492. [CrossRef]
61. Hong, Z.; Mello, A.; Yoshida, T.; Luan, L.; Stern, P.H.; Rossi, A.; Ellis, D.E.; Ketterson, J.B. Osteoblast proliferation on hydroxyapatite coated substrates prepared by right angle magnetron sputtering. *J. Biomed. Mater. Res.* **2010**, *93*, 878–885. [CrossRef]
62. Maximov, M.; Maximov, O.-C.; Craciun, L.; Ficai, D.; Ficai, A.; Andronescu, E. Bioactive Glass—An Extensive Study of the Preparation and Coating Methods. *Coatings* **2021**, *11*, 1386. [CrossRef]
63. Gomez-Vega, J.M.; Saiz, E.; Tomsia, A.P.; Marshall, G.W.; Marshall, S.J. Bioactive glass coatings with hydroxyapatite and Bioglass®particles on Ti-based implants. 1. Processing. *Biomaterials* **2000**, *21*, 105–111. [CrossRef]
64. Cañas, E.; Orts, M.J.; Boccaccini, A.R.; Sánchez, E. Microstructural and in vitro characterization of 45S5 bioactive glass coatings deposited by solution precursor plasma spraying (SPPS). *Surf. Coat. Technol.* **2019**, *371*, 151–160. [CrossRef]
65. Mahato, A.; De, M.; Bhattacharjee, P.; Kumar, V.; Mukherjee, P.; Singh, G.; Kundu, B.; Balla, V.K.; Nandi, S.K. Role of calcium phosphate and bioactive glass coating on in vivo bone healing of new Mg–Zn–Ca implant. *J. Mater. Sci. Mater. Med.* **2021**, *32*, 55. [CrossRef]
66. Cattini, A.; Bellucci, D.; Sola, A.; Pawłowski, L.; Cannillo, V. Microstructural design of functionally graded coatings composed of suspension plasma sprayed hydroxyapatite and bioactive glass: Microstructural Design Of Functionally Graded Coatings. *J. Biomed. Mater. Res.* **2014**, *102*, 551–560. [CrossRef]
67. Berbecaru, C.; Alexandru, H.V.; Stan, G.E.; Marcov, D.A.; Pasuk, I.; Ianculescu, A. First stages of bioactivity of glass-ceramics thin films prepared by magnetron sputtering technique. *Mater. Sci. Eng. B* **2010**, *169*, 101–105. [CrossRef]
68. Dhinasekaran, D.; Kaliaraj, G.S.; Jagannathan, M.; Rajendran, A.R.; Prakasarao, A.; Ganesan, S.; Subramanian, B. Pulsed laser deposition of nanostructured bioactive glass and hydroxyapatite coatings: Microstructural and electrochemical characterization. *Mater. Sci. Eng. C* **2021**, *130*, 112459. [CrossRef]
69. Garrido, B.; Dosta, S.; Cano, I.G. Bioactive glass coatings obtained by thermal spray: Current status and future challenges. *BoletÍN Soc. Española CerÁMica Vidr.* **2022**, *61*, 516–530. [CrossRef]
70. Comesaña, R.; del Val, J.; Quintero, F.; Riveiro, A.; Arias-González, F.; Boutinguiza, M.; Lusquiños, F.; Pou, J. Laser Cladding and Laser Direct Glass Deposition of Bioactive Glass and Glass-Ceramics. In *Bioactive Glasses and Glass-Ceramics*; Baino, F., Kargozar, S., Eds.; Wiley: Hoboken, NJ, USA, 2022; pp. 311–340. [CrossRef]
71. Khanmohammadi, S.; Ojaghi-Ilkhchi, M.; Farrokhi-Rad, M. Evaluation of bioglass and hydroxyapatite based nanocomposite coatings obtained by electrophoretic deposition. *Ceram. Int.* **2020**, *46*, 26069–26077. [CrossRef]
72. Safavi, M.S.; Walsh, F.C.; Surmeneva, M.A.; Surmenev, R.A.; Khalil-Allafi, J. Electrodeposited Hydroxyapatite-Based Biocoatings: Recent Progress and Future Challenges. *Coatings* **2021**, *11*, 110. [CrossRef]
73. Taranu, B.-O.; Ianasi, P.; Rus, S.F.; Bucur, A.I. Simultaneous Precipitation and Electrodeposition of Hydroxyapatite Coatings at Different Temperatures on Various Metal Substrates. *Coatings* **2022**, *12*, 288. [CrossRef]
74. Say, Y.; Aksakal, B. Enhanced corrosion properties of biological NiTi alloy by hydroxyapatite and bioglass based biocomposite coatings. *J. Mater. Res. Technol.* **2020**, *9*, 1742–1749. [CrossRef]
75. Jaafar, A.; Hecker, C.; Árki, P.; Joseph, Y. Sol-Gel Derived Hydroxyapatite Coatings for Titanium Implants: A Review. *Bioengineering* **2020**, *7*, 127. [CrossRef]
76. Azzouz, I.; Faure, J.; Khlifi, K.; Cheikh Larbi, A.; Benhayoune, H. Electrophoretic Deposition of 45S5 Bioglass®Coatings on the Ti6Al4V Prosthetic Alloy with Improved Mechanical Properties. *Coatings* **2020**, *10*, 1192. [CrossRef]
77. Jamari, J.; Ammarullah, M.I.; Santoso, G.; Sugiharto, S.; Supriyono, T.; van der Heide, E. In Silico Contact Pressure of Metal-on-Metal Total Hip Implant with Different Materials Subjected to Gait Loading. *Metals* **2022**, *12*, 1241. [CrossRef]
78. Fan, X.; Chen, J.; Zou, J.; Wan, Q.; Zhou, Z.; Ruan, J. Bone-like apatite formation on HA/316L stainless steel composite surface in simulated body fluid. *Trans. Nonferrous Met. Soc. China* **2009**, *19*, 347–352. [CrossRef]
79. Beig, B.; Liaqat, U.; Niazi, M.F.K.; Douna, I.; Zahoor, M.; Niazi, M.B.K. Current Challenges and Innovative Developments in Hydroxyapatite-Based Coatings on Metallic Materials for Bone Implantation: A Review. *Coatings* **2020**, *10*, 1249. [CrossRef]
80. Schlesinger, M.; Paunovic, M. (Eds.) *Modern Electroplating*, 5th ed.; Wiley: Hoboken, NJ, USA, 2010.

81. Zanca, C.; Cordaro, G.; Capuana, E.; Brucato, V.; Pavia, F.C.; Carrubba, V.L.; Ghersi, G.; Inguanta, R. Galvanic Deposition Of Hydroxyapatite/Chitosan/Collagen Coatings On 304 Stainless Steel. *Chem. Eng. Trans.* **2021**, *86*, 6. [CrossRef]
82. Mendolia, I.; Zanca, C.; Ganci, F.; Conoscenti, G.; Pavia, F.C.; Brucato, V.; La Carrubba, V.; Lopresti, F.; Piazza, S.; Sunseri, C.; et al. Calcium phosphate/polyvinyl acetate coatings on SS304 via galvanic co-deposition for orthopedic implant applications. *Surf. Coat. Technol.* **2021**, *408*, 126771. [CrossRef]
83. Zanca, C.; Mendolia, I.; Capuana, E.; Blanda, G.; Carfì Pavia, F.; Brucato, V.; Ghersi, G.; la Carrubba, V.; Piazza, S.; Sunseri, C.; et al. Co-Deposition and Characterization of Hydroxyapatite-Chitosan and Hydroxyapatite-Polyvinylacetate Coatings on 304 SS for Biomedical Devices. *KEM* **2019**, *813*, 153–158. [CrossRef]
84. Zanca, C.; Patella, B.; Capuana, E.; Lopresti, F.; Brucato, V.; Carfì Pavia, F.; La Carrubba, V.; Inguanta, R. Behavior of Calcium Phosphate–Chitosan–Collagen Composite Coating on AISI 304 for Orthopedic Applications. *Polymers* **2022**, *14*, 5108. [CrossRef]
85. Zanca, C.; Carbone, S.; Patella, B.; Lopresti, F.; Aiello, G.; Brucato, V.; Carfì Pavia, F.; La Carrubba, V.; Inguanta, R. Composite Coatings of Chitosan and Silver Nanoparticles Obtained by Galvanic Deposition for Orthopedic Implants. *Polymers* **2022**, *14*, 3915. [CrossRef]
86. Blanda, G.; Brucato, V.; Pavia, F.C.; Greco, S.; Piazza, S.; Sunseri, C.; Inguanta, R. In Vitro Corrosion and Biocompatibility of Brushite/Hydroxyapatite Coatings Obtained by Galvanic Deposition on 316LSS. *J. Electrochem. Soc.* **2018**, *165*, G1–G10. [CrossRef]
87. Blanda, G.; Brucato, V.; Carfì, F.; Conoscenti, G.; La Carrubba, V.; Piazza, S.; Sunseri, C.; Inguanta, R. Chitosan-Coating Deposition via Galvanic Coupling. *ACS Biomater. Sci. Eng.* **2019**, *5*, 1715–1724. [CrossRef]
88. Blanda, G.; Brucato, V.; Pavia, F.C.; Greco, S.; Piazza, S.; Sunseri, C.; Inguanta, R. Galvanic deposition and characterization of brushite/hydroxyapatite coatings on 316L stainless steel. *Mater. Sci. Eng. C* **2016**, *64*, 93–101. [CrossRef]
89. Chen, B.; Liang, C. Preparation of hydroxyapatite coating by the use of a sacrificial Mg anode method. *Ceram. Int.* **2007**, *33*, 701–703. [CrossRef]
90. Inguanta, R.; Ferrara, G.; Piazza, S.; Sunseri, C. A new route to grow oxide nanostructures based on metal displacement deposition. Lanthanides oxy/hydroxides growth. *Electrochim. Acta* **2012**, *76*, 77–87. [CrossRef]
91. Inguanta, R.; Piazza, S.; Sunseri, C. A Route to Grow Oxide Nanostructures Based on Metal Displacement Deposition: Lanthanides Oxy/Hydroxides Characterization. *J. Electrochem. Soc.* **2012**, *159*, D493–D500. [CrossRef]
92. Battaglia, M.; Piazza, S.; Sunseri, C.; Inguanta, R. Amorphous silicon nanotubes via galvanic displacement deposition. *Electrochem. Commun.* **2013**, *34*, 134–137. [CrossRef]
93. Patella, B.; Russo, R.R.; Aiello, G.; Sunseri, C.; Inguanta, R. Vertical standing copper nanowires for electrochemical sensor of nitrate in water. In Proceedings of the 2020 IEEE International Conference on Flexible and Printable Sensors and Systems (FLEPS), Manchester, UK, 16–19 August 2020; pp. 1–4. [CrossRef]
94. Inguanta, R.; Ferrara, G.; Piazza, S.; Sunseri, C. Fabrication and characterization of metal and metal oxide nanostructures grown by metal displacement deposition into anodic alumina membranes. *Chem. Eng. Trans.* **2011**, *24*, 199–204. [CrossRef]
95. Papaderakis, A.; Mintsouli, I.; Georgieva, J.; Sotiropoulos, S. Electrocatalysts Prepared by Galvanic Replacement. *Catalysts* **2017**, *7*, 80. [CrossRef]
96. Available online: https://www.legor.com/en-us/introduction-the-galvanic-process (accessed on 1 January 2020).
97. *Power Diffraction File*; International Centre for Diffraction Data: Newtown Square, PA, USA, 2007.
98. Downs, R.T.; Hall-Wallace, M. The American Mineralogist Crystal Structure Database. *Am. Mineral.* **2003**, *88*, 247–250.
99. Therese, G.H.A.; Kamath, P.V. Cathodic reduction of different metal salt solutions Part I: Synthesis of metal hydroxides by electrogeneration of base. *J. Appl. Electrochem.* **1998**, *28*, 539–543. [CrossRef]
100. Nobial, M.; Devos, O.; Mattos, O.R.; Tribollet, B. The nitrate reduction process: A way for increasing interfacial pH. *J. Electroanal. Chem.* **2007**, *600*, 87–94. [CrossRef]
101. Azar, Z.; Khalil-Allafi, J.; Etminanfar, M.R. Electro-crystallization of hydroxyapatite coatings on Nitinol rotating disk electrode. *Mater. Res. Express* **2019**, *6*, 055401. [CrossRef]
102. Nur, A.; Setyawan, H.; Widjaja, A.; Lenggoro, I.W. Electrochemical Processes for the Formation of Hydroxyapatite Powders. *Bull. Chem. React. Eng. Catal.* **2014**, *9*, 168–174. [CrossRef]
103. Ebrahimi, S.; Sipaut, C.S. Synthesis of Hydroxyapatite/Bioglass Composite Nanopowder Using Design of Experiments. *Nanomaterials* **2022**, *12*, 2264. [CrossRef] [PubMed]
104. Botelho, C.M.; Lopes, M.A.; Gibson, I.R.; Best, S.M.; Santos, J.D. Structural analysis of Si-substituted hydroxyapatite: Zeta potential and X-ray photoelectron spectroscopy. *J. Mater. Sci. Mater. Med.* **2002**, *13*, 1123–1127. [CrossRef] [PubMed]
105. Aminian, A.; Solati-Hashjin, M.; Samadikuchaksaraei, A.; Bakhshi, F.; Gorjipour, F.; Farzadi, A.; Moztarzadeh, F.; Schmücker, M. Synthesis of silicon-substituted hydroxyapatite by a hydrothermal method with two different phosphorous sources. *Ceram. Int.* **2011**, *37*, 1219–1229. [CrossRef]
106. Kathavate, V.S.; Pawar, D.N.; Bagal, N.S.; Deshpande, P.P. Role of nano ZnO particles in the electrodeposition and growth mechanism of phosphate coatings for enhancing the anti-corrosive performance of low carbon steel in 3.5% NaCl aqueous solution. *J. Alloys Compd.* **2020**, *823*, 153812. [CrossRef]
107. Jüttner, K. Electrochemical impedance spectroscopy (EIS) of corrosion processes on inhomogeneous surfaces. *Electrochim. Acta* **1990**, *35*, 1501–1508. [CrossRef]
108. Hinderliter, B.R.; Croll, S.G.; Tallman, D.E.; Su, Q.; Bierwagen, G.P. Interpretation of EIS data from accelerated exposure of coated metals based on modeling of coating physical properties. *Electrochim. Acta* **2006**, *51*, 4505–4515. [CrossRef]

109. Thanh, D.T.M.; Nam, P.T.; Phuong, N.T.; Que, L.X.; Anh, N.V.; Hoang, T.; Lam, T.D. Controlling the electrodeposition, morphology and structure of hydroxyapatite coating on 316L stainless steel. *Mater. Sci. Eng. C* **2013**, *33*, 2037–2045. [CrossRef]
110. Yazdani Samani, F.; Rabiee, S.M.; Jamaati, R.; Bagherifard, S. Effect of shot peening on electrophoretic deposition of bioactive glass coating on AISI 316L stainless steel. *Ceram. Int.* **2023**, *49*, 17468–17478. [CrossRef]
111. Leikin, J.B.; Paloucek, F.P. (Eds.) *Poisoning and Toxicology Handbook*, 4th ed.; CRC Press/Taylor & Francis Group: Boca Raton, FL, USA, 2008.
112. Shibata, H.; Yokoi, T.; Goto, T.; Kim, I.Y.; Kawashita, M.; Kikuta, K.; Ohtsuki, C. Behavior of hydroxyapatite crystals in a simulated body fluid: Effects of crystal face. *J. Ceram. Soc. Jpn.* **2013**, *121*, 807–812. [CrossRef]

coatings

Article

EIS and LEIS Study on In Vitro Corrosion Resistance of Anodic Oxide Nanotubes on Ti–13Zr–13Nb Alloy in Saline Solution

Bożena Łosiewicz [1,*], Agnieszka Stróż [1], Julian Kubisztal [1], Patrycja Osak [1] and Maciej Zubko [1,2]

1 Institute of Materials Engineering, Faculty of Science and Technology, University of Silesia in Katowice, 75 Pułku Piechoty 1A, 41-500 Chorzów, Poland
2 Department of Physics, Faculty of Science, University of Hradec Králové, Rokitanského 62, 500 03 Hradec Králové, Czech Republic
* Correspondence: bozena.losiewicz@us.edu.pl; Tel.: +48-32-3497-527

Abstract: This work concerns the search for new ways to modify the surface of the biomedical Ti–13Zr–13Nb alloy for applications in regenerative medicine and personalized medicine. Obtained for the first time, oxide nanotubes (ONTs) layers of first-generation (1G) on a Ti–13Zr–13Nb alloy were produced by anodizing in 0.5% HF electrolyte at 20 V for 120 min. The physico-chemical characterization of the obtained bamboo-inspired 1G ONTs was conducted using TEM and ATR-FTIR methods. In vitro corrosion resistance of the 1G ONTs and comparative Ti–13Zr–13Nb substrate in saline solution at 37 °C was conducted by open-circuit potential, Tafel curves, anodic polarization curves, and EIS methods. LEIS and SVET study of local corrosion resistance was also carried out. It was found that surface modification by anodizing of the Ti–13Zr–13Nb alloy under proposed conditions allowed to obtain porous ONTs highly resistant to pitting corrosion. The obtained results give a new insight into the relationship between the morphological parameters of first-generation oxide nanotubes and in vitro corrosion resistance of the Ti–13Zr–13Nb alloy in saline solution at the macro- and microscale.

Keywords: anodizing; corrosion resistance; oxide nanotubes; Ti–13Zr–13Nb alloy

1. Introduction

The human body is a very demanding environment for engineering materials due to the need to use implant materials that must be highly resistant to corrosion, especially pitting, in the environment of tissues and body fluids [1–7]. Titanium and its alloys are currently one of the most commonly used biomaterials in modern medicine. The self-passive oxide layer formed naturally in the air or the passive oxide layer obtained as a result of forced passivation on the surface of the titanium-based implants provides their protection against harmful environmental factors of the human body, thanks to which these biomaterials can be commonly used in implantology, e.g., for the production of short- and long-term implants, such as stents, orthodontic wires, dental implants, knee and hip endoprostheses, plates, bone screws, etc. [1,2]. Insufficient resistance of metallic biomaterials to corrosion may become a source of elements that have a toxic effect on tissues or cause carcinogenic reactions in the body [8].

Titanium alloys used in implantology contain toxic elements such as Al, V, and Ni, the release of which can lead to many diseases such as Alzheimer's, metallosis, neuropathy, or a broadly understood allergic reaction [8]. To eliminate the harmful effects of toxic elements, new alloys based on biocompatible elements such as niobium, zirconium, molybdenum, and tantalum have been developed at the turn of the last few years, which act as stabilizers of the β structure in titanium [9–21]. Therefore, emphasis is placed on the production of new titanium-based alloys with mechanical properties similar to the Ni-Ti alloy and greater corrosion resistance, not containing harmful elements such as Ti–Nb–Zr, Ti–Mo–Zr, Ti–Nb–Ta, Ti–Mo, Ti–Zr, or Ti–Ta [1,2,9–21]. In addition, the surface of titanium and titanium alloys is additionally modified by various methods, from mechanical, through plasma, to electrochemical, in order to increase their corrosion resistance and biocompatibility [1–6,9,10,12,22–26]. Electrochemical methods such as electrochemically assisted deposition (ECAD) [5,24], electrophoretic deposition (EPD) [6], or anodizing [7,9–12,14–18,25–33], the latest development of which is plasma electrolytic oxidation (PEO), have a great potential for development, as they enable a wide modification of the chemical composition, surface morphology, and are cheap and easy to implement for large-scale production.

One of the electrochemical methods most often used to modify the self-passive oxide layer on titanium and its alloys, which ensures the possibility of producing self-assembled oxide nanotubes (ONTs), is anodizing [7,9–12,14–18,25–33]. Oxidation of titanium in an electric field in an aqueous electrolyte occurs according to the commonly accepted Guntershulze–Betz model [34]. The process of electrochemical production of a matrix of ONTs on the titanium and its alloys can be divided into several steps, which depend on the current density. In the first and shortest step, a compact, barrier oxide layer is produced. The beginning of the second step is associated with the activation of the barrier oxide layer by fluoride ions, which, in dissolving the oxide layer, cause the formation of randomly distributed pores. The start of this step is associated with an increase in the current intensity, which is caused by reducing the thickness of the oxide layer at the bottom of the pores. This leads to further deepening of the pores, which, after some time, begin to branch, overlapping each other competing for the flow of current. Under optimal conditions, the current flows evenly between the pores leading to the self-assembly of the porous layer, which begins to take the shape of the ONTs matrix. This is the third step of the process, characterized by a relative stabilization of the current. After reaching the equilibrium, the oxide layer thickness, both at the bottom of the nanotubes and in the space between the nanotubes, is constant as the result of the titanium oxidation rate equaling the dissolution rate of the formed oxide layer by fluoride ions. This process leads to a gradual increase in the length of the ONTs, which can be vividly described as the movement of the barrier layer into the titanium or its alloy, with a relatively small change in the position of the upper surface of the nanotubes associated with slow, chemical dissolution. Sometimes, the increase in the length of ONTs is distinguished as the fourth step of the whole process, but most authors treat it as a further part of the third step. The growth rate of the ONTs gradually decreases during the anodizing process. Since this rate is the same as the chemical dissolution rate taking place on the top surface of the layer, further anodizing does not elongate the formed ONTs [34,35].

The properties of ONTs layers are mainly determined by the chemical composition of the electrolyte and its temperature, the applied anodizing voltage or current density, and the substrate material [7,9–12,14–18,25–35]. An increase in the anodizing voltage increases the pore size by merging small pores into large ones, thereby reducing the number of pores. Increasing the surface roughness of the implant at the nanoscale contributes to facilitating the adhesion of osteogenic cells. Due to the similarity of the structure of ONTs layers to the structure of bone tissue in chemical and morphological terms, the porous surface of the implant improves the osseointegration as a result of faster growth of bone tissue and the formation of a stronger bond between the bone and the implant surface [32]. Obtaining ONTs with a high level of pore organization has become a very useful technology to functionalize the surface of surgical implants, production of biomedical sensors, or drug carriers [7].

In this paper, we continue our study on surface functionalization of the Ti–13Zr–13Nb alloy using anodizing [10,14–18,25,26]. The subject of the research was tuning the surface properties of this alloy towards biomedical applications by producing first-generation (1G) ONTs layers under new anodizing conditions in hydrofluoric acid solution. Due to the potential use for long-term implants, the newly developed 1G ONTs layers were subjected for the first time to in vitro corrosion resistance characteristics in a biological environment. Considering that usually used direct current (DC) and alternating current (AC) methods do not provide full information on the corrosion behavior of biomaterials because they allow the measurement of only the average current values for the entire electrode surface in contact with the electrolyte, both electrochemical impedance spectroscopy (EIS) and localized electrochemical impedance spectroscopy (LEIS) were applied at the macro- and microscale, respectively. Both EIS and LEIS were based on the same principles that a small sinusoidal voltage disturbance was applied to the electrode under test and the current response was measured, allowing the impedance to be calculated. The difference was that in the EIS method, the bulk current was measured, and in the case of the LEIS method, a microprobe scanning close to the electrode surface was used, which allowed the measurement of the local current in the electrolyte and the calculation of the local impedance.

2. Materials and Methods

2.1. Substrate Preparation

The substrate material used for anodizing was a commercial Ti–Zr–Nb alloy with a chemical composition (wt.%): 74—Ti, 13—Zr, and 13—Nb (BIMO TECH, Wrocław, Poland). This bi-phase ($\alpha + \beta$) alloy was obtained as a bar 20 mm in diameter and 1000 mm long. The bar was in the annealed state and before the tests, it was plastically processed in the two-phase range with air cooling in accordance with the specifications of ASTM F1713-08(2021)e1 [36]. Tested samples prepared in the form of discs 3 mm thick were embedded in duracryl, and then wet-ground on abrasive papers with silicon carbide of various grits of gradation of 600, 800, 1200, and 2500 (Buehler Ltd., Lake Bluff, IL, USA) using a metallographic grinding and polishing machine Forcipol 202 (Metkon Instruments Inc., Bursa, Turkey). Then the samples were polished using polishing cloths and suspensions. After polishing, the samples were placed in an ultrasonic cleaner USC-TH (VWR International, Radnor, PA, USA) with ultrapure water (Milli-Q Advantage A10 Water Purification System, Millipore SAS, Molsheim, France) for 20 min to remove unwanted impurities. Then the samples were degreased with acetone (Avantor Performance Materials Poland S.A., Gliwice, Poland). The sample cleaning procedure was repeated twice.

2.2. Anodizing Conditions of Ti–13Zr–13Nb Alloy

For the purposes of anodizing, electrodes were made from the prepared alloy samples. Electrical contact was provided by an insulated copper wire that was attached to the back side of the alloy samples with epoxy resin. The back side of the alloy samples and the side walls were protected using a two-component epoxy resin, which was chemically resistant. The native oxide layer was removed from the electrode surface immediately before anodizing using depassivation in 25% v/v HNO_3 (Avantor Performance Materials Poland S.A., Gliwice, Poland) for 10 min. The electrodes were then cleaned with Milli-Q water in an ultrasonic bath for 20 min and placed in an electrochemical cell.

Anodizing was carried out in an aqueous solution of 0.5% hydrofluoric acid (ACS-grade HF, 48%, Sigma-Aldrich, Saint Louis, MI, USA) at a temperature of 22(1) °C at a voltage of 20 V for 120 min using the PWR800H high-current power supply (Kikusui Electronics Corporation, Yokohama, Japan). The distance between the sample (anode) with a geometric area of 0.64 cm^2 and the platinum foil (cathode) with an area of 16 cm^2 was constant and amounted to 25 mm. After the anodizing process, the electrodes were rinsed in Milli-Q water and dried in air at ambient temperature.

2.3. TEM Measurements

The structure of 1G ONTs on Ti–13Zr–13Nb alloy was examined by high resolution electron microscopy (HREM) method using a JEOL JEM-3010 Transmission Electron Microscope (TEM, JEOL Ltd., Tokyo, Japan) operating at an acceleration voltage of 30 kV, equipped with a Gatan 2 k $\times$ 2 k Orius^{TM} 833 SC200D CCD camera. During TEM studies, the microstructure of ONTs without the presence of a substrate was observed. For this purpose, the anodized Ti–13Zr–13Nb alloy was immersed in isopropanol and sonicated for 60 min in an ultrasonic bath. Then, the ONTs layer detached from the substrate was deposited on a copper mesh covered with an amorphous carbon foil standardized to prepare a sample for TEM observations.

2.4. ATR-FTIR Measurements

To determine the functional groups of the tested materials, the Attenuated Total Reflectance-Fourier Transform Infrared Spectroscopy (ATR-FTIR) method was used. ATR-FTIR absorption spectra were recorded using an IR Trace-100 spectrophotometer (Shimadzu, Kyoto, Japan) equipped with ATR attachment with a diamond crystal for testing solid-states. All measurements were conducted in the spectral region of 4000–400 cm^{-1} at 100 scans per sample at 21(1) °C. The radiation was split into two beams, one of which ran along a path of constant length, and the other was generated by an interferometer with a moving mirror moving at a constant speed. The changing path length difference of the two beams caused mutual interference, resulting in an interferogram. The use of the Fourier transform allowed to transform the interferogram from the time domain to the frequency domain receiving a spectrum. The infrared (IR) beam penetrated the sample to a depth of several microns before being reflected. During the measurements, the crystal and the material under study stayed in contact.

2.5. DC and AC Measurements in Saline

In vitro tests of the corrosion resistance of the studied material were conducted in a saline solution at the temperature of 37(1) °C using the method of open-circuit potential (OCP), polarization curves, and electrochemical impedance spectroscopy (EIS). A solution of 0.9% NaCl pH = 7.4(1) was deaerated for 30 min with argon; 4% NaOH and 1% $C_3H_6O_3$ were used to adjust the pH. For preparation of the saline solution, analytically pure reagents (Avantor Performance Materials Poland S.A., Gliwice, Poland) and ultrapure water were used.

All electrochemical tests were performed using the Autolab/PGSTAT12 computer controlled electrochemical system (Metrohm Autolab B.V., Utrecht, The Netherlands). Electrochemical measurements were carried out in a three-electrode system consisting of a working electrode (WE) in the form of the tested material, a counter electrode (CE) as a platinum mesh, and a reference electrode (RE) placed in the Luggin capillary against which all potential values were measured. The RE was a saturated calomel electrode (SCE) with a potential of 244.4(1) mV. The method of preparation of WE was described in the work [7]. After stabilization of open-circuit potential (Eoc) for 2 h, the polarization curves j = f(E) were recorded using the potentiodynamic method in the potential range ±50 mV in relation to the E_{OC} with the electrode polarization rate v = 1 mV s^{-1}. The obtained polarization curves were the basis for determining the corrosion resistance parameters.

Subsequently, the EIS spectra were recorded at the E_{OC} in the frequency range from 20 kHz to 1 mHz using 10 frequencies per decade. A sine wave with an amplitude of 10 mV as an excitation signal was applied. Kramers–Kronig relations (K–K test) were used to assess the correctness of the obtained EIS data [37]. The analysis of the experimental EIS spectra was carried out based on the equivalent electrical analogs using the EQUIVCRT program with Boukamp's circuit description and the method of complex nonlinear least squares (CNLS) with modulus weighting [38]. The Fisher–Snedecor test F for the confidence level of α = 0.01 corresponding to 99% probability was used to check the significance of the parameters of the equivalent electrical circuits used [37]. The χ^2 (chi-square) test was used for verification of the EIS fit quality.

Susceptibility to pitting corrosion was tested using anodic polarization curves in the potential window from E_{OC} minus 150 mV to 9.4 V at v = 1 mV s^{-1}. To accurately visualize and interpret the obtained results, electrochemical noise was eliminated using the Savitzky–Golay smoothing algorithm and OriginPro 2018 software (OriginLab, Northampton, MA, USA).

Each type of measurement was repeated thrice, and the values of the determined parameters were given as mean values with standard deviation (SD).

2.6. Scanning Electrochemical Measurements in Saline

To analyze local changes in the corrosion resistance of the Ti–13Zr–Nb alloy before and after electrochemical oxidation, the scanning vibration probe technique (SVET) and LEIS were used. The SVET and LEIS techniques allowed to determine the distribution of the ion current density (j) and the impedance modulus (|Z|), respectively, above the surface of the sample in an aqueous solution of 0.9% NaCl with a conductivity of 16.32 mS cm^{-1}. Distribution maps of a specific parameter over the material surface (j or |Z|) were recorded using a PAR Model 370 Scanning Electrochemical Workstation (Princeton Applied Research, Oak Ridge, TN, USA). SVET and LEIS measurements were carried out in a four-electrode system: WE, CE as a platinum grid in the SVET technique and a platinum ring applied to the probe in the LEIS method, RE as SCE, and SVET or LEIS probe. All electrodes were immersed in a saline solution contained in a glass cell known as the TriCell. For both techniques, the probe–sample distance was determined using a video camera to be approximately 150 μm

The SVET technique measured the localized current in a saline above a sample by virtue of the IR drop in the electrolyte [39]. The connection mode of a sample under potentiostatic control with an isolated potentiostat was used, which reduced noise (Figure 1a). A potential gradient between the WE and the CE in the saline was directly proportional to the current and the conductivity of the electrolyte (Figure 1b). A probe immersed in the electrolyte was perpendicular to the WE. The probe was mounted on a precision vibrating stage, which vibrated in the vertical plane. The potential of the probe was directly proportional to its position in the potential gradient. The amplitude of the AC potential on the probe was therefore directly proportional to the current. The probe was scanned in the saline electrolyte such that the distribution of current on the surface of a sample might be mapped in the X and Y planes.

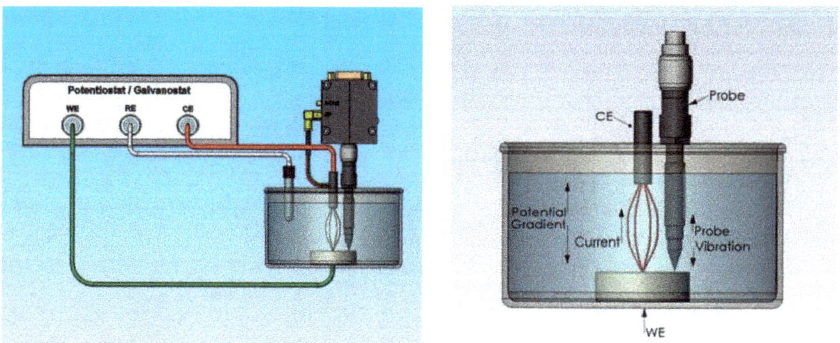

(**a**) (**b**)

Figure 1. Scheme of the SVET setup with Scanning Electrochemical Workstation Software M370, Version 1.20: (**a**) Sample under potentiostatic control with isolated potentiostat; (**b**) TriCell with a four-electrode configuration.

The maps of the ion current density distribution were recorded at a fixed anode current density in the passive range of 60 μA cm^{-2}, which corresponded to a potential of about 2 V vs. SCE on a potentiodynamic curve and a fixed microprobe vibration amplitude of 30 μm. The scanned area was 512 × 512 μm with a step of 16 μm. Using a gold electrode with a diameter of 200 μm immersed in the same solution as the tested materials, a calibration was performed, which enabled the conversion of the voltage recorded by the SVET microprobe into the values of the ion current density.

The LEIS technique was used to monitor corrosion mechanisms and kinetics. The maps of the impedance modulus distribution were measured at the E_{OC}, considered approximately as the corrosion potential (E_{cor}) of the material and the selected frequency, i.e., 500 mHz. More details about the LEIS technique are available in [40,41].

3. Results and Discussion

3.1. TEM Characterization of 1G ONTs on Ti–13Zr–13Nb Alloy

The 1G ONTs layers were produced by anodizing Ti–13Zr–13Nb alloy at 20 V for 120 min in 0.5% HF solution and then detached from the substrate. Figure 2a,b show examples of TEM images of the surface morphology of the formed ONTs layer at different scales. The on-top general view of the obtained ONTs layer in a selected micro-region revealed a uniform distribution of densely packed ONTs with single and very smooth walls and a high degree of self-organization. In the TEM image of a single oxide nanotube, its regular shape can be observed without any fraying at the top of ONT. The cross-section of a single oxide nanotube has a cylindrical shape. The obtained bamboo-like ONTs showed a vertical arrangement. No bundles of 1G ONTs growing locally were observed as was the case with 2G [17] and 3G [18] ONTs on the Ti–13Zr–13Nb alloy.

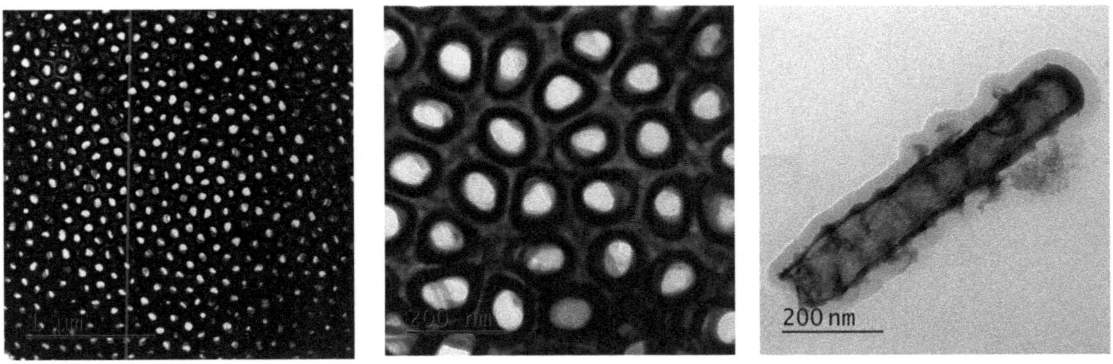

(a) (b) (c)

Figure 2. TEM microscopic image of 1G ONTs produced by anodizing Ti–13Zr–13Nb alloy at 20 V for 120 min in 0.5% HF solution and then detached from the substrate: (**a**) On-top general view of the ONTs layer; (**b**) On-top general view of the ONTs layer at higher magnification; (**c**) Single oxide nanotube.

On the basis of TEM microscopic images, the morphological parameters of the produced 1G ONTs were determined, such as the average inner and outer diameter of the oxide nanotube and its length. Under the proposed anodizing conditions, the ONTs with an average inner diameter of 70(8) nm and an average outer diameter of 90(13) nm were formed. The average length of the ONTs was 0.91(7) μm. The obtained results are consistent within the limit of error with the values of morphological parameters determined on the basis of FE-SEM images in selected local areas of the 1G ONTs surface, formed under comparable conditions [15]. The mechanism of multi-step formation of 1G ONTs layer on the Ti–13Zr–13Nb alloy surface based on activity of fluoride ions in aqueous inorganic solutions was explained in detail in the previous work [18]. For comparison, 2G ONTs with the inner diameter of 61(11) nm, outer diameter of 103(16) nm, and length of 3.9(2) were formed on the same substrate surface in a 1M $(NH_4)_2SO_4$ + 2% NH_4F solution at 20 V for 120 min [17]. On the other hand, 3G ONTs on the Ti–13Zr–13Nb alloy formed in 1M $C_2H_6O_2$ + 4%NH_4F at 50 V for 80 min were characterized by the inner diameter of 218(39) nm, outer diameter of 362(44) nm, and length of 9.7(6) μm [14]. These results show that anodizing conditions, and in particular the type of electrolyte, have a significant impact on the morphological parameters of the obtained ONTs.

The proposed ONTs can be saturated with antibiotics and bactericidal/bacteriostatic substances, which will make it possible to omit the oral route of dosing the drug. The drug will act directly in the bone tissue, inhibiting the growth of bacteria already at the implantation site. Medicinal substances will be released from the inside of the ONTs in the place of the newly inserted implant. Innovative implants with a layer of ONTs on their surface can therefore be proposed as intelligent drug carriers in drug delivery systems, especially for personalized medicine [7]. They will also allow for painless application, faster healing, and significant acceleration of bone tissue regeneration [1,2,32].

3.2. ATR-FTIR Characterization of Ti–13Zr–13Nb Alloy before and after Anodizing

Measurements of transmittance in the fundamental IR range were based on the phenomenon of total internal reflection of light from the interface of two materials with different refractive indices. The ATR-FTIR spectroscopy, as one of the most accurate spectroscopic methods, allowed to assign particular functional groups to specific areas with characteristic absorption bands. The non-anodized Ti–13Zr–13Nb alloy along with the 1G ONTs layer was investigated by ATR-FTIR spectroscopy in the range of 4000–400 cm^{-1} (Figure 3).

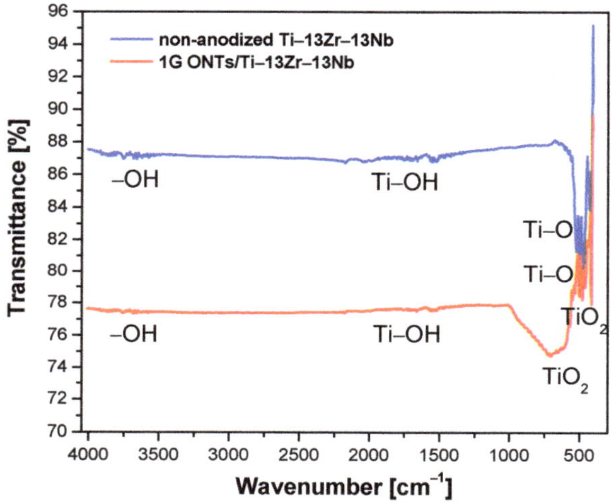

Figure 3. The ATR-FTIR absorption spectrum of the non-anodized Ti–13Zr–13Nb alloy and with 1G ONTs layer formed at 20 V for 120 min in 0.5% HF solution.

Several characteristic bands for TiO_2 are visible in both obtained ATR-FTIR spectra in Figure 3. The broad TiO_2 peak observed at 3896–3577 cm^{-1} is related to the stretching vibrations of the hydroxyl group –OH, which represents water as moisture present on the surface of the material or in the vicinity of the tested sample [42]. Both the ATR-FTIR absorption spectrum recorded for the non-anodized Ti–13Zr–13Nb alloy and the 1G ONTs layer produced show a band at the wavelength of 1630 cm^{-1}, which corresponds to bending modes of water Ti-OH [43]. The peak in the range of 873–558 cm^{-1} visible in the ATR-FTIR absorption spectrum for the 1G ONTs/Ti–13Zr–13Nb sample indicates the presence of TiO_2 in the anodic oxide layer on the alloy substrate surface. The peak at 580 cm^{-1} in both obtained spectra corresponds to vibrations of the Ti-O bond [44]. This is a peak characteristic of TiO_2 present on the surface of the material [45,46]. Bands in the range of 495–453 cm^{-1} correspond to the frequency of TiO_2 in the rutile phase. Bands at 451 and 410 cm^{-1} correspond to the frequencies of the anatase nanocrystalline phase and/or the rutile phase [47]. The results obtained using ATR-FTIR spectroscopy confirmed the presence of the anodic ONTs layer on the Ti–13Zr–13Nb alloy surface. They are also in good agreement with the results obtained using the grazing incidence X-ray diffraction (GIXD) for the obtained 1G ONTs layer, which confirmed the presence of both α-Ti phase and β-Ti phase from the substrate and titanium oxides such as TiO_2 (rutile), TiO_3, and TiO [15].

3.3. *Assessment of In Vitro Corrosion Resistance Measurements in Saline Solution*

3.3.1. Open-Circuit Potential Measurements

The preliminary assessment of the effect of anodizing the Ti–13Zr–13Nb alloy in the proposed conditions on the corrosion resistance in vitro in saline solution was carried out based on the measurement of open-circuit potential. The conditions of the in vitro electrochemical measurements with the potentiostat open loop corresponded to the in vivo conditions in the human body. The E_{OC} value was determined as the potential difference measured between WE and RE without using an external current source in the electrochemical system for 2 h as shown in Figure 4.

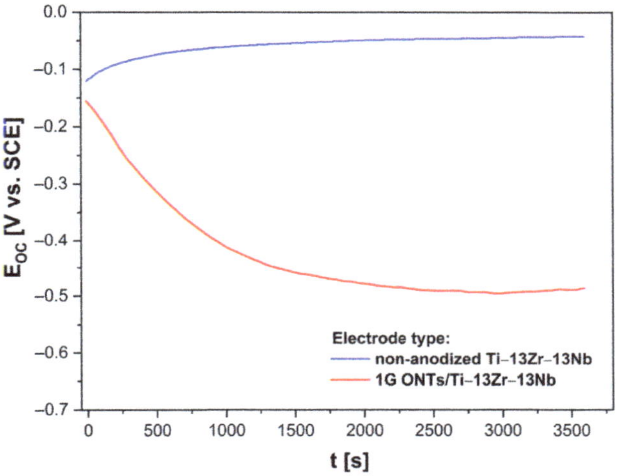

Figure 4. Dependence of open-circuit potential (E_{OC}) on immersion time (t) for the Ti–13Zr–13Nb electrode without and with 1G ONTs layer in saline solution at 37 °C.

After 2 h of immersion, a stable E_{OC} value for both tested electrodes was reached. After this time, the rate of Eoc change was slower than 1 mV min^{-1}. For the non-anodized Ti–13Zr–13Nb electrode, the stable E_{OC} was −42 mV(8) V, while for the 1G ONTs/Ti–13Zr–13Nb electrode, the E_{OC} shift towards cathode potentials was observed. The significant decrease in the E_{OC} value to −486 mV(24) V for the anodized Ti–13Zr–13Nb electrode indicates that the initiation of electrochemical corrosion will occur more easily on the porous surface of ONTs. Such a character of E_{OC} changes initially indicates a greater thermodynamic tendency to the corrosion of porous 1G ONTs/Ti–13Zr–13Nb electrode. Figure 4 also shows the influence of the Ti–13Zr–13Nb alloy anodizing on the course of the E_{OC} = f(t) curve. The observed changes in E_{OC} reflect various trends in the metal | solution interface variability with increasing immersion time. As shown in Figure 4, the Eoc value for the non-anodized Ti–13Zr–13Nb electrode gradually shifted in time towards the anodic potentials. This phenomenon resulted from a spontaneous formation and thickening of an ultrathin oxide layer on the electrode surface with time [22]. The self-passive oxide layer protected the Ti–13Zr–13Nb electrode from dissolving in the electrolyte. In the case of the 1G ONTs/Ti13Zr-13Nb electrode, in the initial phase of the measurements, the E_{OC} rapidly shifted towards the cathode potentials, and then the rate of its changes gradually decreased until it finally reached a stable value.

3.3.2. Analysis of Tafel Curves

Figure 5 shows the Tafel curves recorded in a narrow range of potentials ±50 mV relative to the E_{OC} for the Ti–13Zr–13Nb electrode before and after anodizing in saline solution at 37 °C. The obtained log | j | = f(E) dependences were the basis for determining the corrosion resistance parameters, which are summarized in Table 1.

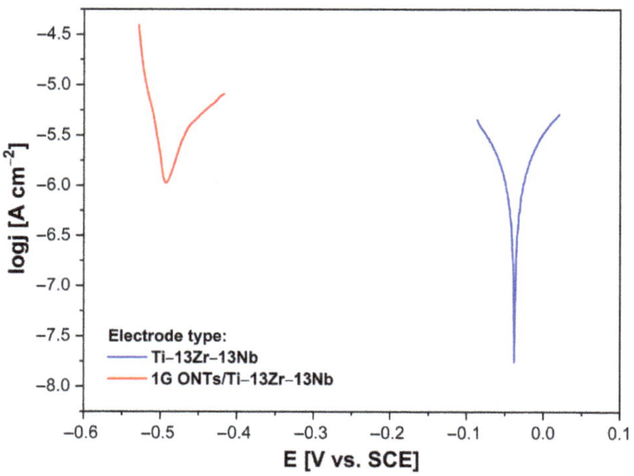

Figure 5. Tafel curves for the Ti–13Zr–13Nb electrode before [10] and after anodizing in saline solution at 37 °C. The polarization scan rate was v = 1 mV s^{-1}.

Table 1. In vitro corrosion resistance parameters determined based on the Tafel curves for the Ti–13Zr–13Nb electrode before and after anodizing in saline solution at 37 °C (see Figure 5).

Electrode Type	E_{cor} (V)	j_{cor} (A cm^{-2})	b_c (V dec^{-1})	b_a (V dec^{-1})	R_p (Ω cm^2)	CR at E_{cor} (mm yr^{-1})
Ti–13Zr–13Nb [10]	−0.038(8)	$3.3(7) \times 10^{-7}$	−0.018(4)	0.012(2)	460(92)	0.003(1)
1G ONTs/Ti–13Zr–13Nb	−0.497(26)	$4.9(9) \times 10^{-6}$	−0.033(8)	0.032(7)	50(7)	0.004(1)

The values of corrosion resistance parameters, such as corrosion potential (E_{cor}), corrosion current density (j_{cor}), and cathodic (b_c) and anodic (b_a) Tafel slope were determined numerically by fitting the parameters of the Butler–Volmer Equation (1) to the experimentally obtained dependence j = f(E) [48]:

$$j = j_{cor}\left\{\exp\left[\frac{2.303(E - E_{cor})}{b_a}\right] - \exp\left[-\frac{2.303(E - E_{cor})}{b_c}\right]\right\} \quad (1)$$

Polarization resistance (R_p) and general corrosion rate at E_{cor} were calculated using Equations (2) and (3), respectively, according to ASTM G102:89(2015)e1 [49]:

$$R_p = \frac{B}{j_{kor}} \quad (2)$$

$$CR = K_1 \frac{EW}{\rho} j_{kor} \quad (3)$$

where B—Stern–Geary coefficient, K_1—conversion constant, EW—equivalent weight, and ρ—density.

The EW for the Ti-13Zr-13Nb electrode was calculated at 12.5 assuming the thermodynamically stable forms of Ti^{4+}, Zr^{2+}, and Nb^{5+} based on the Pourbaix diagram of Ti-H_2O, Zr-H_2O, and Nb-H_2O system, respectively [50]. CR is expressed in mm yr^{-1} assuming in the Equation (3) the value of K_1 equal to 3.27×10^{-3} mm g μA^{-1} cm^{-1} yr^{-1} and taking into account j_{cor} expressed in μA cm^{-2}.

The potentiodynamic polarization curves presented in Figure 5 illustrate the course of anodic and cathodic reactions in the corrosion process of the non-anodized and anodized Ti-13Zr–13Nb electrodes in saline solution. One of the most important electrochemical parameters that quantify the changes in the corrosion resistance of the materials under study is the E_{cor} (Table 1). This thermodynamic quantity has values similar to the E_{OC}. The E_{cor} can be used as a comparative parameter for the assessment of the corrosion resistance of various materials in the same corrosive environment. Figure 5 shows the E_{cor} shift towards the cathodic potentials for 1G ONTs/Ti–13Zr–13Nb electrode in comparison with the Ti–13Zr–13Nb electrode before anodizing. Such electrochemical behavior indicates a lower corrosion tendency for the Ti–13Zr–13Nb electrode covered with a self-passive oxide layer with stronger barrier properties [10,14,22,51,52].

The j_{cor} value for the porous 1G ONTs/Ti–13Zr–13Nb electrode is ca. 15 times higher as compared to the j_{cor} determined for the non-anodized substrate (Table 1). The obtained values are directly proportional to the CR at the E_{cor}. These results indicate faster dissolution rate of the anodic 1G ONTs layer as compared to the self-passive oxide layer formed spontaneously. However, it should be emphasized that the material consumption of both tested electrodes is on the same order of 10^{-3} mm yr^{-1} (Table 1).

The Tafel slope for the cathodic (b_c) and anodic (b_a) branch were determined based on the Equations (4) and (5), respectively [48]:

$$b_c = -\frac{2.3RT}{\alpha nF} \tag{4}$$

$$b_a = \frac{2.3RT}{(1-\alpha)nF} \tag{5}$$

In the above equations, R is assigned to the gas constant equal to 8.314 J K^{-1} mol^{-1}, T denotes the temperature in K, α means the cathodic transfer coefficient, n is the number of electrons involved in the reaction, and F is the Faraday constant equal to 96,500 C mol^{-1}.

For both tested electrodes $b_c > b_a$, i.e., the rate determining step of the corrosion process is the slower anodic reaction (Table 1). Based on the results obtained, it is possible to propose the corrosion mechanism of the investigated materials, which is consistent with the passive dissolution under anaerobic conditions in a neutral aqueous solution [3,53].

The course of the charge transfer reaction can be described by the following general reaction in which H_2O acts as the oxidation agent:

$$Me + n \cdot H_2O \leftrightarrow Me(OH)_n + \frac{n}{2} \cdot H_2 \tag{6}$$

The $Me(OH)_n$ products in Reaction (6) can be metal oxides, hydroxides, or hydrated oxides. Gaseous hydrogen is also evolved as a product of this reaction.

The Reaction (6) is coupled with the water reduction consuming electrons from the oxidation reaction. The Reactions (7) and (8) also proceed in a neutral solution:

$$O_2 + 2H_2O + 4e^- \rightarrow 4OH^- \tag{7}$$

$$2H_2O + 2e^- \rightarrow H_2 + 2OH^- \tag{8}$$

Based on the b_c and b_a values in Table 1, one can see that the rate of anodic reaction described by Equation (6) is slower as compared to the rate of the reduction reactions represented by Equations (7) and (8). It can be concluded that anodizing does not change the nature of the electrochemical processes occurring on the Ti–13Zr–13Nb electrode in saline solution. However, a strong decrease in the R_p value is observed for the Ti–13Zr–13Nb electrode covered with the 1G ONTs layer compared to the non-anodized electrode with an amorphous native oxide layer (Table 1) [10]. The corrosion resistance of such electrodes is dependent on both the structure and thickness of the oxide layer [15].

3.3.3. EIS Study on Mechanism and Kinetics of Electrochemical Corrosion in Saline Solution

EIS method was used for characterization of the interfacial properties of the Ti–13Zr–13Nb | oxide layer | saline solution system. The experimental Nyquist diagrams for the Ti–13Zr–13Nb electrode before [10] and after anodizing recorded at E_{OC} in saline solution at 37 °C are shown as symbols in Figure 6a and Figure 6b, respectively. The corresponding phase angle Bode diagrams are presented in Figure 7. Symbols in Figures 6 and 7 are experimental data and continuous lines are CNLS fit. The CNLS fitting procedure used the concept of electrical equivalent circuits in which, instead of capacitors, the constant phase element (CPE) was applied. The impedance of CPE ($\hat{Z}_{CPE}$) was defined by the Equation (9) [10]:

$$\hat{Z}_{CPE} = \frac{1}{T(j\omega)^{\phi}} \quad (9)$$

where T is the capacitance parameter expressed in F $s^{\phi-1}$ cm^{-2} dependent on the potential of electrode potential, and ϕ relates to the angle of rotation of purely capacitive line on the complex plane plots of $\alpha = 90\,°(1-\phi)$. The CPE is a leaking capacitor with nonzero real and imaginary components. For $\phi = 1$, T is equal to the capacitance of the double layer (C_{dl}), and purely capacitive behavior is observed. According to Equation (9), pure capacitance is for ϕ equal to 1, infinite Warburg impedance for ϕ equal to 0.5, pure resistance for ϕ equal to 0, and pure inductance for]ϕ equal to 1 [37].

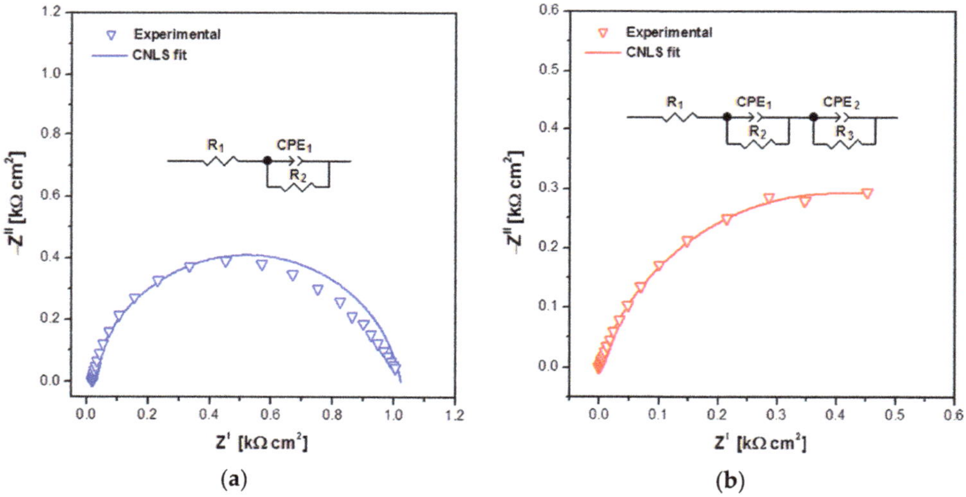

Figure 6. Nyquist diagram for the Ti–13Zr–13Nb electrode in saline solution at 37 °C with the equivalent electrical circuit model for the pitting corrosion process: (**a**) Non-anodized [10]; (**b**) After formation of 1G ONTs layer.

The maximum value of φ is slightly less than 90° in the case of the non-anodized and anodized Ti–13Zr–13Nb electrodes (Figure 7). For the non-anodized electrode, only one time constant is present in the electrical circuit (Figures 6a and 7a). Such impedance behavior characterizes titanium and its alloys coated with a thin oxide layer in a biological milieu [3,6,10,14,52]. Two time constants are visible in the electrical circuit for the Ti–13Zr–13Nb electrode with anodic 1G ONTs layer (Figures 6b and 7b). The obtained results are in accordance with our previous study on the EIS behavior of the autoclaved alloy with a sintered HAp/$nSiO_2$/Ag hybrid coating [6] and the Ti–13Zr–13Nb alloy with formed ONTs layers of second- and third-generation [10,14]. The experimental high values of $|Z|_{f\to 0}$ are typical for materials with capacitive behavior and high corrosion resistance [37].

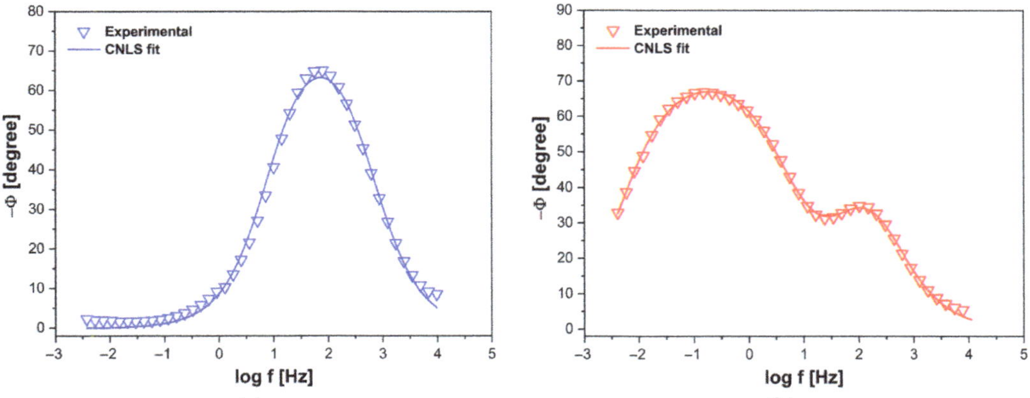

Figure 7. Phase angle Bode diagram for the Ti–13Zr–13Nb electrode in saline solution at 37 °C: (**a**) Non-anodized [10]; (**b**) After formation of 1G ONTs layer.

The EIS experimental data on the protective properties of a self-passive oxide layer on the surface of the non-anodized Ti–13Zr–13Nb electrode were approximated using the equivalent electrical circuit model for the pitting corrosion process which is called one-CPE model (Figure 6a) [10,37]. This model with four adjustable parameters as R_1, CPE_1-T, CPE_1-ϕ, and R_2 displays only one semicircle on the Nyquist plot [3,6,10,14,37,52]. In this model, R_1 is the solution resistance, CPE_1-T denotes the CPE_1 capacitive parameter, CPE_1-ϕ is the CPE_1 exponent associated with the C_{dl}, and R_2 is the charge transfer resistance through the interface of electrode | oxide layer | electrolyte. The construction of such equivalent electrical circuits as well as the physical meaning of the individual circuit parameters have been described in detail in earlier work [10,37].

In the case of the Ti–13Zr–13Nb electrode with the layer of 1G ONT, to approximate the experimental EIS data, the equivalent electrical circuit model for the pitting corrosion process illustrated in Figure 6b was used. This two-CPE model is described by seven adjustable parameters, R_1, CPE_1-T, CPE_1-ϕ, R_2, CPE_2-T, CPE_2-ϕ, R_3, and displays two semicircles on the Nyquist plot [6,10,37]. In this model, presence of a two-layered structure of the passive oxide film on the surface of the metallic electrode is assumed. The semicircle at high frequencies (HF) refers to the outer oxide layer with a porous structure (ONTs) and is described by the circuit parameters such as R_1, CPE-T_1, CPE-$ϕ_1$, and R_2. R_1 is the solution resistance, CPE_1-T is the capacitance of porous ONTs layer, CPE_1-ϕ is the CPE_1 exponent, and R_2 corresponds to the resistance of the ONTs layer. The remaining parameters of the circuit as CPE_2-T, CPE_2-ϕ, and R_3 describe the second semicircle at low frequencies (LF) which refers to the inner-barrier oxide layer directly adjacent to the alloy substrate and showing strong protective properties. CPE_2-T and CPE_2-ϕ are the CPE_2 capacitive parameter and the CPE_2 exponent, respectively, related to the barrier oxide layer. R_3 is the resistance of this barrier oxide layer.

Figures 6 and 7 illustrate the CNLS-fitted data marked as continuous lines which were obtained using the described electrical equivalent circuits. The very good quality of the CNLS fit is observed. All CNLS-fit parameters determined using the one-CPE equivalent electrical circuit model shown in Figure 6a for the Ti–13Zr–13Nb electrode before anodizing are summarized in Table 2. Table 3 presents all CNLS-fit parameters determined using the two-CPE equivalent electrical circuit model shown in Figure 6b for the Ti–13Zr–13Nb electrode with 1G ONTs layer.

Table 2. The parameters with standard deviations determined by approximation of the experimental EIS data for the non-anodized Ti–13Zr–13Nb electrode in saline solution at 37 °C [10] and the one-CPE equivalent electrical circuit model for the pitting corrosion process (see Figure 6a).

Electrode Type	R_1 (Ω cm^2)	CPE_1-T (F cm^{-2} s $^{\phi-1}$)	CPE_1-ϕ	R_2 (kΩ cm^2)
Non-anodized Ti–13Zr–13Nb	20.69(2)	0.73(5) $\times$ 10^{-5}	0.877(8)	1.01(1)

Table 3. The parameters with standard deviations determined by approximation of the experimental EIS data for the anodized Ti–13Zr–13Nb electrode in saline solution at 37 °C and the two-CPE equivalent electrical circuit model for pitting corrosion process (see Figure 6b).

Electrode Type	R_1 (Ω cm^2)	CPE_1-T (F cm^{-2} s $^{\phi-1}$)	CPE_1-ϕ	R_2 (Ω cm^2)	CPE_2-T (F cm^{-2} s $^{\phi-1}$)	CPE_2-ϕ	R_3 (Ω cm^2)
1G ONTs/ Ti–13Zr–13Nb	0.57(1)	5.08(34) $\times$ 10^{-3}	0.805(9)	1.46(3)	1.87(2) $\times$ 10^{-2}	0.821(1)	916(12)

The value of R_2 = 1.01(1) kΩ cm^2 is obtained for the Ti–13Zr–13Nb electrode before anodizing (Table 2), which is close to the R_p parameter determined based on the polarization curves near the E_{OC} (Table 1). The obtained results prove the correctness of the performed EIS tests. In physical and chemical terms, the parameters R_2 and R_p mean the same and refer to the ongoing corrosion process according to Equation (6). In the case of the Ti–13Zr–13Nb electrode with 1G ONTs layer, the charge transfer resistance associated with the outer oxide layer, R_3 of 916(12) Ω cm^2, is over 627 times higher in comparison with R_2 (Table 3). This phenomenon may result from the fact that the local electrolyte concentration inside the oxide nanotubes strongly increases compared to the pH in the volume of the electrolyte, which intensifies the destructive processes at the bottom of the oxide nanotubes. The deviation of CPE-ϕ parameter from 1 can be related to physico-chemical or geometrical inhomogeneities [37].

3.3.4. LEIS and SVET Study of Local Corrosion Resistance in Saline

Figure 8 shows the distributions of the local values of the impedance modulus $|Z|$ over the surface of the Ti–13Zr–Nb alloy before and after electrochemical oxidation in 0.9% NaCl solution. The average value of impedance irregularities determined for non-anodized Ti–13Zr–Nb and after electrochemical oxidation are approximately 1.6 kΩ and 13 Ω, respectively. On this basis, it can be concluded that the surface of the non-anodized Ti–13Zr–Nb alloy is characterized by a relatively heterogeneous distribution of impedance values compared to the Ti–13Zr–Nb alloy with 1G ONTs layer. The cause of local impedance fluctuations (of the order of several kΩ cm^2) is probably the non-uniform thickness of the natural passive layer formed on Ti–13Zr–Nb in the initial state. From a thermodynamic viewpoint, Ti and its alloys can react quickly with oxygen and produce stable metal oxides on the surface. Native oxide films formed in the air are ultrathin [3,22,51]. The thickness of the native oxide film exposed to air increases over time. Its composition and microstructure are dependent on the temperature and pH of the environment. The native oxide film on the surface of Ti and its alloys often becomes inhomogeneous as it grows, which may be the cause of a relatively heterogeneous distribution of local impedance values. Therefore, in order to increase the stability of the native oxide layer on the titanium and its alloys, ensuring high corrosion resistance of implants, forced passivation is often used in clinical practice. It should be noted that the average value of the impedance modulus determined for the non-anodized Ti–13Zr–Nb is over 8 times higher compared to Ti–13Zr–Nb after electrochemical oxidation. For Ti–13Zr–Nb in the initial state, it is 20.3 kΩ cm^2, and for Ti–13Zr–Nb after electrochemical oxidation, it is equal to 2.5 kΩ cm^2. This effect is related to the formation of a porous layer of ONTs as the outer part of the passive layer. The analysis of the impedance distribution maps recorded at E_{OC} confirms and complements the results o btained with the EIS method (Figures 6 and 7).

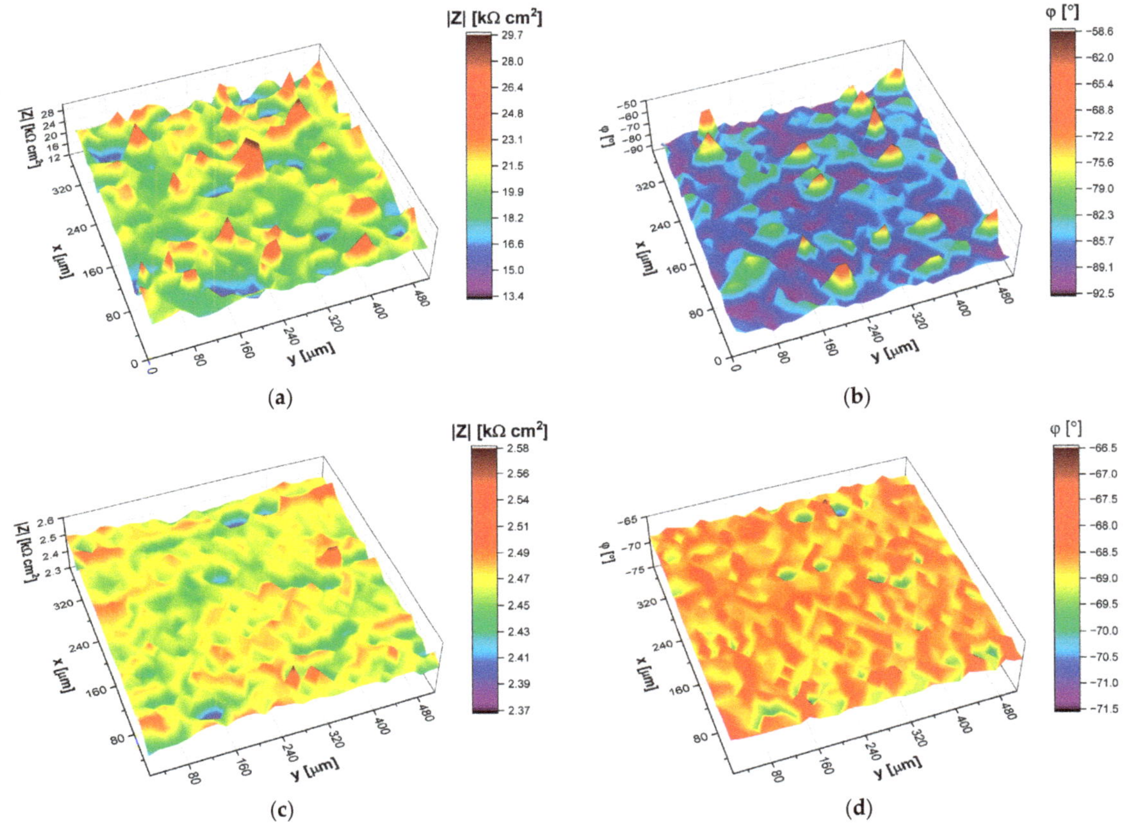

Figure 8. LEIS maps recorded at E_{OC} in saline solution at 37 °C for the Ti–13Zr–13Nb: (**a**) | Z | for the non-anodized alloy; (**b**) φ for the non-anodized alloy; (**c**) | Z | for the anodized alloy; (**d**)]φ for the anodized alloy.

The scanning vibrating electrode (SVET) technique also known as scanning vibrating probe (SVP) technique was used to measure the localized current flowing in a saline electrolyte above the tested samples. Figure 9 presents distributions of local values of ion current density (j) over the surface of the non-anodized and anodized Ti–13Zr–Nb alloy in a 0.9% NaCl solution.

The obtained SVET maps show that the average value j determined for the Ti–13Zr–Nb alloy in the initial state is about 2 times higher compared to the anodized sample. For the non-anodized Ti–13Zr–13Nb alloy, the average j is 81 μA cm^{-2}, and for the Ti–13Zr–Nb substrate covered with a layer of 1G ONTs, it is 41 μA cm^{-2}. The observed difference in the anodic current values indicates that the electrochemical oxidation process is faster/easier on the non-anodized surface of the Ti–13Zr–Nb alloy compared to the material on which the oxide layer has already been formed. It proves that it is easier to oxidize the fresh surface of the Ti–13Zr–Nb alloy than to thicken the already existing oxide layer. The average value of current density irregularities determined for the Ti–13Zr–Nb alloy before and after anodizing are about 10 μA cm^{-2} and 9 μA cm^{-2}, respectively. On this basis, it can be concluded that the surface of the Ti–13Zr–Nb alloy, both in the initial state and after electrochemical oxidation, is characterized by a relatively homogeneous distribution of ion current density values.

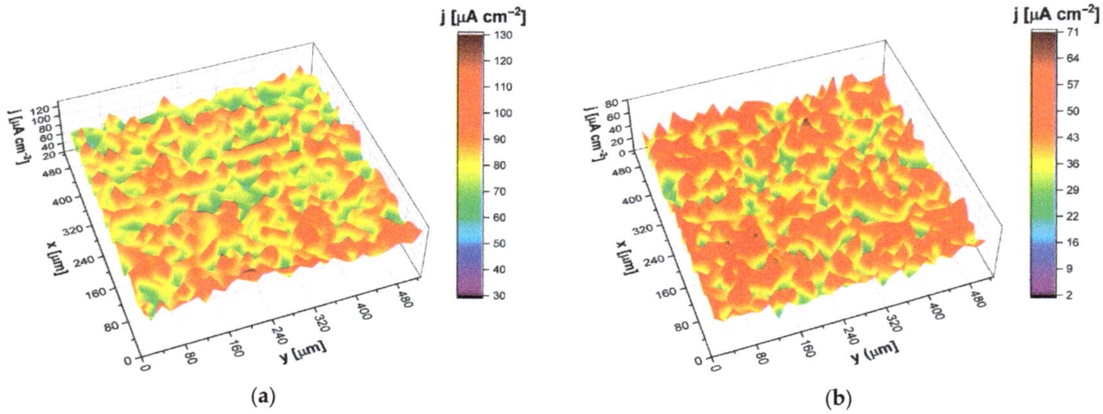

Figure 9. Ion current density maps for the Ti–13Zr–13N alloy in saline solution at 37 °C: (**a**) Non-anodized; (**b**) Anodized. X and Y coordinates denote the length of the mapping in μm.

3.3.5. Study of Anodic Polarization Curves on Susceptibility to Pitting Corrosion in Saline

The susceptibility of the Ti–13Zr–13Nb alloy without and with the 1G ONTs layer to pitting corrosion in the saline environment was determined on the basis of recording the anodic polarization curves obtained by the potentiodynamic method. A continuous change of the WE potential was recorded with the electrode polarization rate v = 1 mV s^{-1} with the simultaneous recording of the current flowing through the interface electrode | solution. The measurement was carried out from the potential 150 mV as more negative in relation to the E_{OC} towards the anodic potentials to the value of 9.4 V, recording the cathode–anode transition. Comparison of anodic polarization curves in the form of log | j | =f(E) for the non-anodized and anodized Ti–13Zr–13Nb electrodes is shown in Figure 10.

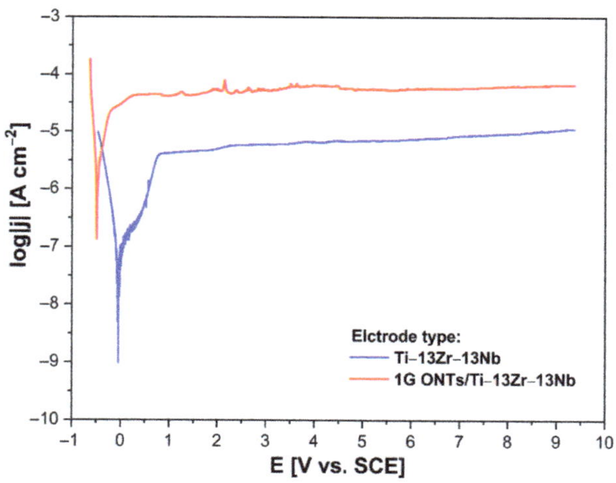

Figure 10. Anodic polarization curves for the non-anodized and anodized Ti–13Zr–13Nb electrode in saline solution at 37 °C.

Based on the obtained potentiodynamic characteristics in a very wide range of potentials, a similar course for both tested electrodes and their passive anodic behavior can be stated. However, one can observe that the formation of the 1G ONTs layer by anodizing of the Ti–13Zr–13Nb electrode causes the shift of the anodic polarization curve towards cathodic potentials. This is due to the decrease in the corrosion resistance of the porous 1G ONTs layer in comparison to the smooth and ultrathin oxide layer formed spontaneously on the surface of Ti–13Zr–13Nb. The lower passive current densities of the order of 10^{-5} A cm^{-2} were also observed for the non-anodized Ti–13Zr–13Nb electrode, which indicate the slower anodic dissolution according to Equation (6) in the case of the self-passive layer. Importantly, on both recorded anodic polarization curves, no increase in the measured passive current density is observed up to 9.4 V, which would be associated with the initiation of pitting corrosion in saline solution. Determination of the breakdown potential of the oxide layers on the Ti–13Zr–13Nb electrode above the applied anodic limit was impossible due to the limitations of the potentiostat and the stability of RE. The obtained anodic characteristics confirm the excellent corrosion resistance in vitro of both the non-anodized and anodized Ti–13Zr–13Nb electrodes in saline solution containing aggressive chloride ions, which is required for long-term implants.

4. Conclusions

Based on the conducted studies using TEM and ATR-FTIR methods, it was found that the surface modification of the biomedical Ti–13Zr–13Nb alloy via anodizing in a 0.5% HF electrolyte at 20 V for 120 min allowed to produce nanotubular oxide layers.

The obtained 1G ONTs with a rutile structure have the average inner diameter of 70(8) nm, average outer diameter of 90(13) nm, and average length of 0.91(7) μm. EIS study of the anodized Ti–13Zr–13Nb electrode revealed the impedance behavior typical for titanium and its alloys covered with a porous oxide layer, which was confirmed by local corrosion resistance parameters obtained by LEIS and SVET methods. The slight decrease in corrosion resistance of the anodized Ti–13Zr–13Nb electrode in comparison with the self-passive Ti–13Zr–13Nb electrode was revealed. However, no susceptibility to pitting corrosion up to 9.4 V was found in potentiodynamic studies for both the non-anodized and anodized Ti–13Zr–13Nb electrode, confirming their excellent in vitro corrosion resistance in saline solution. Thanks to the ability to control the diameter and length of the nanotubes, the porous surface of the 1G ONTs layer can also be used as an intelligent drug carrier and accelerate the process of osseointegration.

Author Contributions: Conceptualization, B.Ł.; methodology, B.Ł., A.S. and J.K.; software, B.Ł. and J.K.; validation, A.S., J.K., P.O. and M.Z.; formal analysis, A.S., J.K., P.O. and M.Z.; investigation, A.S., J.K., P.O. and M.Z.; resources, B.Ł.; data curation, A.S., J.K., P.O. and M.Z.; writing—original draft preparation, B.Ł., A.S., J.K. and P.O.; writing—review and editing, M.Z.; visualization, A.S., J.K. and P.O.; supervision, B.Ł.; project administration, B.Ł.; funding acquisition, B.Ł. All authors have read and agreed to the published version of the manuscript.

Funding: This research received no external funding.

Institutional Review Board Statement: Not applicable.

Informed Consent Statement: Not applicable.

Data Availability Statement: Not applicable.

Conflicts of Interest: The authors declare no conflict of interest.

References

1. Jakubowicz, J. *Ti-Based Biomaterials: Synthesis, Properties and Applications*; MDPI AG: Basel, Switzerland, 2020. [CrossRef]
2. Anene, F.; Jaafar, C.A.; Zainol, I.; Hanim, M.A.; Suraya, M. Biomedical materials: A review of titanium based alloys. *Proc. Inst. Mech. Eng. Part C J. Mech. Eng. Sci.* **2020**, *235*, 3792–3805. [CrossRef]
3. Łosiewicz, B.; Osak, P.; Maszybrocka, J.; Kubisztal, J.; Stach, S. Effect of autoclaving time on corrosion resistance of sandblasted Ti G4 in artificial saliva. *Materials* **2020**, *13*, 4154. [CrossRef] [PubMed]
4. Aniołek, K.; Łosiewicz, B.; Kubisztal, J.; Osak, P.; Stróż, A.; Barylski, A.; Kaptacz, S. Mechanical properties, corrosion resistance and bioactivity of oxide layers formed by isothermal oxidation of Ti-6Al-7Nb alloy. *Coatings* **2021**, *11*, 505. [CrossRef]

5. Osak, P.; Maszybrocka, J.; Kubisztal, J.; Ratajczak, P.; Łosiewicz, B. Long-Term Assessment of the In Vitro Corrosion Resistance of Biomimetic ACP Coatings Electrodeposited from an Acetate Bath. *J. Funct. Biomater.* **2021**, *12*, 12. [CrossRef]
6. Dudek, K.; Dulski, M.; Łosiewicz, B. Functionalization of the NiTi Shape Memory Alloy Surface by HAp/SiO_2/Ag Hybrid Coatings Formed on SiO_2-TiO_2 Glass Interlayer. *Materials* **2020**, *13*, 1648. [CrossRef]
7. Łosiewicz, B.; Stróż, A.; Osak, P.; Maszybrocka, J.; Gerle, A.; Dudek, K.; Balin, K.; Łukowiec, D.; Gawlikowski, M.; Bogunia, S. Production, Characterization and Application of Oxide Nanotubes on Ti–6Al–7Nb Alloy as a Potential Drug Carrier. *Materials* **2021**, *14*, 6142. [CrossRef]
8. Costa, B.C.; Tokuhara, C.; Rocha, L.A.; Oliveira, R.C.; Lisboa-Filho, P.N.; Pessoa, J. Vanadium ionic species from degradation of Ti-6Al-4V metallic implants: In vitro cytotoxicity and speciation evaluation. *Mater. Sci. Eng. C* **2019**, *96*, 730–739. [CrossRef] [PubMed]
9. Ossowska, A.; Zieliński, A.; Supernak, M. Formation of High Corrosion Resistant Nanotubular Layers on Titanium Alloy Ti13Nb13Zr. *Solid State Phenom.* **2011**, *183*, 137–142. [CrossRef]
10. Smołka, A.; Dercz, G.; Rodak, K.; Łosiewicz, B. Evaluation of corrosion resistance of nanotubular oxide layers on the Ti13Zr13Nb alloy in physiological saline solution. *Arch. Metall. Mater.* **2015**, *60*, 2681–2686. [CrossRef]
11. Luz, A.R.; de Souza, G.B.; Lepienski, C.M.; Siqueira, C.J.M.; Kuromoto, N.K. Tribological properties of nanotubes grown on Ti-35Nb alloy by anodization. *Thin Solid Film.* **2018**, *660*, 529–537. [CrossRef]
12. Stróż, A.; Goryczka, T.; Łosiewicz, B. Electrochemical formation of self-organized nanotubular oxide layers on niobium (Review). *Curr. Nanosci.* **2019**, *15*, 42–48. [CrossRef]
13. Schneider, S.G.; Nunes, C.A.; Rogero, S.P.; Higa, O.Z.; Bressiani, J.C. Mechanical properties and cytotoxic evaluation of the Ti-13Nb-13Zr alloy. *Biomecânica* **2000**, *8*, 84–87. [CrossRef]
14. Stróż, A.; Łosiewicz, B.; Zubko, M.; Chmiela, B.; Balin, K.; Dercz, G.; Gawlikowski, M.; Goryczka, T. Production, structure and biocompatible properties of oxide nanotubes on Ti13Nb13Zr alloy for medical applications. *Mater. Charact.* **2017**, *132*, 363–372. [CrossRef]
15. Stróż, A.; Dercz, G.; Chmiela, B.; Łosiewicz, B. Electrochemical synthesis of oxide nanotubes on biomedical Ti13Nb13Zr alloy with potential use as bone implant. *AIP Conf. Proc.* **2019**, *2083*, 030004. [CrossRef]
16. Smołka, A.; Rodak, K.; Dercz, G.; Dudek, K.; Łosiewicz, B. Electrochemical Formation of Self-Organized Nanotubular Oxide Layers on Ti13Zr13Nb Alloy for Biomedical Applications. *Acta Phys. Pol.* **2014**, *125*, 932–935. [CrossRef]
17. Stróż, A.; Dercz, G.; Chmiela, B.; Stróż, D.; Łosiewicz, B. Electrochemical formation of second generation TiO_2 nanotubes on Ti13Nb13Zr alloy for biomedical applications. *Acta Phys. Pol.* **2016**, *130*, 1079–1080. [CrossRef]
18. Łosiewicz, B.; Skwarek, S.; Stróż, A.; Osak, P.; Dudek, K.; Kubisztal, J.; Maszybrocka, J. Production and Characterization of the Third-Generation Oxide Nanotubes on Ti-13Zr-13Nb Alloy. *Materials* **2022**, *15*, 2321. [CrossRef]
19. Lee, T. Variation in Mechanical Properties of Ti-13Nb-13Zr Depending on Annealing Temperature. *Appl. Sci.* **2020**, *10*, 7896. [CrossRef]
20. Davidson, J.A.; Kovacs, P. New Biocompatible, Low Modulus Titanium Alloy for Medical Implants. U.S. Patent No. 5,169,597, 8 December 1992.
21. Lee, M.; Kim, I.-S.; Moon, Y.H.; Yoon, H.S.; Park, C.H.; Lee, T. Kinetics of Capability Aging in Ti-13Nb-13Zr Alloy. *Crystals* **2020**, *10*, 693. [CrossRef]
22. Szklarska, M.; Dercz, G.; Simka, W.; Łosiewicz, B.A.c. impedance study on the interfacial properties of passivated Ti13Zr13Nb alloy in physiological saline solution. *Surf. Interface Anal.* **2014**, *46*, 698–701. [CrossRef]
23. Osak, P.; Maszybrocka, J.; Zubko, M.; Rak, J.; Bogunia, S.; Łosiewicz, B. Influence of Sandblasting Process on Tribological Properties of Titanium Grade 4 in Artificial Saliva for Dentistry Applications. *Materials* **2021**, *14*, 7536. [CrossRef] [PubMed]
24. Osak, P.; Maszybrocka, J.; Kubisztal, J.; Łosiewicz, B. Effect of amorphous calcium phosphate coatings on tribological properties of titanium grade 4 in protein-free artificial saliva. *Biotribology* **2022**, *32*, 100219. [CrossRef]
25. Stróż, A.; Maszybrocka, J.; Goryczka, T.; Dudek, K.; Osak, P.; Łosiewicz, B. Influence of Anodizing Conditions on Biotribological and Micromechanical Properties of Ti–13Zr–13Nb Alloy. *Materials* **2023**, *16*, 1237. [CrossRef] [PubMed]
26. Stróż, A.; Luxbacher, T.; Dudek, K.; Chmiela, B.; Osak, P.; Łosiewicz, B. In Vitro Bioelectrochemical Properties of Second-Generation Oxide Nanotubes on Ti–13Zr–13Nb Biomedical Alloy. *Materials* **2023**, *16*, 1408. [CrossRef] [PubMed]
27. Durdu, S.; Cihan, G.; Yalcin, E.; Altinkok, A. Characterization and mechanical properties of TiO_2 nanotubes formed on titanium by anodic oxidation. *Ceram. Int.* **2021**, *47*, 10972–10979. [CrossRef]
28. Ossowska, A.; Olive, J.-M.; Zielinski, A.; Wojtowicz, A. Effect of double thermal and electrochemical oxidation on titanium alloys for medical applications. *Appl. Surf. Sci.* **2021**, *563*, 150340. [CrossRef]
29. Ossowska, A.; Zieliński, A.; Olive, J.-M.; Wojtowicz, A.; Szweda, P. Influence of Two-Stage Anodization on Properties of the Oxide Coatings on the Ti–13Nb–13Zr Alloy. *Coatings* **2020**, *10*, 707. [CrossRef]
30. Handzlik, P.; Gutkowski, K. Synthesis of oxide nanotubes on Ti13Nb13Zr alloy by the electrochemical method. *J. Porous Mater.* **2019**, *26*, 1631–1637. [CrossRef]
31. Stępień, M.; Handzlik, P.; Fitzner, K. Electrochemical synthesis of oxide nanotubes on Ti6Al7Nb alloy and their interaction with the simulated body fluid. *J. Solid State Electrochem.* **2016**, *20*, 2651–2661. [CrossRef]
32. Wu, S.; Wang, S.; Liu, W.; Yu, X.; Wang, G.; Chang, Z.; Wen, D. Microstructure and properties of TiO_2 nanotube coatings on bone plate surface fabrication by anodic oxidation. *Surf. Coat. Technol.* **2019**, *374*, 362–373. [CrossRef]

33. Zhang, S.; Qin, J.; Yang, C.; Zhang, X.; Liu, R. Effect of Zr addition on the microstructure and tribological property of the anodization of Ti-6Al-4V alloy. *Surf. Coat. Technol.* **2018**, *356*, 38–48. [CrossRef]
34. Guntherschulze, A.; Betz, H. Neue Untersuchungen Uber die elektrolytische Ventilwirkung: I. Die Oxydschicht des Tantals. *Z. Fur Phys.* **1931**, *68*, 145–161. [CrossRef]
35. Kowalski, D.; Kim, D.; Schmuki, P. TiO_2 nanotubes, nanochannels and mesosponge: Self-organized formation and applications. *Nano Today* **2013**, *8*, 235–264. [CrossRef]
36. *ASTM F1713-08(2021)e1*; Standard Specification for Wrought Titanium-13Niobium-13Zirconium Alloy for Surgical Implant Applications (UNS R58130). ASTM: West Conshohocken, PA, USA, 2021.
37. Lasia, A. *Electrochemical Impedance Spectroscopy and Its Applications*; Springer Science + Business Media: New York, NY, USA, 2014; ISBN 978-1-4614-8932-0. [CrossRef]
38. *User Manual for Frequency Response Analysis (FRA) for Windows*, version 4.9; Eco Chemie, B.V.: Utrecht, The Netherlands, 2001.
39. Łosiewicz, B.; Popczyk, M.; Szklarska, M.; Smołka, A.; Osak, P.; Budniok, A. Use of Scanning Vibrating Electrode Technique to Localized Corrosion Evaluation. *Solid State Phenom.* **2015**, *228*, 353–368. [CrossRef]
40. Łosiewicz, B.; Popczyk, M.; Smołka, A.; Szklarska, M.; Osak, P.; Budniok, A. Localized Electrochemical Impedance Spectroscopy for Studying the Corrosion Processes in a Nanoscale. *Solid State Phenom.* **2015**, *228*, 383–393. [CrossRef]
41. Łosiewicz, B.; Kubisztal, J. Effect of hydrogen electrosorption on corrosion resistance of Pd80Rh20 alloy in sulfuric acid: EIS and LEIS study. *Int. J. Hydrogen Energy* **2018**, *43*, 20004–20010. [CrossRef]
42. Pezhooli, N.; Rahimi, J.; Hasti, F.; Maleki, A. Synthesis and Evaluation of Composite TiO2@ZnO Quantum Dots on Hybrid Nanostructure Perovskite Solar Cell. *Sci. Rep.* **2022**, *12*, 9885. [CrossRef]
43. León, A.; Reuquen, P.; Garín, C.; Segura, R.; Vargas, P.; Zapata, P.; Orihuela, P.A. FTIR and Raman Characterization of TiO_2 Nanoparticles Coated with Polyethylene Glycol as Carrier for 2-Methoxyestradiol. *Appl. Sci.* **2017**, *7*, 49. [CrossRef]
44. Ghann, W.; Kang, H.; Sheikh, T.; Yadav, S.; Chavez-Gil, T.; Nesbitt, F.; Uddin, J. Fabrication, Optimization and Characterization of Natural Dye Sensitized Solar Cell. *Sci. Rep.* **2017**, *7*, 41470. [CrossRef]
45. Al-Amin, M.; Dey, S.; Rashid, T.; Ashaduzzaman, M.; Shamsuddin, S. Solar Assisted Photocatalytic Degradation of Reactive Azo Dyes in Presence of Anatase Titanium Dioxide. *Int. J. Latest Res. Eng. Technol.* **2016**, *2*, 14–21.
46. Zhu, Y. Insight into the Structure and Formation of Titanium Oxide Nanotubes. *Funct. Mater. Lett.* **2008**, *1*, 239–246. [CrossRef]
47. Petrović, B.; Kisić, D.; Jokanović, B.; Rakočević, Z.; Nasov, I.; Trajkovska Petkoska, A.; Jokanović, V. Wetting Properties of Titanium Oxides, Oxynitrides and Nitrides Obtained by DC and Pulsed Magnetron Sputtering and Cathodic Arc Evaporation. *Mater. Sci.-Pol.* **2019**, *37*, 173–181. [CrossRef]
48. Bard, A.J.; Faulkner, L.R. Electrochemical Methods. In *Fundamentals and Applications*, 2nd ed.; Wiley: New York, NY, USA, 2001.
49. *ASTM G102:89(2015)e1*; Standard Practice for Calculation of Corrosion Rates and Related Information from Electrochemical Measurements. ASTM: West Conshohocken, PA, USA, 2015.
50. Pourbaix, M. *Atlas of Electrochemical Equilibria in Aqueous Solutions*, 2nd ed.; NACE International, National Association of Corrosion Engineers: Houston, TX, USA, 1974; ISBN 10: 0915567989.
51. Freitag, M.; Łosiewicz, B.; Goryczka, T.; Lelątko, J. Application of EIS to study the corrosion resistance of passivated NiTi shape memory alloy in simulated body fluid. *Solid State Phenom.* **2012**, *183*, 57–64. [CrossRef]
52. Osak, P.; Łosiewicz, B. EIS study on interfacial properties of passivated Nitinol orthodontic wire in saliva modified with Eludril® Mouthwash. *Prot. Met. Phys. Chem. Surf.* **2018**, *54*, 680–688. [CrossRef]
53. Brunette, D.M.; Tengvall, P.; Textor, M.; Thomsen, P. Titanium in Medicine. In *Material Science, Surface Science, Engineering, Biological Responses and Medical Applications*; Springer: Berlin, Germany, 2001; pp. 145–162, ISBN 978-3-642-56486-4.

MDPI AG
Grosspeteranlage 5
4052 Basel
Switzerland
Tel.: +41 61 683 77 34

Coatings Editorial Office
E-mail: coatings@mdpi.com
www.mdpi.com/journal/coatings

www.ingramcontent.com/pod-product-compliance
Lightning Source LLC
Chambersburg PA
CBHW061505210926
51850CB00008B/1831
9783725840649